SEWER MANAGEMENT SYSTEMS

The Wiley Series in Infrastructure Management and Design

Infrastructure Systems: Mechanics, Design, and Analysis of Components
Demeter G. Fertis

Sewer Management Systems
Thomas J. Day

SEWER MANAGEMENT SYSTEMS

THOMAS J. DAY

JOHN WILEY & SONS, INC.

New York · Chichester · Weinheim · Brisbane · Singapore · Toronto

Unless otherwise noted, all charts, pictures and data have been provided courtesy of American Sigma, Medina, New York. Used with permission.

This book is printed on acid-free paper. ♾

Published simultaneously in Canada.

Library of Congress Cataloging-in-Publication Data:

Day, Thomas J.
Sewer management systems / Thomas J. Day.
p. cm.— (The Wiley series in infrastructure management and design)
Includes index.
ISBN 0-471-31799-3 (alk. paper)
1. Sewage disposal plants—Management—Automation. 2. Sewerage—Automation. I. Title. II. Series.
TD746 .D39 2000
628.2—dc21 99-042171

Printed in the United States of America.

10 9 8 7 6 5 4 3 2 1

This book would not have been possible without the efforts and encouragement of the one to whom it is dedicated, James J. Nicolo, J.D., P.E., who always supported, defended, and encouraged all the work that led to this writing. At no time in our association did Jim ever waver or question my integrity or actions. For his unshakable faith in me, I dedicate this writing to him, for I will never be able to repay him for all he did for me. To him I am grateful; I hope I never let him down.

ACKNOWLEDGMENTS

No project is ever accomplished alone. This book is no exception. The many associations I have made have formed and shaped me; I have grown from them all and am very grateful to each of them, mentioned here or not.

To the reviewers, I am extremely grateful. Raymond Staniec, P.E., was the chief reviewer, and his critical eye for detail is greatly appreciated and acknowledged. James J. Nicolo J.D., P.E., as usual, gave time and effort to another of my projects; his comments are appreciated. Craig Bradstock, Dan Kiesch, and Heyward Johnson, with whom I worked for many years, did much detail work for me; I acknowledge their work and thank them. Robert Lendzinski, P.E., with whom I worked with closely for many years, taught me much about wastewater process: I thank him as well. Len Bernstein, P.E., and Mike Kyser, P.E. (Badger Meter), who signed my recommendation for the fellowship that directed me into the privatization field, I thank for their time and never-ending efforts.

On the corporate level, I am deeply indebted to Dick Wissenback and the entire American Sigma team, who provided me with an incredible resource pool and made much possible that would have eluded me without their assistance. I learned much from them and hope to continue learning from Dick and this incredible group in Medina, New York.

ABOUT THE AUTHOR

Thomas J. Day's diverse background and education led him across many territories and administrations in implementing varied sensitive and critical utility projects. The author of over 30 technical papers, publications, and presentations, he has been responsible for many time-dependent and complex projects that had no precedence or base on which to draw, requiring out-of-the-box thinking and new concepts. He is the recipient of the distinguished Water Environment Federation's Water Research Center's Travel Fellowship, along with other letters of commendation and appreciation for his contributions to the water industry.

He holds a certificate in paralegal studies, an associate degree in management, a bachelor's degree in electrical engineering, and a master's degree in computer and software engineering, and has continued postgraduate work. For over 11 years he was the in-house consultant to the Philadelphia Water Department in the Operations Division. Today, he is director of technology for American Water Services, Inc., a private operator of water and wastewater utility systems.

CONTENTS

INTRODUCTION

This book is written with the assumption that the reader knows what and where the sewers are. It does not matter if the seeker is a Ph.D. or a high school dropout; what matters is that the reader has a genuine desire to enter an area relatively new to engineering, where there are more unknowns than knowns in the everyday workplace. The book assumes that the reader has some experience in the area, but simply from a comfort viewpoint and not a knowledge base. There is always comfort if there is an established rapport with the material in general.

The book opens with an overview of sewer maintenance and management. A short history and orientation for perspective are provided. Chapters 2 through 5 outline the implementation of a basic surveillance network and the process of sample taking. The elements of Chapter 2 provide a baseline understanding of flow and how to measure it. The idea is to present introductory concepts that can be built into full models. Chapter 3 describes the methods for laying out a system and understanding what is required from a site analysis.

Chapter 4 presents more baseline definitions. It then introduces structures and features of the sewer infrastructure that are of particular interest. Chapter 5 enters the external word of the infrastructure by describing factors that affect it but over which people have little or no control. The often overlooked subject of rain analysis is defined.

Chapters 6 and 7 discuss the selection and analysis process. A step-by-step method that can be utilized in making system purchase decisions is provided. This is adaptable to any system. Included are issues on data communications, utility services, and sequencing.

Chapter 8 shows how to apply the data generated to tangible, real-world tasks. It is from the use of the data that the expense of the system is truly justified. Included are sections on tides and temperature issues. Chapter 9 discusses additional functions that may be desirable after the system is up and running or that may be designed in if there is enough foresight and engineering talent available. Included is information on algorithm development for warnings and features for automatic sewer control.

Chapter 10 demonstrates the return on investment provided by the system. It shows how to use the system as a funding source and not just a funding pit. This chapter will assist system operators in getting their projects through and making the best use of the data generated.

The text is clearly structured to take the reader methodically through each step in the construction and operation of a sewer monitoring system. Advanced readers will find they can skip chapters containing familiar information.

I once heard someone describe a sewer as a living, breathing thing. My many years of experience in utilities enable me to validate this statement. A sewer is its own entity and requires an attention sequence that is not generally adaptable from the world of plant maintenance. It is possible to bring some of those factors forward, but the logistics of street maintenance, dealing with the public, governmental influences, and the years of matching and mismatching sections of the system produce something not generally envisioned by the system designers. These circumstances call for specialized texts and precise experience in the management of such systems.

For the living and breathing entity that you are entrusted with, I wish you well. I hope this contribution can make some small improvement to your already outstanding work in sewer infrastructure management. In the sewer business, the reward for good work is more work. May yours become easier as a result.

CHAPTER 1

Evolution of Infrastructure Management: Modern Automation for the Sewer Systems

1.1 THE HIDDEN RESOURCE: THE INFRASTRUCTURE

The purpose of taxes, levies, and fees imposed by any government jurisdiction is to fund services required to maintain and continue the health and safety, legal obligations, and goodwill of that political jurisdiction. This is consistent across all governments in the modern world, whether in the United States, Europe, the Far East, or anywhere else. There are degrees and rates of levies, but populations paying for government services are a fact almost everywhere in the world today. Other than legislative bodies and tangible governmental administrative services, such as police and social services, the money goes to the infrastructure, that which underpins the very fabric of the community, is the vehicle for its continuance, and serves as the vessel for its growth.

It is buried. It is hidden. It is behind and within. It is generally overlooked by the public, unless a problem arises. It is an asset that utilities are currently wrestling to price and give worth. It is generally not discussed without some political motivation. It is used every day to the betterment of the community that it serves. It cannot be purchased directly from a store, manufacturer, or commodity exchange. It is unique to every community. It is the infrastructure.

Roads, parking lots, utility services, storage facilities, and all the other hidden or otherwise unobserved systems that provide for the well-being and convenience of any given community fall under the category of infrastructure. These elements develop during the growth of a town, city, or country, are erected in sections over many years, are designed and retrofitted, and pass from one management scheme to another, one political administration to the next. But there is one common thread that ties all these things together. The infrastructure serves the public.

Infrastructure is the frame on which a town, city, and nation grows. It is the foundation for commerce and trade. It is the services used by populations as they grow and prosper. It also has one unique characteristic not otherwise found in any business environment: It is demanded and expected by the public and by businesses. In this regard, the infrastructure of any community is a unique and highly individualized frame by which government renders service to its citizens. Despite this great importance to the community, its citizens, and its businesses, it is seldom seen or directly observed.

1.2 WHAT MAKES AN INFRASTRUCTURE UNIQUE?

The growth of cities and towns all over the world is driven by the available technology to fuel economies both local and national. Businesses grow and the populations form around them. The businesses change forms and owners. The areas become established as business centers, residential sections, or commercial zones. They rise and fall based on general market needs and demands. Some of these areas are as old as the country itself. Some are relatively new and only a few years old. The one commonality that binds the community and whose effectiveness generally is a major contributor to its ultimate success or its final demise is the infrastructure that serves it.

The infrastructure controlled by a local government or its entity is the single most distinctive part of any community. This is because its survival and continuance is predicated on the existing particular economic climate or culture as much as the intended need and available technology. In private business, intended need and available technology usually generate the market. Market forces generally have less impact on infrastructure maintenance than the political climate or culture of an area do, although these factors all play direct parts. If building, living, or growing occurs in that area, the infrastructure must be there. If an area is dying, it must continue to be supported until the death knell is sounded. Even then, the infrastructure may have to be supported if it connects two prospering areas. Roads, utility services, and other public property thread within a given community at all levels. This interconnection is what sets the needs of the infrastructure apart from those of private or general business.

It also requires unique approaches to appease each of the groups and interested parties that believes it knows best how to manage the infrastructure.

1.3 THE BEGINNINGS OF THE ROLE OF GOVERNMENT IN MANAGEMENT OF UTILITIES

Most nations genuinely try to provide the best care for citizens, regardless of their governmental structure. So if you accept that the goal of government is to provide services for its people, the infrastructure is a legislative responsibility.

The ancient Egyptians, Greeks, and Romans carefully documented their efforts to provide these services to their citizens. Many of the Roman infrastructures are still in existence. Aqueducts and sewers throughout Europe, the baths in Bath, England, and the massive ruins in Italy all attest to the Roman government's attempt to provide services to the people.

As the planet became more populated and diverse from the seventeenth through the nineteenth centuries, more services were required. The various kingdoms through Europe and Asia and the colonies of the Americas required that the shifting populations be provided with services and fixtures to assist in their growth. Governments funded infrastructure programs as a way of bringing settlement and stability to these new communities. In all cases, water and wastewater services were required in order to stop the spread of disease among these new, highly concentrated populations.

The twentieth century brought with it the most phenomenal technological marvels that the world had ever known. It also brought a nightmare of expansion and waste problems. Governments all over the world were faced with technological challenges that were never thought of before—nuclear waste from power plants, toxic solids found in water and in wastewater plants, residuals from telecommunications systems. The United Kingdom and Europe chose to turn many of these services over to private enterprise and divested the government of these issues and concerns. The feeling was to let those with the technological know-how deal with them and the governments would pay for the service. The British government, to take one example, sold power, telecommunications, and other strictly governmental services outright to private entities in the mid- to late 1980s. The United States still wrestles with these issues.

So the question now is: Is a government capable of running infrastructure assets in a modern world in the best interest of its citizens? Should it be involved at all, or has the evolution of these services crossed that boundary?

1.4 ISSUES WITH INFRASTRUCTURE MANAGEMENT BY GOVERNMENT

The consensus is that business, commercial, and public interests want the infrastructure run like a well-oiled business enterprise. This is a very desirable concept. Herein the problems begin. The main problem is that governments exist to govern and legislate, not run businesses. Therefore, the only mechanisms available to governments to perform these services are the same as those used to run courts, councils, police departments, and permitting. But is the procurement officer who purchases stationery qualified to purchase pumps, motors, and engineering services? It is probably unfair to put a person in that position. But, in the beginning, this is the only available mechanism for procuring the goods and services needed to operate infrastructure systems.

Citing this type of shortcoming, legislatures developed agencies that can and do specialize in the specific areas of the infrastructure that require unique technical and managerial attention. Specialists within these groups develop an intuitive knowledge of their area because they are in the position of seeing all the needs—or at least are poised to see all the needs—of their area of responsibility. They are provided with budgets and given goals and objectives based on the needs of public health and safety, business growth and commercial climates, political interests, available technology, and the perpetuation of the agency. These are difficult constraints, as a shortcoming in one may be enough to cause the agency to fail.

However, the need to generate a profit is not paramount, as it would be in a private enterprise. Public protection must supersede any profit motivation. Costs are always a consideration, but the protection of the public always—and rightly—outweighs the profit motive. Budgets are sacrificed for public protection, as are relationships among political, business, and commercial parties. The balance is difficult to maintain and requires high-level negotiating skills. The agency must constantly try to show that it simultaneously provides full service, seeing to all the individual needs of the tax- or ratepayers, and is cost-effective.

1.5 INFRASTRUCTURE AND THE WATER UTILITY

Let us now turn our attention to the management of water utilities. Water utilization, for the initial part of this discussion, can be divided into water source, filtration, conveyance, use, collection, treatment, and return. In many instances, the return point and the initial source are not the same, as when water comes from a well, is used, and then is returned to a stream. The focus here is on the latter part of the utilization chain: collection, process, and return.

A collection system is a matrix delivery system that gets wastewater from the user to a treatment plant for processing. These matrices are varied and unique to any given community and developed as the community did. Treatment is generally performed by a publicly owned treatment works (POTW)[1] or a sewage treatment works (STW)[2] that processes wastewater for return to a receiving stream or river. Now many sewage plants are becoming private entities. In the United Kingdom, they are all private. In France and Sweden, most are privately run but still influenced by the government. In the United States, most still fall under the jurisdiction of some government entity or utility formed to service many neighboring communities.[3]

The world community now provides many models for public, private, and joint partnership ownership and operations in a variety of sizes and styles of infrastructure systems, including treatment plants. There is no

1. This is the term used by the U.S. Environmental Protection Agency (EPA).
2. This is the term used in the United Kingdom.
3. At the time of this writing, approximately 3% of all wastewater processing facilities in the United States are under some form of private management. This percentage varies slightly by source.

Figure 1-1 Sewer infrastructures can be large or small. Many times, the general public does not recognize them as sewers until there is an issue. Here, the immense size of an outfall in general public view is compared to the size of the technician.

stable or universal model to which an agency can turn to find a baseline pattern. Each agency, in an ongoing evolutionary process, must blend the many solid models that currently exist and develop the best fit for a particular community's requirements.

1.6 THE NEED FOR MORE INFORMATION

It has become extremely complicated to manage a water utility of any sort. Water is not only a necessity of life but also is used in every conceivable facet of a community, town, city, and nation. As the technology of these entities evolves, so does the complexity of the residuals they generate. Many of these residuals are within some water waste stream, known as *wastewater.* Therefore, the ability to deal with wastewater demands intimate knowledge of its structure and points of origin.

Intelligent decisions are based on the proper gathering, presentation, and interpretation of facts, hundreds of millions of dollars have been invested over the last 20 years in automation, computers, smart transmitters, data acquisition devices, and software systems and subsystems to make the wastewater process more effective and cost-efficient. The next 20 years promise to see an exponential increase in treatment systems and the money to purchase them.[4] Discharge permitting systems have been implemented by regulatory agencies to maintain water quality in receiving streams. Environmental interest groups continually review the regulations and have them revised where necessary.

Despite advances in wastewater processes and automation, many difficulties still exist in the overall handling of wastewater. It is not that a mistake was made in the decision to focus on plant automation; that was a logical choice, as a plant is confined to a readily manageable area and can be dealt with locally. Rather, the problem is that the evolution of automation and advanced treatment systems has not yet caught up to all areas of the wastewater processing system. While a tremendous amount of effort has been expended on the automation, operation, and maintenance of sewage treatment works, because of the priorities government assigned in its distribution of funding during the 1970s and 1980s, virtually none has been devoted to the conveyance system to get the wastewater to the plant. The delivery matrix is the root of the process of wastewater processing. In general, wastewater plant automation has performed very well. The next step in development is called *sewer infrastructure management.*

The interest in sewer management systems appears to result from regulatory interest in point pollution sources and infiltration and inflow issues. These issues are a concern of regulators in most countries and, in the 1970s, drew the attention of the Environment Protection Agency (EPA) in the United States and the National Rivers Authority in the United Kingdom. Both took steps to attempt to correct the situations. The diversity of the efforts had to do with available technology, size differences between the nations, and the age of the infrastructure. The general belief was that a properly managed sewer system would deliver a better-quality sewage in a manner easier for the treatment works to process. It simultaneously addressed the need to protect both public health and safety and the property of the citizens who pay the water bills.

4. Any number of sources provide very high estimates for the cost of wastewater improvements. The estimates vary so much that it makes little sense to quote them. All sources agree, however, that the figure is very high.

The collection system, known generically as *the sewers,* is part of the return of water from use. Other parts are the treatment plants and the return (river, bay, or ocean). What happens to one part tends to affect the others; this is why it is difficult to separate them.

A considerable body of work already exists on two of the three parts: treatment and return. It now makes good sense to discuss the collectors and the concepts called *sewer infrastructure management.*

1.7 THE DEVELOPMENT OF SEWER INFRASTRUCTURE MANAGEMENT CONCEPTS

The management of entities that do not have fixed physical boundaries tends to be difficult for many reasons. Streets, utility services, and other public services that are spread out have special considerations that the best plant management practices generally do not address. The first and foremost of these reasons is strictly logistics. Considerable time may be lost moving from one problem site to another due to traffic, weather, and other unpredictable occurrences. These delays and setbacks cost money, directly or indirectly. Managers have wrestled with practical methods of minimizing the effects of these logistical delays but have achieved only minimal success.

Let us take a step back first, for the specific needs for modern management of a sewer system cannot be fully realized until its assets are adequately described. A sewer cannot be constructed, repaired, or moved without extensive excavation work. It is usually out of sight or at least not in view of the general public. Little aesthetic or other physical, tangible benefit is associated with the construction of new sewers other than their necessity. It is difficult to justify the need for a sewer program of any type to a layman, as the structure you are attempting to maintain is usually well buried and invisible. It is difficult to display the system other than by secondary means including maps, pictures, or similar methods; in fact, sewers are rarely seen, except when a problem occurs. This invisibility makes them distinct from bridges and roads, which are visible to the public, although the population as a whole may use the sewers more often.

A defining feature of a sewer infrastructure is that the entire basis for the operation of the treatment plant relies on what the collection matrix can deliver. In most instances, this connection is not noted as part of either a collector system or a treatment works plan. The need for constant dialog

Figure 1-2 Repairs being made to sewers hidden behind the wall of the watershed. This is a heavily used recreational area. The existence of the sewer infrastructure is virtually unknown to the public unless there is an accidental discharge. (Author photo.)

between plant personnel and the collector system team is therefore sometimes minimized. However, usually neither side really understands how much the matrix can deliver in relation to the plant's hydraulic capacity. The plant personnel know how much the facility can process in a given time frame. What they are not sure of is how much will be coming down the matrix and when. The plant can only be ready for the unexpected. The entire sewer system can be unpredictable during certain events and under certain conditions. The transfer of information regarding the collection matrix can be difficult because of the limited amount of field information available and the different information formats used by each part of the system. It is thus ideal to develop information based on current field conditions and make that information available not only to the collector systems personnel but to the plant people as well. This allows at least early warning of impending conditions within the system and increases lead time for reaction. This is the beginning of proactive infrastructure management for the sewer system.

It is difficult to perform any proactive maintenance without historical information. Most of this type of information is developed over the years from field complaints and determining potential trouble spots from maintenance records. Many municipalities have developed original methods of handling these situations given their administrative constraints. Still, more effective management tools are necessary to provide prompt, reliable field information to necessary personnel so that efficient and cost-effective solutions can be developed.

There are two key components to any infrastructure management system. The first is that people must make the proper decisions based on their experience. The second is to realize that the only purpose of the infrastructure management system is to support the efforts of the personnel involved in the decision making and not to replace intuitive or otherwise experience-based decision-making capacities. Proper sewer infrastructure management underpins all these efforts and provides for better service in the long run to the citizens of any given community.

Additional benefits to the operating agency in the use of any sewer management technique include capital planning, parts and supplies allocation, and the management of customer complaints. Proper management allows an optimal flow of information, a method of checks and balances, a tracking system to provide repairs based on complaints, and replacement of structures as a result. Short- and long-term projects can be developed based on need arising from customer input and not strictly from an engineering assessment.

1.8 THEORY AND PRACTICE OF INFRASTRUCTURE MANAGEMENT CONCEPTS

Remember that any governmental management systems approach is a three-part concept. One part is the support equipment, services, and software that make up the product. The second and most important is the human operator, who makes decisions based on the information generated. The third is the public's satisfaction that the money generated from rates is being well invested. The aim is to develop a system that satisfies the public and meets the needs of the utility.

A goal of any infrastructure management system is to provide as much decision-related information to as many people as possible involved with wastewater. In theory, this can be performed readily using instruments, soft-

Figure 1-3 The hidden infrastructure becomes quite visible after certain natural events. This sewer was pushed out of a hillside along a jogging path after a severe winter and rapid snow meltdown. Such unavoidable events bring the hidden infrastructure to the public notice. (Author photo.)

ware, and networks. In practicality, this is very difficult due to the human issues pertinent to sharing information, political environments, and areas of responsibility. Practically, there may be a limit to the amount of information that can be made available.

The installation of an infrastructure management system is geared to improving efficiencies based on factors that are usually out of the control of the operating agency. The effects of weather, traffic, political climate, regulatory mandates, and lawsuits are the most common sources of issues that the operators have to deal with. Weather is unpredictable; amounts of precipitation and the way it inhibits maintenance crews in the field cannot be accurately assessed prior to an event. Traffic can cause problems and delay the arrival of maintenance crews and equipment to a worksite. Political circumstances cause problems for the operating agency, as political hot spots tend to get more attention than less-known trouble sites. Regulatory mandates usually take priority, and sometimes their good intentions are not readily apparent. Lawsuits can hinder, change, or shape the methods used in work implementation, applications, locations, and task prioritization. The management system must therefore be poised to provide timely and accurate information to address these issues.

The first kind of assistance that the management system can provide is support to the maintenance crews. Theoretically, prompt, accurate information can protect them at the worksites, provide site-specific information concerning the problem, and target their efforts more efficiently. A properly installed and maintained electronic monitoring system provides current and historical location information that the maintenance forces can use not only to clear the problem as reported but also to determine its root. Short- and long-term trend analysis can produce a body of information that points to the cause and suggests corrections to a variety of irregularities in the collector matrix. By matching these trends with rainfall time and duration series, planning and other proactive and predictive procedures can be conceived and implemented to improve the efficiency of the collection matrix.

The practical side of any systems implementation for the sewers relies heavily on what information is available. However, workers may fear they will no longer be needed and may lose their jobs if large amounts of new information become available. This is simply not true. A system can tell where a problem is. It may even suggest potential remedies just by the nature of the data display. People still need to perform the work, however, and that does not change. It therefore makes practical sense to bring in the maintenance forces and explain to them that their jobs become more secure and necessary with the system's implementation. They will appear more efficient to the general public, as they will probably no longer randomly pop manholes to determine blockage locations. They can go right to the site and start setting up to work. They can take the time to really understand why the problem is occurring and bring potential solutions to their supervisors. Therefore, assessing and merging theory and practice means making workers aware of the infrastructure system. The maintenance forces are essential to the success of any base system. They need to be brought on board and made part of the team. Remember, the maintenance forces can make or break the system implementation, especially in its inception.

Another benefit is the ability to provide long-term historical information to engineering groups for planning, construction, compliance, and other significant or capital-intensive activities. In the past, engineering groups commonly embarked on a short-term study for their needs or relied on a desktop model. Neither approach can provide the longitudinal information required for accurate decision making. In theory, there is a central depository of information on flows, levels, water quality, and other parameters for each location of interest within the sewer infrastructure matrix. The depository is available to everyone who requires the information. Practically, however, data management within a government agency tends to be segregated. An operations division retains most of the information and the engineering group probably contracts with a service firm for the short-term results, which it usually keeps for itself. This approach tends to be self-defeating and very expensive in the long run. It also provides less than optimal results. Local-area network (LAN) technology helps to bridge this gap, along with inter-unit cooperation. By providing system access to the engineering, planning, or related group, the operations division can provide high-quality data in the format needed for the study while retaining possession of and responsibility for the database. It also encourages the engineering group to have all short-term data retrieved in a format that can be stored in the select database and limits the need for the battery of short-term studies each time issues arise. In a more practical sense, it provides greater access for everyone and a reduced cost to the ratepayers.

Often, in the treatment works, simulations and testing are required to determine current capabilities. Plant stress testing may be required as a method of determining plant capacities, whether for utility knowledge or for a compliance effort. Hydraulic modeling is another simulation effort used in planning and compliance efforts. The loading to the plant facility is, in fact, the sewer system. A history of its activities is important to determine any changing conditions that would affect the outcome of the tests and simulations.

Regulatory compliance is a two-edged sword. It helps the environment but generally hurts the pocketbook. This is because the basis for its comparison is its weighing of pollution loadings verses water quality. Also, compliance issues have costs, not least for the manpower needed to compile reports and documents. However, regulation is necessary and does provide some assurance that the watershed is being protected over the long term.

Many compliance programs are data-submission endeavors, and simply providing information with integrity is an act of good faith, reassuring the public that all that can be done for the good of the watershed is being done. Data are delicate in light of the judgments that can be rendered from its interpretation. The agency needs data to prove both the good and the bad. The good is how well the agency has done in protecting the environment, the efficiencies of its processes, and the implementation of its procedures. The bad is what needs to be done to make specific improvements, reassess processes, and reinvestigate in-place procedures. In some of the cases where specific improvements appear necessary, the information may correctly show that the technological limits of the plant or the collector system matrix and/or one of its processes have been reached and no more can be done. Regulators need to assess the information and determine if their requests are within both technological limitations and human abilities. This is best achieved with proper data presentations.

New directives from well-meaning regulation administrators must be carried out. Regulators are concerned with a better environment in general and the quality of the watershed specifically. Their directives and initiatives are meant to provide long-term relief to the environment from manmade pressures. In reality, many decisions by regulators are made without regard for actual conditions or the difficulty of the challenge. Many political and business reasons contribute to such decisions. Nevertheless, they must be taken into account, and accurate data is the way to provide regulators with concrete and definitive explanations of why a program should or should not be continued or enforced. In practicality, sewer infrastructure management systems probably reduce the cost of compliance programs as the ability to produce hard data tempers the more aggressive or cutting-edge programs for which there may be no apparent targets or results. Also useful is the ability to provide regulators with instant hard facts about actual system response, so they can form more realistic opinions of the system and the matrix they are regulating.

An area of modern concern is the legal issues generated by the proliferation of lawsuits against operating agencies. In the past, the need for legal defense in utility operation was rare, but today agency lawyers need reports supported by data—including rain, flow, level, and water quality information—for their use in defending the utility, agency, or authority from negligence claims. In many instances, a city is not liable for damage if a storm exceeds the design capacity of the sewer system. Therefore, if the system was designed for a 10-year storm event and strategically placed rain gauges documented a 25-year storm, the lawyers may successfully argue that there is no agency liability based on the design criteria. Practically, the accusing public generally wins something regardless of the amount of information provided. The more information provided to an attorney, however, the more easily the utility can be defended and frivolous claims dismissed. Further, questionable claims can be mitigated so the least possible exposure to the public treasury can be anticipated. Remember that costs of lawsuits, frivolous or not, all add to the rate structure and cost the ratepayers in the long run. The utility could probably justify the cost of a monitoring system based on the indirect cost savings from dismissal of frivolous claims. This is why here the theoretical and practical are so close and provide a measurable savings for the operating agency.

Modern management of sewer systems is much more involved than it was just ten years ago. The biggest reasons are regulatory compliance endeavors and rapid changes in technology. A more ominous reason is the constant budget reviews and restrictions implemented in the age of downsizing and outsourcing. Information to justify programs, personnel, physical resources, and other necessities for the day-to-day and long-term operations of the sewer infrastructure is critical to its longevity. It is easier for managers to develop and justify a budget or program if they have hard data in a format understood by the administration. The theory is that the data speaks for itself, and all the manager has to do is to package it and make an administrative presentation. However, political factors and issues such as environmental hot spots generally dictate the programs. In the real world, what managers can aim for is to leverage the most necessary programs with real, reliable, continuous data from the field while satisfying political interests.

1.9 MERGING THE FACTIONS FOR THE COMMON GOOD

The need to protect the environment and the need to ensure the longevity of the infrastructure are why we have to have modern infrastructure management systems. Regardless of political or social overtones, protection of the environment—in this case, the watersheds—is a common and most noble endeavor. Longevity of the infrastructure is usually a goal of the general public, although it is generally hidden from everyday view. However, the public is aware of its necessity and certainly does appropriate funds for its existence. But the public is usually interested only in results, not how they are achieved. Infrastructure managers *are* concerned with the methods used to achieve those results. Managers can combine these two needs, using environmental protection, the ultimate protection of the watershed, as a path for programs to maintain the infrastructure. The key to all of this is hard, accurate, reliable, and well-presented data. The method for obtaining data is infrastructure management.

The factions involved are spread throughout the entire utility structure. The plants require specific information to operate. The sewer infrastructure system is generally complaint driven, with limited preventive maintenance. System owners, managers, and operatives also get all the blame for personal property claims regardless of whether the plant closed the gates or not. The engineering group is required to develop long-term replacement plans for the infrastructure with limited insight on the system response. The various management teams are expected to develop standard operating procedures based on intuitive knowledge of system response, usually provided from the engineering group. Legal units defend the utility from claims and have limited access to anything but past cases and inquiry reports. The many factions' needs can all be met with reliable, centralized information.

Good infrastructure management means providing as much information as possible to as many of the various divisions within a utility—engineers, management teams, the legal unit—as possible. A good manager realizes that these factions all work as a system, protecting the environment and serving the public at large. By satisfying all their data needs, the potential for factional conflict decreases and the system operates with relative efficiency. Remember that infrastructure management involves both the people who operate the system and the tools they possess to make decisions. The higher the quality of both parts, the better the systems operate.

One of the many shortcomings common in the past was the inability to make data available to these divisions in realistic fashion. In fairness, this was the problem in most industries, not only wastewater. This has been overcome with computer networks and their ability to transfer extremely large and complex data files literally in milliseconds from one part of the system to another. The continued effort to unify software and file formats can help the various divisions see their common problems, issues, and goals. Therefore, the resolution of this one issue indirectly allows for the potential solution of many more.

Crossing the data bridge is a vital step in the development of an effective sewer management system. As the concepts of sewer infrastructure management systems mature, the collector systems will be able to provide the same breadth and width of data on their systems as plant people can with their ongoing automation and more amicable computer platforms.

Figure 1-4 Runoff from weather events can cause a variety of problems in the sewer system. In this case, timbers block the natural flow of the watershed. Such entry into the sewer infrastructure can require immediate action to protect public property. (Author photo.)

1.10 THE INITIAL STAGES OF SEWER INFRASTRUCTURE MANAGEMENT

The concepts presented in this text are relatively new. The aim is to provide an introduction to all the major areas that any operating agency, regulatory body, interest group, or other party may require to become acquainted with the inception, development, and implementation of modern sewer management techniques. Such introductions are always difficult, as there is little background on which to draw and a small base on which to begin construction.

Many times in this text, the development of a concept leads to the description of an implementation.[5] This appears to be the best method for

5. T. Day et al., *Permanent Large Scale CSO Monitoring Systems,* Sewers of the Future, Houston, Texas, September 1995. Many of the concepts used were developed as part of the investigation.

Figure 1-5 The importance of watershed issues for sewer management systems is evident here. This sensitive watershed area is also used for recreational purposes. Note the toilet paper hanging from the trees and bushes. Modern sewer management techniques assist in reducing such conditions. (Author photo.)

acquainting the new student with this area of study and actual methods of implementation. Such developed concepts provide baseline methodologies that can be readily implemented as a matter of procedure.

1.11 SUMMARY

Infrastructure management of the sewer using modern management tools and a central repository of the information can make the tasks of diverse groups easier and more cost-effective. This is not to say that people are going to be laid off because of automation. They certainly will not be; their skill sets will need to be either changed or redirected. It also means that the people involved will be working smarter, producing more, and possibly merging their goals for more homogeneous solutions. This is the evolutionary process of infrastructure management. It will allow the various parts of a utility to operate more like a single system serving the public. Hard, verifiable data become the key in system management.

CHAPTER

2

Basic Instrumentation and Fundamental Flow Measurement

2.1 AN OVERVIEW OF INSTRUMENTATION

It is generally understood that direct observation of hidden assets is, at best, quite difficult. This is due to the very nature of their work, which is to perform without being noticed by the population. Therefore, the only way to observe the conditions within an infrastructure is through some relational means—in this case, in situ instrumentation.

The goal of any instrument is to identify a direct, repeatable relationship between the condition that presents itself, though difficult to observe, and some relational operator readily seen by interested parties. This relationship is then presented or described in a manner that provides the appearance of the actual condition and makes it available as necessary. It does not matter if the relationship is linear or nonlinear; it only matters that the condition is repeatable—that is, for a given set of conditions and responses, a relationship can be generated.

Infrastructure management tools were originally adapted from process equipment and marine measuring devices. As interest grew and the cost of sewage processing increased, devices were made more specific to the infrastructure. The reason for this is simple. In plant manufacturing, regardless of how hazardous the chemical or compound being produced is, the substance in the process loop is always known in one form or another. In marine work, seawater has basic characteristics, even with pollution and other influences. Again, the content of seawater is known, more or less. This is not the case with sewage. As the compounds move from place to place, from one part of the system to another, they are combined with residuals and debris from multiple locations—runoff, manufacturing, or residential. It is difficult to determine the relative concentrations within any given waste stream. Therefore, the parameters used to manufacture the protective cases of process manufacturing or marine equipment may or may not be applicable or reliable in an infrastructure device. A longer discussion of this comes later, but it is correct to note here that the environment of the sewers is unique and hazardous, requiring special protection and other considerations that are not directly applicable to its historically adapted devices.

Today, many reliable and accurate products are designed and constructed specifically for sewer work.[1] The basic equipment for understanding the infrastructure's hydraulics has come a long way in a short period. The water quality equipment is following quickly but is based on market demand, which is still being assessed, and the potential outcome is still being determined. However, as the need for water quality data grows, so will that portion of the market as well. Cost and return-on-investment issues are still in the evaluation stage.

This section addresses the basic equipment with which one can provide a relational view of the hydraulics within an infrastructure. The potential assets and applicability of each item are presented so readers can make intelligent and logical decisions for their own particular systems.

2.2 INTRODUCTION TO INFRASTRUCTURE INSTRUMENTATION

Infrastructure instrumentation is designed to provide relational representations of the system's hydraulic conditions at the moment of measurement. This is performed by noting changes in the measurement relative to some predetermined reference.[2] A reference is a point of stability against which changes are measured. It is the baseline for accuracy in the instru-

1. T. Day, *Modern and Advanced Flow Measurement: Equipment for Wastewater Process and Applications,* Chesapeake Water Environment Association, Ocean City, Maryland, July 1996.

2. "Sensing Instrumentation," in WEF MOP #21, *Instrumentation in Wastewater Treatment Facilities,* 1993, ch. 4.

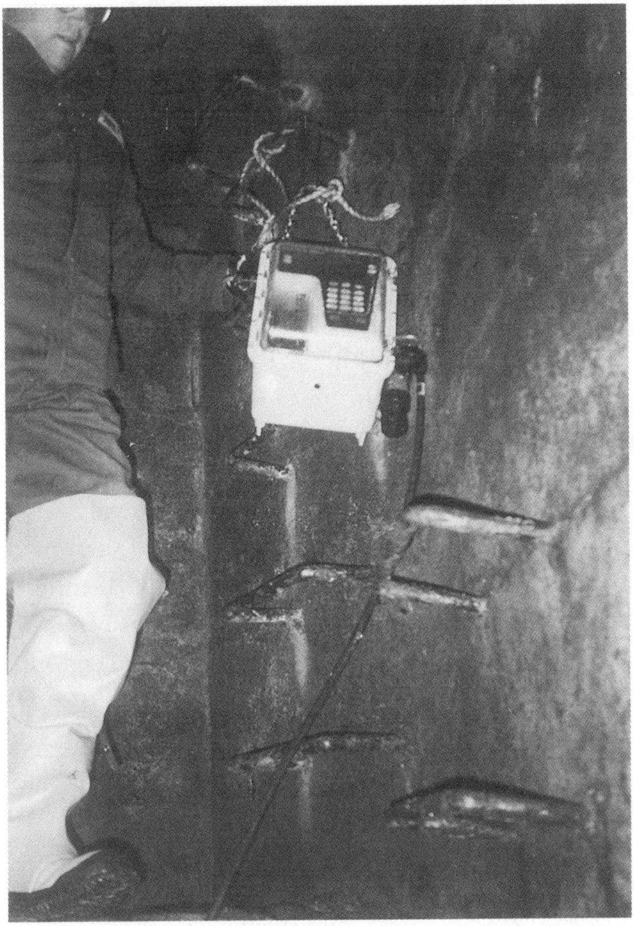

Figure 2-1 An excellent variety of portable and readily installable instruments is available for providing short- and long-term information about the sewer infrastructure. Their ease of deployment allows them to be placed in areas where issues and concerns require information and remedial action. (Courtesy of American Sigma. Used with permission.)

ment. If the reference drifts or otherwise changes without proper compensation, the relational representation has no meaning. Therefore, infrastructure instrumentation must accurately reflect changes in reference.

The sewer system is considerably more hazardous than many other environments for which instrumentation exists, because with sewers so many factors are unknown. The collector systems are randomly used as disposals and outlets for the undesirables of a society. They are well bounded and confined, and dilution has little effect on the total concentration being delivered to the sewage mixture. On the way to the sewage treatment works, these undesirables combine at random, and chemical reactions take place within the sewers. What comes out at the influent head is therefore different from what was originally sent in. This is the initial consideration in choosing acceptable sewage monitoring equipment. It also makes the job of selecting suitable coatings and elements difficult. Many manufacturers of sewer instruments do not reveal the materials they use, as advanced or unique materials give them strategic advantages in marketing and longevity. Sometimes, depending on the applicability of the device, size and shape—along with material construction—provide strategic advantages in applications and therefore, the market itself. Shape is also a big issue with fouling and debris gathering. It is important to check manufacturer and supplier warranties, applications, and approvals to be certain that the proprietary coatings and materials will withstand the environment.

An important consideration in any sewage flow is the amount of solids, in suspended, colloidal, settlable, and floatable forms, that is transported with the sewage as a function of total deliverable material to the treatment works. Solids clog and destroy instrumentation placed in the system. The transport function is random; there are, however, particulars for each area. Some areas are heavy in debris and others are light. Some carry large solids as floatables, others are strictly suspended. Some storm systems are generally clear after the first flush, as all they carry is runoff from a particular drainage basin. Solids concentration and total loadings are an important characteristic of sewage. Any shape and profile exposed to the direct flow in the sewer has the potential for collecting debris and, therefore, the potential for either inaccurate readings or no readings at all. Thus it is also important to be sure that an instrument's size and shape are appropriate for the application.

Another consideration is that if the content of the mixture is changing constantly on its way to the treatment works, what is its condition at any point along the way? This is the second quandary. Treatment plants can determine, within reasonable limits, concentrations and compositions within broad boundaries at the head of the plant. If they could not, biological treatment would not be possible. So there are some reasonable broad boundaries that can be put to the sewage content particular to a given area, eliminating the potential for illegal dumpers and other unknowns. The conditions and the reactions on the way to the plant cause much of the debate in material selection; they make the job of choosing suitable coatings and elements difficult. Fortunately, most manufacturers understand this and develop their products accordingly.

A modern instrument can be divided into three parts: sensor, signal transmission medium, and the electronics package. The sensor (also called an element or probe) sits in or near the condition to be monitored, and detects changes relative to the reference condition imposed. The signal transmission medium carries the raw signal generated by the condition from the sensor to the electronics package and is, in most cases, cabling of either twisted-pair copper wire or fiber; however, radio and other methods are being researched and tested to get the signal from the sender outward. The electronics package contains the reference condition, evaluates and records changes from the reference as seen by the sensor, and provides some kind of output or indication of change. Everything added to the device beyond these descriptions is still pertinent to one of these three parts. A sensor can be labeled an element, probe, or other term indicating an extremity of the actual device. It sits in or near the condition of interest, depending on the application. A datalogger or data communications function is not necessarily part of the base instrument but an addition. It is a secondary consideration and should be separate from the measurement technology.

The output—the processed reference change as computed by the electronics package—can be a signal, indicator, data point in a computer file, or any combination thereof. A signal output is a variance of a range of an electrical signal, scaled so it references the conditions of application. Signals are usually 4–20 mA or 1–5 V in range. The initial reading is usually 4 mA or 1 V, generally calibrated as zero. This is called a *live zero.* These are the references of choice, as in physics there tends to be a spike or jump in electrical potential when electrical references are moved from true zero to the first value of interest. This is known as *bounce* and is damped in a lot of electronics but can readily show up in oscilloscope readings. Live zeroes have no bounce associated with them, which is why they have become accepted. Until the age of microcomputers, these signals were generally considered linear, or forced linear for displays or recording, even if variances occurred. This was simply due to the limits of the electronics of the day. In this age of microcomputers, signals need only be repeatable, as equations can be written into electronic memories to convert repeatable signals into usable values.

The linearized output is applied to a circuitry that geneates a reference voltage or current scaled to the desired conditions. The electronic reference either drives a display, is applied to a datalogging device, or is otherwise recorded. Because the conditions generated are linear at the output, any number of suitable devices can provide reasonable and accurate records or indications of the conditions being referenced.

Another method for determining references is the pulse count. In days gone by, this was the mainstay for determining flows, rainfall, and other exact quantities or totals where required. A potential is imposed over a contact closure. When changing conditions reach some threshold, the contact closure opens and a pulse is recorded. Each pulse is scaled to some relative quantity, say 1,000 gallons or 0.01 inches of rain. The number of pulses is recorded and summed over a time period. The total number of pulses is multiplied by the scalar. This provides a quantity for the user. Pulse counts went by the wayside with the advent of digital counters and microelectronics. However, they are still used in rain gauges and other simple instrument applications. For this reason, they are discussed here.

A faster-growing and more modern method is to use data transmissions protocols, which provide a direct, converted reading to a computer or datalogger equipped to read the protocol. The recommended standards of RS-232, RS-488, fieldbus, or other specific Internet protocols provide high-speed data transfer far superior to any electronic signal or pulse reference. Discussion of each of the recommended standards is beyond the scope of this text, but note that the protocols must match throughout the entire data transmission. One must be either consistent in one's choice of protocol or willing to purchase protocol converters within the data transmission area of preference.

Direct computer interfacing has many advantages. Not only can the data generated by the instrument package be transmitted directly, without further scaling or referencing, but a variety of additional support information can be transmitted with the data packets, such as location, time and date, and specific conditions within the instrument. Disadvantages include the need for additional circuits, which increases the size of the electronics package, strains on battery life on non-AC-powered locations, and increased demands on software. Variances in the standards may also cause irregularities in the total interfacing, including irregular readings or, possibly, failure of the device. Computer interfacing may not be required in temporary or similar applications, so the needs of the user should be properly evaluated to determine if an interface is proper for the installation.

It is important to remember that sewer instrumentation is no different from any other type of measuring or control device. It simply has unique characteristics, like every industry. Sensors are located in, at, or near the condition (usually a flow) and changes from a reference indicate measurement. Electronics packages convert these changes into information usable by the operators. However, they must be made capable of residing in the hazardous and very changeable environment of the sewer.

The key to good instrumentation is to find high-quality components and systems that are matched, rather than attempting to integrate devices not necessarily made to work together. Using equipment that is matched is different from matching equipment. Matched pieces of equipment have exactly the same internal protocols, data sets, and qualities, which allows them to seamlessly integrate as a single cohesive system. Matching equipment is integrating devices the best way possible considering their features, in order to meet a need. Such equipment may or may not be of high quality; therefore it may become advisable to remove devices of lower quality and begin matching with more applicable devices.

The quality of the matched equipment and the matching device is always of concern and should be carefully examined. Quality is generally apparent or verifiable through references and specific field testing. Matched systems for sewer infrastructure management are difficult to obtain, as most are currently custom manufactured. This is due to the nature of the individual collector systems into which they must be introduced. However, there are some matched systems available that meet the hazardous location requirements of sewer environments. Check the references well and be sure that the manufacturer produces all of the devices you require. If not, be certain that the aftermarket products you require will directly blend in seamlessly, or as seamlessly as possible, with the core system of your choice.[3]

Matched systems cannot be constructed from dissimilar parts by the definition introduced here. Many of today's components are constructed to be parts of such systems and are made for integration into custom, com-

3. "Testing and Quality Assurances" in WEF MOP #21, *Instrumentation in Wastewater Treatment Facilities,* 1993, 33.

plex, or otherwise nonstandard matching systems. This is because, as noted above, each infrastructure system is unique and has changed forms many times over the years the community has existed. Devices with electrical signals or computer interfaces are examples of systems that blend into larger systems. If there is a need to construct a large or custom system, matching these features will make the integration less complex.

These guidelines give a baseline and parameters to introduce actual hardware into the discussion. It is important to give baselines so one is not just throwing hardware into a sewer, wondering why the results do not appear to correlate. The definitions and parameters introduced above provide one way of establishing the platform from which the sewer infrastructure system can be built.

The following sections describe the types of instruments necessary to develop a modern sewer infrastructure system. Although no recommendation is made as to brands, it is advisable to obtain as many components from a single manufacturer as possible. Mismatched pieces of equipment produce less than desirable results and will result in added expense. Even the integration of high-quality components into a cohesive system will yield less favorable results than the deployment of same-source equipment. Unless you are intimately familiar with all third-party instruments and the software to drive them, it pays to obtain as much from a single source as possible. Experience shows that deployment and integration is smoother with same-make instruments and software, as they are matched. Another argument for single-source integrated systems is whether outsourcing will be used. If the decision is made to retain maintenance of the system in-house, a uniform parts supply is generally easier to obtain through the municipal procurement channels. If the choice is to outsource, bid development is made easier by the ability to develop a better baseline for equipment and parts so proper pricing can be determined.

2.3 INFRASTRUCTURE INSTRUMENTATION: THE HARDWARE AND SOFTWARE

This part of the discussion covers the measurement of flows through measuring the associated level and velocities, where applicable, and rain measurement in its relation to wet-weather hydraulics. Not all technologies and variations of products can be presented because of the rapid changes in this area; however, the sections below discuss most popular kinds of in-

Figure 2-2 Flow being measured using level and velocity. Cable fastens to banding, which is installed to mount sensors. (Courtesy of American Sigma. Used with permission.)

strumentation and will allow readers to make reasonable judgments concerning the best instrumentation for their own application. It is important to understand completely all products currently offered so that the best possible fit can be obtained between a particular infrastructure and the instruments used in it.[4]

Each device has individual routines and functions, all in software form, that drive the instruments. Internal software is intended to transpose the relational reference into a usable value. Internal software should not be modified by the user under any circumstances without the manufacturer's instructions. Usually, the internal software contains algorithms developed by the manufacturer that are unique to the measurement process. They may be patented or otherwise covered by trade laws. External software brings the transposed relational relationships forward for use. This

4. T. Day, *An Assessment of State-of-the-Art Flow Measurement Techniques for Sanitary Sewers and Overflows,* (paper presented at the U.S. Environmental Protection Agency Sanitary Sewer Conference, Washington, DC, April 1995).

software comes in many forms; specifics can be requested from the manufacturer.

The lack of software standards causes great variances in the type and specification of coding used in computerized applications. The Institute of Electrical and Electronic Engineers wrestles with potential standards that can provide uniform analysis and practical consistency for such software, but the task is formidable and not likely to be resolved soon.[5]

It is therefore necessary for the user to determine the type of data required and the form in which it should be made available to the user. This information should be given to vendors so the best match, based on current product offerings and cost of specific modifications, can be made on behalf of the customer.

2.4 THE MEASUREMENT OF FLOW AND ITS RELATION TO PRECIPITATION

Flow in a sewer infrastructure, in its steady state, can be relatively predictable in its behavior and characteristics. The introduction of precipitation to the equation brings the flow out of steady state and into a condition that, generally, is unpredictable. Understanding and introducing enough parameters to generate predictability, to some degree, of the flow and flow rate produces valuable information concerning the delivery of sewage under adverse conditions to the sewage treatment works. A corollary interest is in the effects on the watershed and surrounding properties when all the flow does not get to the works, for whatever reason.[6]

The condition where all the flow does not get to the treatment plant can be controlled or uncontrolled. In the case of a combined sewer overflow, the discharge is generally regulated by the overflow weir. In the case of a sanitary sewer overflow, the known point of discharge produces the same effect. A damaged sewer line tends to be uncontrolled because the extent of the damage is unpredictable and usually unknown at the time of discharge. Damaged lines have dry-weather effects as well.

The conditions of first flush after a prolonged dry spell are also of issue.

5. The IEEE subcommittee on software specification currently reviews such matters.

6. *Urban Pollution Management,* Foundation for Water Research, Buckinghamshire, United Kingdom. One of the many reasons the UPM was developed and published was to provide a uniform, stable assessment instrument with which to assess this relationship.

The buildup of solids in the storm system, washdown from street catchments, and introduction of nonpoint pollution from runoff all have significant and measurable effects on the sewer infrastructure as a whole. It is possible to generate relationships based on cause and effect and repeatable characteristics by measuring flow and relating it to characteristic precipitation.

This introduction discusses how flow is measured and how rain is described in relation to flow. This shorthand account provides the background necessary to understanding the methods utilized by the equipment described to generate the relational representation and the reference conditions imposed. A full discussion is included later; however, a basic understanding of the parameters is essential at this point in order to comprehend the functions of the equipment.

Flow is measured in two parts: level and velocity. Level, the depth of water within a geometrically fixed conduit, is the easiest to gauge. Level is a representation of the volume of water at that point, or of the geometric cross section of the conduit, whichever is applicable to the device. Because the conduit is fixed, the volume or cross section can be determined using geometric relationships of the depth to the wetted perimeter of the conduit. The wetted perimeter is that portion of the conduit where the water is in contact with the walls during its sojourn.

Velocity is the rate at which the water is moving through the geometrically fixed conduit. Layers of velocities within a profile or cross section usually contribute to the total movement of the water.

The factors generated from the level calculations and the contribution made by the velocity profile produce a flow or flow rate. The flow or flow rate, depending on the scalars used, in a sewer system is affected by both dry-weather and wet-weather conditions, sets of hydraulic responses that occur in the infrastructure and affect the flow or flow rate when rain is or is not falling. Wet-weather conditions are a set of hydraulic responses that occur in the infrastructure and affect the flow or flow rate as a result of the time and duration of a rainfall.

2.5 RAINFALL MEASUREMENT: RAIN GAUGES

A rain gauge is a simple device. It is the only device discussed in this section that is not, either entirely or in part, directly exposed to the sewer environment. It provides the basic information about how much rain fell in a given area, which directly affects the flow or flow rate in the sewers. It is the trigger for knowing the difference between wet- and dry-weather events. It is also probably the most neglected device for infrastructure management and the device that requires the least amount of service.

There are two popular devices for measuring rainfall: the tipping bucket and the electronic weight. Both devices measure rainfall in 0.01-inch increments. This is the required increment for most permit and compliance reporting today. Both devices mount easily and readily. The devices may be equipped with a heater assembly as an option for melting snowfall, in areas where applicable, to determine a water equivalent of snow precipitation. Both types can be purchased with their own instrument package or can be easily integrated into a third-party system, being considered matching components. Devices purchased independently generally have their instrument package manufactured by a third party; therefore, integration into another system is fairly straightforward.

The electronic weight is the more accurate device, but it requires more maintenance than the tipping bucket. The principle is that the instrument collects the rainwater and weighs it, converting weight into inches. This type of gauge is recommended only for scientific studies where accuracy is extremely critical for success of the endeavor, as the device requires constant calibration and is temperature sensitive. The high frequency of maintenance makes this device unsuitable for infrastructure management systems.

The tipping bucket system is the more popular and has a long history of success in sewer infrastructure management. The device consists of two 1-cm^3 buckets mounted on a seesaw-type rocker. Each bucketful indicates 0.01 inches of precipitation. When the bucket is full, it tips the rocker, tripping a switch that breaks a voltage or current loop. The electronics package counts the number of times and the interval at which the loop is broken, thus providing information about the intensity and duration of a rain event can be made. Tipping buckets are durable and require little attention in the field. They are not affected by temperature changes, as the electronic weight system is. They also quickly integrate into other instrument packages that provide the electrical loop and the means of counting the breaks in the loop. Most manufacturers provide a calibration bottle with a metered amount of water that is timed for delivery into the buckets. It is recommended that the calibration procedure of the manufacturer be used to verify the accuracy of the tipping bucket. Utilizing the manufacturer's techniques and procedures also provides standards for uniformity and justification to regulatory agencies.

The tipping bucket system has one minor drawback. During very intense storms, water is delivered to the buckets faster than they can receive them and tip. Therefore, some is lost by splashing. However, this problem is mostly limited to events such as 50- and 100-year storms—that is, storm events of a given intensity and duration; statistical and prediction techniques can be used in conjunction with long-term data to calculate the intervals at which such storms can be expected to occur in a coverage area. Each manufacturer of tipping buckets provides a correction curve for such occurrences, and this curve can usually be coded into the electronics package fairly quickly; therefore, this drawback is minor. It is wise, however, to verify the manufacturer's correction curve or other method used for corrections when the storm intensity exceeds the capacity of the device.

Rain gauges should be displayed in strategic places along interceptors, sensitive watershed areas, or other regions of interest. Because rain is a fleeting quantity, it is desirable to install as many rain gauges as possible to best determine coverage for an area. Their coverage areas should also be weighted against one another. Weighted areas are known as *Theissan areas* and prove valuable in determining coverage over the long term. A more accurate description of the fleeting characteristics of rainfall follows, and the uses of Theissan area distribution appear in a subsequent section.

Rain gauges should be installed at fairly uniform elevations and be located away from overhangs and surrounding buildings. Air currents and trees also affect the quantities of rain recorded. The National Weather Service offers parameters for rain gauge installation, as does every engineering school in the United States and Canada. Most are more theoretical than practical due to property right-of-ways and other local situations. Also, most of the recommendations are not fixed in a formal methodology that is repeatable for similar conditions; they are strictly recommendations. Therefore, reasonable judgment must be employed in placing gauges.

Maintenance involves the placement of debris screens within the funnel portion of the rain gauge. Other than periodic calibrations, the only maintenance required is to keep the device free of debris. Birds' nests, leaves in the autumn, and other natural occurrences interfere with optimal operation. Due to the length of time required to calibrate the instrument, it is recommended that the devices be removed and replaced with a spare unit while the original is brought back to a controlled area where the manufacturers' procedures can be utilized without undue cost.

The rain gauge is the simplest of devices and the most overlooked, but it is the cornerstone instrument in infrastructure management. Tipping buckets tend to be easier to maintain and more tolerant of the environment for the long term. Use the manufacturer's suggested calibration procedures and correction curves where applicable. Generate Theissan areas and weigh them against one another to provide for optimal short- and long-term analysis.

2.6 LEVEL MEASUREMENT: PRESSURE

The first parameter of concern within the infrastructure itself is the level or height of water within a conveyance. From the engineering point of view, this eventually translates into the hydraulic grade line (HGL), used in computer modeling and evaluation studies. From the operations point of view, *level* refers to the various heights that the water attains as a response to the system conditions and loadings. Direct observation of the levels within the system is generally impossible. Therefore, a relational means is employed to determine levels within the system.

The pressure or weight of water in an environment is an excellent means for determining levels within the infrastructure. Many of the older texts refer to this as *head* or *head pressure.* The pressure, as read directly, can be converted into a depth based on a relational construct known as the *water column.* The depth of the water in the theoretical water column converts to head (or water depth) using the scientific conversion value of 27.68 in/lb H_2O. There are any number of relational methods to weigh the water in this theoretical water column. No matter what the method, in order to utilize the 27.68 in/lb conversion, the pressure of the water column, or the amount of water directly above the sensor, must be determined. In this case, the theoretical water column is isolated directly above the sensor. This approach dictates that the installation must be free of obstruction directly above the device and for a reasonable distance around it. The fluid property of water allows for this relational measurement and provides for quite accurate levels.

Pressure sensors evolved from a combination of microelectronics and advanced applications of electromagnetics. The simplest pressure gauge and the device used in many integral applications is the strain gauge, which

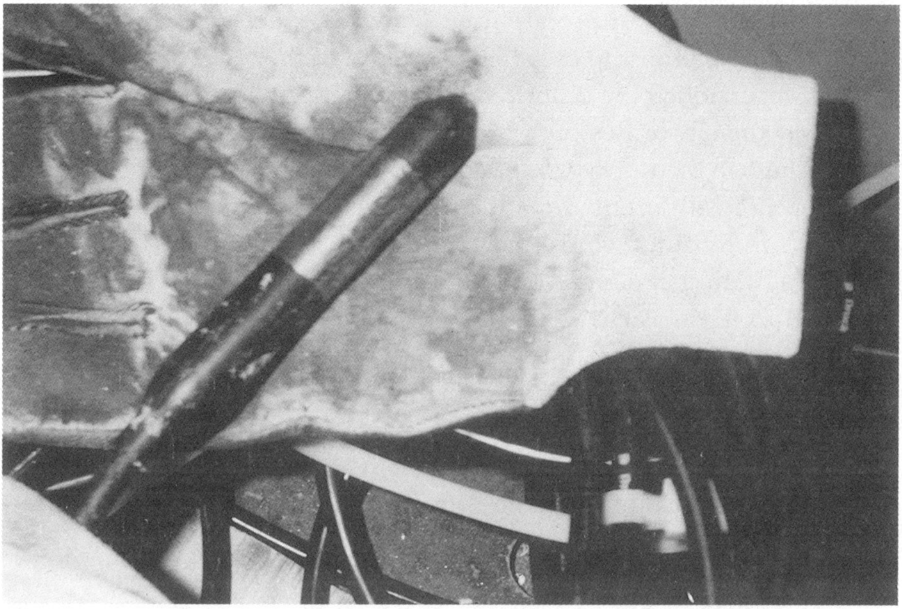

Figure 2-3 A simple level sensor. Note size compared to work glove. (Author photo.)

measures the strain exerted by water pressure on a diaphragm, which moves a metal core through an electromagnetic coil. The diaphragm keeps the coil isolated from the water and dry. The movement of the core through the coil of wire produces an electrical potential, referenced to and as a result of the pressure above the sensor. On one side of the diaphragm is the pressure of the water column as exerted; the second or equalizing pressure is the ambient air pressure that keeps the coil dry and provides the baseline or reference pressure. This is still the most widely used device due to its accuracy and relatively low cost of manufacture, along with its lengthy life span. Advanced manufacturing techniques allow these devices to be made in many sizes and shapes for use in a variety of applications.

The reference for developing the relational operator that results in the depth of water measurement is the difference between the ambient air pressure and the pressure exerted by the strain of the water column against the diaphragm. The difference between these two pressures is the force of the water column and is measured in lbs/in or kg/cm, depending on the preference of units. This standard and accepted conversion is found in most laboratory manuals. The relational operator is converted directly to feet or meters (or the units of preference). The atmospheric reference of zero becomes the relationship by which the depth of the water is referenced.

It is worth noting the distinct difference in the atmospheric pressure between, say the city of Boston, which is at sea level, and the city of Denver, which is high in the Rocky Mountains. The atmospheric reference pressure is considerably different and affects the accuracy of the device if not accounted for. In other words, the zero point at different elevations varies considerably. In past designs (and still common today), the sensor element mounted in the sewer is vented to the ambient atmosphere of the electronics package, usually at the electronics package's elevation. Therefore, the zero is always relational to the height of the vent tube and the location of the instrument package. It is not necessary in these designs to fix the zero point, as the zero point is always the ambient air at the end of the vent tube. This is done to provide a base pressure relationship, described as the relational relationship, to the air at the surface of the water, which is considered zero depth. Blockages in the vent tube produce erroneous results as the reference becomes skewed and, eventually, lost. Also, moisture in the vent tube may affect the electronics package as well as skew the zero reference. This is why more reputable manufacturers provide desiccant cavities to absorb the moisture from condensation and other sources.

Currently, patented techniques are used by some manufacturers to maintain a seal on the vent tube. These make the issue of the relational relationship of a zero point moot. However, this is cutting-edge technology and still in the introductory phase. It is necessary to follow the manufacturer's recommendations and to have the specific tools and equipment required to set the relational reference point.

Pressure transducers are also sensitive to vibration, collisions, shock, heavy flows, and other hazards. This tends to make them unsuitable for direct insertion into fast-moving water for head references. Stilling wells (pieces of perforated pipe in which the sensor is mounted) provide shielding from these hazards while still allowing it to measure the rise and fall of water levels. Note that not only is the diaphragm sensitive to the sewer perils but there also may be considerable electronics located in the sensor element itself, depending on manufacturer and model. A condition that is nearly impossible to prevent is damage from ground hits due to lightning.

Another parameter that has a tremendous effect on the relational reading is the temperature of the water. The subject of water temperature is discussed in great detail in later chapters, as its effects have great impact on the infrastructure as a whole. It is enough to note here that any selection of pressure device should compensate for temperature. There are as many temperature compensation methods as there are algorithms for relational generation. Each manufacturer provides its methodology with the product. The method of compensation should be reviewed to ascertain its applicability to a given infrastructure.

Early devices required that the raw reference signal be converted to depth using a scalar pressure conversion function and send a reference signal to a chart recorder or similar device. The pattern from the chart recorder then had to be converted or read against a scale to interpret the data. Modern electronics in the sensors now provide a reference potential (usually 1–5 V or 4–20 mA) as the pressure conversion is performed in the sensor element. The strain gauge approach is still employed as a basis for pressure level measurement. It needs to be noted, however, that most modern pressure transducers are sophisticated in their applications of the electromagnetic response and that many patents have been issued on design, features, and style of these elements. The innovation of each manufacturer has led to the high reliability and relatively low cost of this form of measurement.

Interesting variations on pressure gauges appear from time to time. One of these is the vibrating wire, which operates on the same principle as a strain gauge except that the diaphragm is connected to a wire reed rather than a magnetic coil. A series of frequencies are generated, spread, and sent over the wire reed from a signal generator to the vibrating wire connected to the diaphragm. A magnetic pickup determines one specific frequency that the vibrating wire is resonating as a result of the pressure of the water against the diaphragm and the tension of the wire. The theory is that the vibrating wire vibrates at only one frequency, which is the frequency equivalent to the pressure being exerted against it. Consider the physics experiment involving two violins, one intoned on one of its notes, an open string, and the other sounds that note of the open string on its own. This is the relationship. The frequency range is not linear but repeatable; therefore, relational equations can be coded to determine the relation of the nonlinear function against predictable results. The repeatable frequency is then convertible to pressure. The relative pressure is directly convertible to depth of water.

These devices are accurate and sensitive. Some are used as barometers to measure changes in atmospheric pressure with great accuracy. Because of their durability, they can be bored directly under wells to determine water levels. Their accuracy and durability suits them to difficult and extreme applications where other technologies fail. In order for these devices to work properly, they must be site calibrated and in thermal equilibrium, as the excitation frequency changes with temperature.

Pressure tape, for some reason, never became popular as a measuring element in spite of its many good qualities. The electromagnetic coils are embedded in a silicone mold that resembles a ruler or yardstick. The strip is placed at or near the invert of a conduit and the aggregate pressure of the water displaces the coil some distance from the magnetic pole. The reference generated is proportional to depth. These devices have yet to grow in popularity in spite of their ease of installation, relatively low maintenance, and improved manufacturing techniques.

In general, the strain gauge maintains the highest share of the pressure transducer market. Its straightforward design and the advanced techniques in manufacturing allow many variations of these devices to be applicable to many products. Also, its accuracy in a small scale or size makes it appropriate for shipping to manufacturers as a subassembly. This is desirable, in many cases, as the instrument manufacturer can concentrate on developing new technology rather than copying old or adapting outdated subassemblies to state-of-the-art devices.

A popular application of this technology is to interface a pressure transducer directly with a flowmeter body. Level (which is translated into a cross-sectional area) is a required parameter for flow using the continuity equation $Q = VA$, where Q is the flow rate, V is the average velocity of the water, and A is the cross-sectional area as determined geometrically from the level. Therefore, it benefits meter manufacturers to accommodate a level device in their equipment. A pressure transducer is usually the easiest and least expensive type to employ. It also allows a meter manufacturer to match ranges the instrument expects to measure optimally with the range of the pressure transducer.

Calibration of a standard strain-gauge type transducer is generally performed electronically using the slope/offset formula $y = Mx + B$. The manufacturer provides slope figures and physical offsets are determined at time of installation. Slope figures are repeatable variations from the straight line. Physical offset is the distance off the bottom or other preferred measuring point at which a sensor has to sit for a good reading. One of the calibration procedures includes taking readings at zero pressure and subtracting the actual zero reading from the reference reading, which is generally offset from the bottom or invert of a pipe. This is an acceptable method of performing this calibration. Variations in electronics, initial calibration from factory conditions, and resistance in the cable contribute to the differences between calculated zero and actual zero. It is also suggested that the zero reading be taken at the point of installation, as ambient atmospheric pressures also affect the reference reading.

A pressure transducer's accuracy is greatly compromised when installed in areas of turbulent or heavy flow. They are sensitive devices, even with stilling well shielding. Pressure transducers are also susceptible to fouling and clogging. Where the potential for rapid flow and/or solid debris capable of colliding with the device exists, another type of device may be a better choice. Bear in mind, though, that anything introduced to a stream of wastewater becomes a target, and proper placement and installation can minimize problems.

The pressure transducer is one of the least expensive devices for measuring head rise in the sewer system. It can be used individually with a secondary data acquisition device or in combination with a flowmeter or sampler. The pressure elements have a relatively low cost of construction and need little maintenance. This makes them suitable for many applications and allows a substantial number of replacement units to be kept on hand. Although sensitive to the environments to which they are exposed and operate, protection for reliable operation is readily accomplished with proper shielding at installation and site choice.

2.7 LEVEL MEASUREMENT: ULTRASONIC

A class of sensors that do not necessarily need to reside in the subject flow are ultrasonic sensors. These have evolved dramatically over the last 20 years into successful and adaptable devices. They are now produced by several reputable manufacturers and can be supplied in a variety of styles. Because of their nonintrusive nature, they are popular where conditions are hazardous or difficult or where pressure sensor insertion is not feasible. Ultrasonics versatile and have dropped in price steadily as their popularity grows and as electronics become more advanced. This class of sensor can also be found in applications other than level; however, head rise is the subject of this section.[7]

7. T. Day, *Overview of Advances in Flow and Level Measurement Technology,* American Water Resource Association, Ft. Lauderdale, FL, September 1996.

Figure 2-4 Ultrasonic head in service measuring level in a flume. The deadband is the length of the cone. (Author photo.)

The concepts that make up ultrasonic level measurement are relatively straightforward. A sound pulse in the ultrasonic range is generated by the excitation of a crystal. Although the frequency is too high to be detected by the human ear, at times a faint click can be detected when the crystal is energized. The range of frequencies used in level measurement generally is between 100 KHz and 1 MHz, though other frequencies are also used. The generated pulse is directed toward the surface of the water and is reflected back to the source. The echo off the surface is received by the ultrasonic head, which is listening for the echo, and level is derived based on the time of sound propagation and temperature of the ambient air.[8]

The calculation for depth of water is actually derived from the free airspace above the liquid, based on the propagation time of the signal. Therefore, actual depth of liquid (l) is determined from the free space (fs) in the conduit plus the offset (of) of the sensor head minus the diameter (d) of the pipe $(l = d - fs - of)$. This follows if one considers that the distance from

8. "Sensing Instrumentation: Sonic, Ultrasonic, and Microwave Sensors," In WEF MOP #21, *Instrumentation in Wastewater Treatment Facilities,* 1993, ch. 4.

the sensor head to the bottom of the channel is always fixed, and that one portion is measured directly, which is the free airspace. The depth of water, which is the consideration of the measurement, is accurately deduced from the difference of the free space and the total internal distance from the bottom of the pipe to the sensor head.

Sound is changed by variations in ambient temperature regardless of the medium of signal propagation. As a result, ultrasonic sensors are adversely affected by temperature changes, which makes them unstable without proper compensation. Most devices have temperature compensating devices, usually a crystal or thermocouple, that correct the ultrasonic travel time equation for variances in ambient temperature. This makes it important to understand the temperature characteristics of the particular infrastructure. In systems where the temperature is relatively consistent and does not experience great seasonal differences, temperature compensation is not as great an issue. In systems where the variation is large, temperature is critical to proper measurement.

A method for temperature compensation that does not use direct temperature measurement is the reference distance approach. The reference distance refers to a known, fixed distance over which a generated sound pulse is reflected, its travel time recorded, and its distance scaled. The electronics knows to listen for a particular pulse related to that specific fixed distance and makes necessary adjustments to the actual level measurement to determine a correct reading. The major advantage to the reference distance approach is there is no compensation device to wear out or corrode. The major disadvantage is that this type of ultrasonic assembly may require a large deadband, which makes it unsuitable for the confined spaces of most sewer monitoring applications. The major advantage to the temperature compensator is that it requires a minimal offset distance in the sensor head, allowing for a greater reading span. Each type of level device has its place in an installation. Deadband is always a problem with ultrasonics, so be certain that the manufacturer has properly compensated for deadband in its product.

A variation of the target approach that is generating interest involves truncating return pulses that fall outside a specific range. By doing this, the reference target can be made to fall in a very small area, which is practical for sewer or other infrastructure applications. This band-pass filtering approach allows for only the most representative signal echoes to return. As long as the reference point remains in the ambient and representative air, a proper relational reference can be made for the temperature variations as they affect the sound pulse.

Ultrasonic devices are generally more expensive than pressure transducers and are susceptible to signal dispersion due to foam and acoustic particulars of the installation. They appear, however, to have a much longer life expectancy than the pressure device and require less maintenance. This is generally due to the fact that they do not have direct contact with the water.

The ultrasonic sensors generally do not require stilling wells in the installation; in fact, they would not operate properly in them. This is due to a phenomenon called *cone expansion.* Ultrasonic pulses expand over distance. The angles of expansion range from 7 to 15 degrees depending on the manufacturer and the cut of the crystal.

Manufacturers that supply infrastructure equipment have addressed the signal deadband and cone expansion issues in innovative ways. One approach, which completely eliminates cone expansion, is to form the crystal around the interior peripheral of the sensor. The crystal then fires into the center of the sensor element and is reflected down. This produces a narrow cone, almost a straight beam, which is applicable to many installations of restricted size. Another solution is an innovative approach in which the signal is traversed horizontally, then reflected downward. This not only keeps the signal confined but also puts the deadband in a region that has no effect on the measuring distance. Therefore, the physical offset of the sensor element itself is the only nonmeasurable space. This means that, if the physical offset is included in the sensor measurement, a surcharged pipe can be successfully measured up to its crown. This is extremely valuable and can prove useful in applications where a pipe surcharges often and exact measurements are required.

Some manufacturers install a second crystal in the event that the first fails. This is a characteristic called *redundant sensing.* The electronics can shift to the reserve set, therefore saving or postponing the replacement process. Anything over two crystals in an element appears to be overly redundant, although multicrystal element heads are manufactured and sold. Electronics and software algorithms can rapidly fire the crystals, receive the echoes in short order, and discriminate between the timing of the pulses. Some designs use the spare crystal as a periodic check on the active crystal and its measurements. The manufacturing processes and precision crystal

Figure 2-5 Two ingenious styles of ultrasonic heads. The particular style and type avails itself to numerous applications. The four-crystal device fires many pulses to eliminate errors. The single element focuses the sound pulse to eliminate cone expansion. (Author photo.)

production in place today can provide for extended life of the sensor head and a better product than was available as little as five years ago.

Since little to no contact with the water occurs, little maintenance is required. Ultrasonics are, however, subject to moisture buildup on the sensor head. Most manufacturers coat their heads to prevent this problem. Alignment is also important, as the devices need to be as perpendicular to the water's surface as possible. Many providers install a level right on the sensor head to ensure that the angle is true to the liquid surface. The installation should be performed at the most tranquil portion of the flow, if possible, as surface variations, depending on their intensity, affect the reading anywhere from minimally to completely.

In some applications, ultrasonics simply do not work because there is too much surface debris or foaming. Surface debris results in uneven surface area and has the potential to disrupt the signal in the same manner as surface turbulence. (Floatables, in fact, cause surface turbulence.)

Areas of high sludge, pumping, or other agitation, which are all susceptible to foaming. Ultrasonic devices are technically limited in these applications and, if foaming of any kind is indicated, alternative solutions should be considered. All devices have technical limitations. It is important to know them as a potential site is examined.

Low-power battery and portable units are now available and perform especially well when system entry is a problem and utility service is a premium. If the dimensions of the measurement point are known with reasonable accuracy, and if the surface water level is known, a portable device can be installed without making a confined-space entry. Although not desirable for long-term applications, these devices lend themselves well for short-term HGL determination, the study of hydraulic response or infiltration, and inflow studies where the need is immediate but temporary.

Ultrasonic level devices offer reliable, low-cost data with minimal maintenance. Most units are AC-powered, but portable battery units are available that offer a wide range of options. Although limited in application by foam and heavy surface debris, they perform in any generic application where level is required. Proper placement and understanding of the devices' technical limits ensure long-term and reliable service.

2.8 LEVEL MEASUREMENT: BUBBLERS

A class of older and very reliable sensors comprises pressure devices known as *bubblers*. They fell from the mainstream with the advances in pressure transducers and ultrasonic level devices but have returned with technical advancements that extend their range of application. Also, the ability to blend bubblers with modern electronics has provided for measurements that otherwise would require more sophisticated methods.

Formerly bubbler systems were costly, but recent advances in electronics have brought their price down remarkably in the last several years, making them competitive with the other devices discussed. They are workhorses and can be used in conditions of heavy silt, foam, sand, and other areas of debris that other technologies find difficult.

To measure depth, an air bubble is blown down a tube in water using a compressor, pump, or other suitable device. The force needed to make the air bubble come out of the tube is the reference pressure, read by a differential pressure cell. The reference pressure is measured in psi, or pounds per square inch. The relational pressure is the force of the water trying to keep the air back. The difference is the relational operator that is converted to depth of water. The differential cell provides an output suitable for transcribing the data generated. It determines this pressure difference (usually in pounds) and a conversion is made to depth of water (using the value of 27.68 in/psi).

A bubbler works well in difficult environments. In forcing air out a tube, it keeps itself free of debris. For this reason, it is well suited to sandy, silt-laden, foamy, and debris-laden areas. Most devices have self-cleaning or purge features that blow higher volumes of air through the tube in order to keep the opening clear.[9]

Most modern bubbler devices are self-contained, provide standard signal outputs, and use plastic tubing that easily installs within a conduit. Some manufacturers even package functional bubbler units as part of their flowmeter at a very reasonable price, which provides for a seamless application of this technology for flow measurement. Bubblers (other than their electronics package) are not usually susceptible to temperature and have a proven track record for reliability. Bubblers continue to prove themselves the device of choice for difficult or unusual measurement applications.

2.9 THE MEASUREMENT OF THE QUANTITY OF FLOW

The measurement of flow is critical for many functions within the infrastructure. From design to process, the amount of flow determines the size, shape, construction material, and location of the infrastructure. Even the quality and content of flow is not as basic to the sizing and direction of a sewer system as the amount the sewer will be carrying.

The measurement of open-channel flow (wastewater flows are generally open-channel) is a difficult science that until recently generated little attention. As the price of water and its handling increased, so did the attention paid to flow measurement. Water processing is now a utility price like electricity and gas. The utility price requires that the volumes being processed be accurately measured.

A serious complication to this issue is the non-uniform shape of the pipes that convey the wastewater. Where the pipes' cross section is circular or rectangular, the issues around measurement are mitigated somewhat. But the infrastructures of older cities and towns feature varied shapes, orig-

9. "Sensing Instrumentation: Bubblers," in WEF MOP #21, *Instrumentation in Wastewater Treatment Facilities,* 1993, ch. 4.

inally designed to assist in grit deposition, cleaning, and materials transport. These unique shapes require an extra level of effort to be sure the sensors make an accurate reading. Also, the combination of shapes as they join within the infrastructure system can produce disturbances that may render even the best devices invalid.

2.10 FLOW MEASUREMENT: PRIMARY ELEMENTS

Our discussion of primary flow elements is limited to an overview because the introduction of modern electronics and technology show that these devices have inherent errors of up to 25%.[10] They are, however, integral to many collection systems, and firsthand knowledge is required to properly utilize and maintain them.

A primary element is a fixed device that is designed to restrict the velocity of water or wastewater to a calculated known and to develop a flow based on the rise and fall of the level within the geometry of the device. The unique geometric shapes provide for the velocity restrictions and usually are named for their inventor. For many years, it was accepted this was the only reasonable way to measure open-channel flow. Advances in modern electronics and development of modern flow measuring equipment make the need for primary elements less necessary, if at all. It is important to know and understand their characteristics, as they did provide long-term solutions and many of these devices are still in place and in service.

The first device to measure flow came from a spillway. This device is known as a *weir*. By knowing the level in or over the weir (depending on the design), an instantaneous flow rate can be observed. Weirs are static devices made from wood, plastic, fiberglass, steel, or other durable material able to withstand the flow of water over it. They are classified as *V*-notch, square, triangular, trapezoidal, or any other geometric shape for which flow rates have been calculated based on its characteristics. Any standard book on open-channel flow provides conversion charts and detailed information on the flow measured by a weir of ordinary shape. The first remote application of instruments on a weir consisted of a float over the notch and a chart recorder drawing a pattern with a pen and arm, which required interpre-

10. This is a variable number based on the testing center. Most errors in primary elements are a result of improper installation rather than a fault of the design itself. Most flumes are, in fact, very accurate and cleverly designed devices.

tation. They were regularly used in storm relief and stream/creek flow calculations.

A number of types of velocity-restricting flumes and nozzles is available today. All of these devices develop a flow based on head rise and restriction of velocity from the geometric shape of the primary element. Kennison, Palmer-Boles, Trapezoidal, Penn/Badger, and other well-known manufacturers and models are still actively used in open-channel/sewage flow measurement. These devices have standard equations based on their noted characteristics—which are, again, available in any civil engineering manual. As modern instrumentation became popular, remote level sensors were added to the flumes and nozzles, and the float was removed. Many of these devices are still manufactured and in service. A number were developed to address a specific application; others were a result of laboratory or university research. Some also were specific to instrument manufacturers, which developed them to provide a matched system for generating flow information for their own instrumentation designs. Each geometry generated a unique performance curve based on the interior diameter of the element, where level correlated to flow.

It is important to remember that the major function of primary elements is the restriction of velocity of all the layers within a flow distribution to a set of known parameters. The unique shapes and designs seen in the multitude of these devices represent ways of accomplishing this goal. It is a difficult task, and the intuitive and creative approaches of inventors shows their true understanding of the unique properties associated with open-channel hydraulics. Some devices, however, are more popular than others, for reasons as simple as tradition or as complex as functionality.

The most popular primary device is the Parshall flume. James Parshall developed a device for measuring open-channel flow based on a differential pressure cell. His original presentation was called the partially filled differential pressure cell. The unique characteristic is the rollover effect of the device, which causes a slight backwash of water to flush the flume free of debris and keep it clean. This flume is discussed at length here because of its vast general acceptance and applications within the industry.

These devices are usually poured-in-place concrete or formed fiberglass, although cast iron was used for many years. A calculated measuring point resides in the influent side of the flume at approximately two-thirds the distance from the throat, where the level is measured and correlated to a flow rate. This point accepts an ultrasonic level sensor readily; pressure trans-

Figure 2-6 A Parshall flume retrofitted with an ultrasonic head. Primary elements, known also as nozzles or flumes, were the only reasonable method of determining open-channel flow prior to the age of modern electronics. (Author photo.)

ducers in stilling wells and measuring tapes have been installed as well. The size of the flume is noted by the measurement of the narrowest portion of the throat. As noted above, the reverse action of the waste stream generates a backwash that assists in keeping solids in suspension, thereby reducing clogs and blocks that would produce false levels and, thereby, false flows. The Parshall flume retains its popularity and maintains a commanding presence in wastewater.

Parshall flumes are subject to an installation difficulty that, once it occurs, is almost impossible to correct and leads to inherent inaccuracy. This is the condition of submergence. The flume is required to be mounted level, with at least 10 diameters upstream straight run and 5 diameters downstream straight run for accuracy. This means that a 12-inch Parshall flume requires 120 inches (10 feet) straight run upstream and 60 inches (5 feet) level downstream. The first shifting of the earth, blockage in the lines, or other distortion of this system renders the nozzle inaccurate. The forces of nature alone are enough to cause the level to be lost. Improper installations are another major problem, as most contractors do not realize the critical nature of maintaining the level of the system for accuracy.

A correction factor can be implemented with the Parshall flume to compensate for the inherent submergence as a result of flawed installation or ground shift. Placing two level sensors to determine the submergence factor and working that back into the flow calculation provides some relief for the situation. The submergence reference must be verified regularly so the correction can be maintained. This is difficult but quite effective if proper regular maintenance and verification procedures are implemented.

All primary element measuring devices appear to have inherent measurement issues. To do the devices justice, they were developed in light of necessity and with the technological know-how of the day, whey they were state of the art. For many years, they were as accurate as possible, given current research and development. Advances in direct velocity measurement devices are now bringing the shortcomings of these devices to light. It is important to know why devices such as Doppler and transit time are superceding the longtime mainstays.

Installation of a primary nozzle or flume assumes several parameters for successful velocity restrictions. The first is a significant upstream and downstream straight run of pipe, up to 20 pipe diameters (10 and 5 in the case of the Parshall flume described above) in both directions, depending on the nozzle. The straight run allows for laminar flow profiles and minimum perturbation of the velocity layering. The second parameter is total gravity flow in most nozzles. Any pumped flow is unexpected and not accounted for in the flow calculations. Gravity flow in general sewer systems is about 3 ft/s. This is sufficient to keep the solids in suspension as far as the treatment works. Pumped flow can be as high as 15 ft/s, which generally falls outside the calculated parameters as outlined by the nozzle designers.

The third parameter is a perfect installation at the recommended level or pitch with no ground movements. This is extremely difficult to obtain. The fourth is periodic maintenance to keep the device is free of debris so as not to perturb the flow profile or artificially inflate the level. The Parshall flume is one of the few designed to minimize internal buildup of debris. The upstream and downstream straight runs must also be regularly inspected and kept free of debris, which can perturb velocity profiles and produce irregularities.

The fifth parameter is infrastructure modifications. Installation of pump and lift stations on the lines is commonplace as areas develop. As noted above, the discharge velocity of a pump station at times may exceed 15 ft/s, which, given that design gravity velocity is generally around 3 ft/s, will create difficult conditions, depending on the size and location of the pump site. Contractor installations from new developments being added to the system may or may not affect the profiles, depending on where the connection pipes are added. It is therefore desirable to ascertain and be constantly aware of modifications and changes to the system to avoid these pitfalls.

Primary element measuring devices today are generally outfitted with electronics to ensure their accuracy. Level devices installed in the location for head rise can detect and record even minor changes and send the reading to electronic storage. A variety of velocity devices have been retrofitted to these flumes in order to better understand the hydraulic characteristics that they generate. The marriage of primary elements and modern electronics provides comfort in two regards. First, there is a mindset that still believes that primary elements are the only proper device for measuring wastewater. Second, the retrofitted elements provide assurances that modern technology is being employed for maximum accuracy.

In spite of their known difficulties, primary elements are widely accepted and utilized in the measurement of water flows. They are as much a part of any infrastructure as the piping. It is probably better to understand them and maintain them than simply discount their value and bypass them. Their usefulness in new installations has yet to be determined in the light of modern electronics and sensor technology.

2.11 FLOW MEASUREMENT: PROPELLER METERS

The propeller meter is a velocity measurement device that uses a calibrated propeller, turned by the forward motion of the water, to generate the parameters for flow. Propeller meters are still used in laboratory settings as a basis for establishing velocity rates of liquids in testing and certification. Their earliest application, unrelated to infrastructure management, was to the speed of ships over water. Although magnetic devices eventually replaced them, propeller meters served as excellent velocity measurement tools for many years.

The device is inserted directly into the water flow. Layers of passing water turn the propeller blades. The speed at which the propeller blades turn is related to the velocity of the passing water flow. Because the device takes up a considerable surface area, it accommodates many layers of water passing over it. Generally, the speed of the propeller is directly calibrated to the speed of the passing water. This measurement is the simplest and the most direct.

An issue with propeller meters in infrastructure management is the amount of debris that wastewater carries. Any intrusion into the flow profile creates an obstruction in the waste stream. This inhibits the accuracy of the device and makes periodic maintenance particularly difficult. It is advisable not to use these devices as permanent fixtures in sewer monitoring due to the high degree of required periodic maintenance.

Propeller meters are, however, quite suitable for streams, creeks, and other clear bodies of water, and in cases where the infrastructure carries clear water from one part of the system to another. They are also applicable as a velocity probe for spot measurements required by the system operators.

Technological innovations are making the propeller meter less desirable. However, economics and need always play the deciding roles when it comes to choices of equipment. Propeller meters are certainly worth considering based on the economy of the situation at hand and the conditions of the application.

2.12 FLOW MEASUREMENT: POINT VELOCITY

A cross section of flow shows any number of independent points of reference for velocity identification. Flow is described in layers; with point velocity measurement, it is possible to choose a single point as the basis for a calculation of flow.

An evolving device in flow measurement technology is the electromagnetic point velocity meter. The original device was developed in the early 1970s to measure ocean currents; adapted later for sewer monitoring applications, it was for many years the mainstream instrument for sewage flow applications. Although it is still widely used, other technologies are becoming prevalent and competing with the electromagnetic device.

Point reference velocity involves measuring velocity at a known point in a profile, working that known point into an assumed distribution of velocities within a cross section, determining a velocity average, and calculating a flow based on the continuity equation $Q = VA$, as discussed above. This technique tends to have the highest success rate on pipes of circular or rectangular cross section that are relatively free of debris and have tranquil flows within them. The main reason for the success of these sewer shapes is the assumed uniform distribution of the velocity profile within the flow volume. If the velocity profile is known with reasonable accuracy, only one or a few points need be known to calculate the flow or flow rate.

The point velocity is generally determined through a small, circular electromagnetic field generated just above the sensor head. Based on Faraday's law of electromagnetic induction, a potential is generated with the passing of a conductor through this field. The conductor, as defined by the application, is a single velocity point or several reference velocity points that are represented in the velocity profile and generate the potential by their passage through the magnetic field. The required parameter, after this, is the level, so the geometric shape for a cross-sectional area can be determined. Level is generally determined by the use of a pressure transducer, but any level device can be employed based on the accommodations and device modifications of the manufacturer. These representative points, worked back into a known flow profile, can be calculated into a flow or flow rate.

Measured profiles of the entire velocity distribution within the conduit at given intervals appear to be essential for proper installation. Profiling of a velocity distribution is recommended in the first-time installation of any type of device. It appears to be a necessity with the electromagnetic point meter, as the shape and distribution of the profile directly affect the accuracy of the result. An average velocity rate is then generated from the combination of the assumed profile, which was based on direct spot measurements, and the measured point velocity generated from the electromagnetic field. The setup software must realign the assumed distribution profile to accommodate any changes in the general distribution that affect the average velocity generated. With little perturbation of the flow profile and equal distribution of the velocities, the electromagnetic device may provide reasonable flow measurements.

Debris is an issue with these instruments, as the electromagnetic field is maintained at a weak level to prolong battery life. As battery technology increases or AC power is more generally employed, the fouling of the sensor in this technology will become less of an issue. Respecting manufacturers' installation procedures and following their periodic maintenance recommendations can minimize debris gathering.

The boundary condition of the measurement area is directly above the sensor, subject to the change in level of the sewer and the placement of the probe in relation to the water's changing surface, and is fixed depending on the strength of the magnetic field. The boundary condition for the cross-sectional area follows the changing level of the water.

Electromagnetic point velocity meters come in several configurations. Two common ones are the handheld velocity device and the in-place meter. The in-place meter can, in fact, be used as a handheld device. This may prove a bit cumbersome, but it is still a viable option when profiling is necessary and access to the handheld device is just not possible.

There is still a large number of installed point velocity devices, given their long exposure on the market. As technology advances and improvements are made to the device, the appeal to the marketplace is expected to continue, especially in creek and stream measurement, where conditions are relatively stable.

2.13 FLOW MEASUREMENT: TRANSIT TIME

A unique application of ultrasonic sound technology is the transit time flowmeter. Transit time technology was first developed by the Westinghouse Corporation in the early 1960s to measure discharge from hydroelectric stations. In the middle 1970s, the Environmental Protection Agency (EPA) funded a program to apply this technology to sewage flow measurement.[11] There was equal interest in the European arena; several products were developed and marketed through England and Western Europe. Today, the transit time flowmeter is a viable mechanism for measuring open-channel flows, especially in very large conduits and other difficult flow applications.

There are several major United States and European manufacturers of this technology for open-channel flow measurement, and each manufacturer has a particular style and implementation that makes its product distinct from its competitors. However, the basic concept is consistent among all the manufacturers. This diversity in the product offerings means that most applications in any given infrastructure are amenable to transit time measuring devices.

In theory, a sound pulse of known frequency, generally, in the ultrasonic range, is angularly transversed through a flow of liquid time of travel or flight is recorded, and another pulse of like frequency is generated and returned to the origin. The time it takes for the pulses to make the journey

11. U.S. Environmental Protection Agency, EPA-600/2–76–243, Project No. 11024 FVQ.

is called *transit time.* The propagation of sound through stagnant water is fixed and identified in any physics text. Therefore, the movement of the water either speeds up or slows down the sound pulse. This relative change in the propagation speed of the sound pulse being affected by the movement of the water is relational not only to the velocity of the liquid but also to the direction in which the water is flowing. A direct reading of the velocity values for the entire travel path of the sound pulse is affected by every incremental value through each of the measured layers it transverses. The effect of the total velocity profile of the area of traversal produces an average velocity reading. Because the boundary condition of the sensors is fixed, the major boundary consideration is the water's surface.

Two manufacturers' installation requirements are presented here to acquaint the reader with transit time technology. One of the installation requirements is to use a single set of acoustic couplers mounted at 20% of the known depth of the conduit. This is known as *chordal velocity measurement.* The other requirement provides for predetermined multiple signal paths based on pipe diameter and capacity.[12] This is known as *multipath measurement.* In very small pipes (24 inches and under), it appears there is little difference in the two design criteria under ideal conditions. This is due to the physical size and mounting requirements of the ultrasonic transducers. Each approach has specific advantages and disadvantages, and their application is entirely site-specific.

Chordal measurement is the integration of the velocities of all the layers in one chord transversing the flow volume and making that value representative of the entire flow profile. The chord is at a fixed position within the pipe, so its mounting characteristics are relatively predictable. This fixed position is determined by measuring the usable depth of the pipe, locating the point that represents 20% of pipe diameter, and mounting a pair of ultrasonic transducers in that position. A deadband of 10% is employed to eliminate measurement bounce. The direct velocity measurement begins at 30% of total depth. Any measurement under 30% is determined using the Manning or other related flow estimation equation. Above 30%, the representative velocity is used in the flow calculation.

The multipath solution is an expansion of the chordal approach. In the multipath installation, the usable depth of the pipe is determined. Then a

12. This material is derived from product literature and used as illustration. The manufacturer of each device should be contacted for updated installation requirements.

certain number of ultrasonic couplers (the number varies by application and manufacturer) is used to establish stages for velocity measurements. Up- and downstream considerations, pipe shape and condition, and hydraulic profiles are all considerations in the number of stages necessary. The more stages installed, the more accurate the profile. The paths are then integrated across the profile to determine a representative velocity for that volume, which is entered into the flow equation. Measurements below the minimum path are estimated using any select flow estimation equation.

There are advantages and disadvantages to each of these designs. Cost is always a consideration and should be weighed against the specific conditions of the field installation. Both devices require AC power for the standard product, which can be a problem in some installation cases. Battery and solar devices have been successfully employed in remote and otherwise difficult utility hookups. However, the battery and solar technologies tend

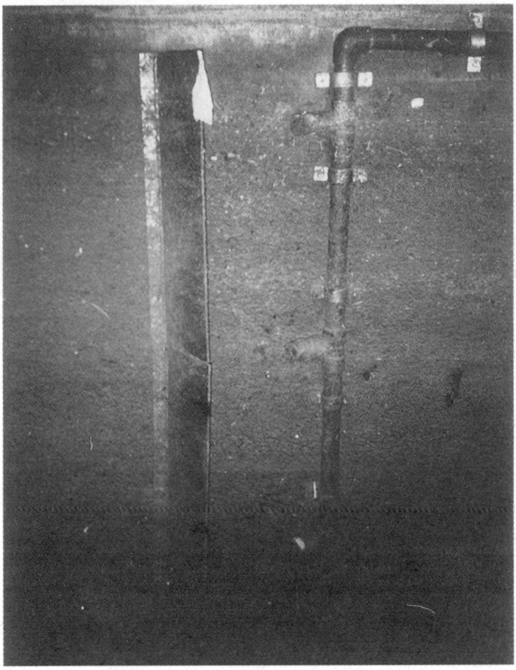

Figure 2-7. Transit time device used in permanent installation. This device, even though coated with debris, still provides accuracies within manufacturers' tolerances. The metal shield keeps solids from building up on the transit time heads. (Author photo.)

Figure 2-8 A unique style of transit time technology. A Doppler sensor is used to verify the transit time signal in the same instrument. (Author photo.)

to be the weakest link in this or any other proven solution. The price of solar cells and battery maintenance should be carefully considered against AC utility delivery.

Multipath technology appears to accurately resolve skewed, turbulent, or variable flow patterns and backwashes within very large conduits by virtue of the number of paths. It also appears to provide a better representation of the average velocity by representing a greater cross section of the flow than a single measurement path. Any increase in the number of paths also increases the aggregate cost of the unit, including the up-front engineering. Chordal devices appear to be reliable in predictable or otherwise reasonably uniform hydraulic conditions and can provide a consistent installation program across the infrastructure by virtue of their specifications. This significantly reduces the up-front engineering costs and provides stability in the installation process.

Flow is calculated, in each case, by determining the cross-sectional area and a representative velocity across a volume of passing water. The cross-sectional area is determined from a level device detecting the head rise within the conduit. Level is measured using any available level device applicable to the specific site. The wetted perimeter is then developed into a cross-sectional area. Conduit shape is programmed into the setup program; most standard devices have functions for common channel configurations. Velocity is integrated, as described previously, based on the manufacturers' choice of sensor configuration. A relational velocity is developed that is representative of the volume being measured. The relational velocity is then worked into the flow calculation. Most of these devices record reverse flow, as the propagation of the signal itself is characteristic of the flow's direction. In cases where level is not sufficient to cover the ultrasonic couplers, Manning or other estimation equations may sufficiently represent the flow rate. It is generally accepted that Manning can sufficiently estimate the flow in a conduit if the level is 10% or below the nominal capacity. Therefore, the use of substitution equations for estimates appears proper under these conditions.

Transit time technology appears to have two significant features that affect its usefulness in any given field location. The first is the conditions of installation, and the second is the aggregate cost of the equipment.

Installation of any transit time device can be a tedious process, as most of the these devices are permanent installations requiring utility services and other fixtures for proper deployment. The ultrasonic transducers require precise alignment, which can be hard to achieve during flow conditions. Laser alignment and precise angle measurements are needed for valid transit time signal readings. Cabling of multipath devices can be cumbersome in specific locations. Entrained air is a nemesis of any device that uses sound pulses and it especially affects transit time devices. Site considerations should note the potential for air entrainment within the flow distribution. For measurement of tidally influenced or very large watershed regions, the permanent installation and known placement locations may be dictated by maintenance or regulatory purposes and not optimal measurement placement.

Because the cost of one transit time meter can equal the cost of up to five portable flow measurement devices, consider the specific aspects of the application before determining whether a transit time device is best suited for the job. If the location is representative and properly profiled and its hydraulic conditions complex, permanent installation is prudent and probably the only reasonable choice.

Successful instrument installation always requires site examination. The factors above can all be readily determined through a proper field inspection and inventory. The hydraulic conditions may not be manifest at the time of inspection but can be deduced from the wear and staining within the sewer. Temporary devices can also be installed to determine hydraulic conditions, both wet and dry.

The transit time flowmeter is a tried-and-true, reliable technology, provided the application warrants its installation. In spite of potentially cumbersome installation requirements, transit time devices have proven to provide traceable, repeatable accuracy and long-term reliability while resisting the hazards that plague the infrastructure.

2.14 FLOW MEASUREMENT: AREA VELOCITY DOPPLER

A technology that has come into the mainstream fairly quickly and established a respectable market share is the area velocity Doppler meter. Advances in microelectronics and improvements in material sciences have made this technology a low-cost and attractive alternative.[13]

The concept of area velocity is the same as volumetric flow measurement. A cross-sectional area is determined using a level-measuring device and the wetted perimeter. Velocity is obtained by determining the shifts in frequency of an array of sound pulses as they reflect off passing material in the waste stream. The combination produces a flow rate. The term *area velocity* is also generally used by manufacturers as a description so as not to confuse it with a flow estimation or velocity-only or level-only device.

The meters use the Doppler shift as a method of determining velocity within a fluid flow. Christian Doppler, in 1865, described the Doppler shift in sound propagation, a principle that has since been applied in many ways. It is the effect one can hear when a siren or train passes. The tone of the siren or the train changes based on the relative position of the listener. This tone change or shift is predictable based on the air temperature, frequency, and relative position of the moving object and listener. In the Doppler meter, a set number of these shifts, measured at given intervals, are integrated; a relational velocity can be derived from the integration. Doppler meters were generally limited to full-pipe velocity devices until several years

13. T. Day, *Demystifying Doppler Technology,* Water Engineering and Management, April 1996.

ago. Advanced signal processing and software applications have allowed Doppler technology to become more readily applicable to a variety of instrumentation.

The two forms of the technology used in sewer systems are usually referred to as *peak Doppler* and *averaging Doppler.* Each of these applications of signal processing has significant advantages and disadvantages in terms of both instrumentation and data interpretation. Perhaps the best way to understand the difference in the two velocity algorithms is to understand the terminology and language of frequency. Frequency can be considered a domain in the same manner that time is so considered. A domain is a sphere of activity or territory being investigated. A time domain is where the increments of change are based on time or the passing of time.

In Doppler technology, the frequency domain is considered. The change in frequency is the important variable. The range is the spectrum of frequency shifts. To understand this, remove time from a graph axis and simply add frequency in its place. This now references the frequency domain and the spectrum being considered.

Peak Doppler takes a spectrum spread of frequency shifts, seeks the highest or greatest shift in the spectrum, and, after filtering, performs a reduction to the value based on the specific algorithm preferences of the

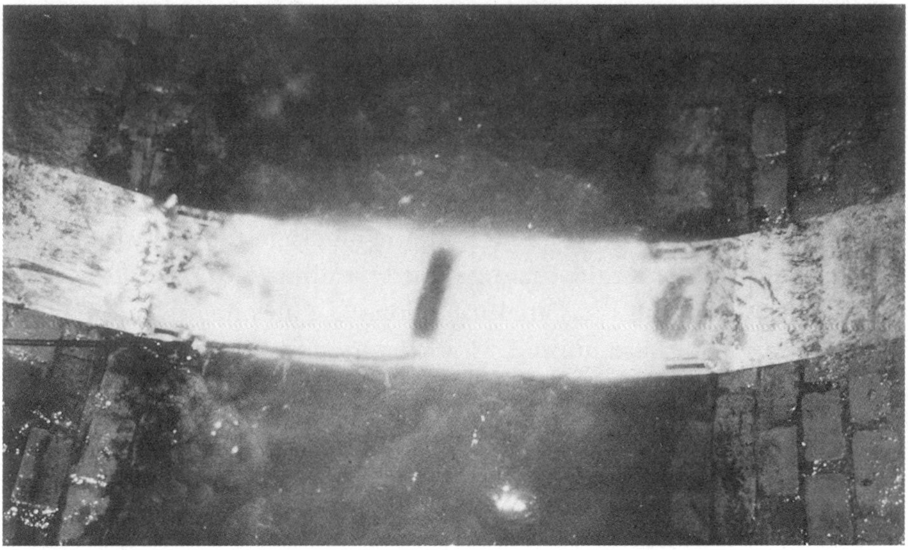

Figure 2-9 Area velocity probe installed in a brick-lined sewer. Band holds sensor in place. (Author photo.)

Figure 2-10 Preparation of area velocity sensors for installation on a large band. Such preparations should be performed on the surface. (Courtesy of American Sigma. Used with permission.)

manufacturer. The manufacturer has set the specific filtering options to reject highs and lows in the spectrum band in order to better utilize the typical frequency shifts, representing conditions and not extraneous conditions or erroneous reading. There is a patent on peak Doppler technology; therefore, its use is limited. An advantage is that this algorithm is the easier of the two to implement and provides fairly accurate results in generally favorable conditions. A downside is that line blockages or other irregularities may cause higher or fluctuating velocity readings and provide less than accurate indications of the actual conditions, as only one or several unrepresentative readings are utilized. However, the manufacturer can provide corrective algorithms in either the instrument or interrogation software for many conditions and specifications.

Averaging Doppler calculates weighted averages based on the frequency shifts, filters for points outside a given range, and generates a value. This approach assigns weights to all the recorded frequency shifts, so the highs and lows are considered in the total calculation. Filtering is performed to remove extraneous or otherwise erroneous data. Some argue that averaging all of the values provides a better picture of the actual conditions, and there is evidence to support this claim. Averaging, however, involves an algorithm that is extremely difficult to implement correctly. Yet many of the difficulties with averaging Dopplers are being overcome as the technology improves.[14]

A flow rate is produced in these devices by generating a cross-sectional area through the use of a level instrument, then taking the velocity (as determined by one of the algorithms discussed above) and calculating a flow rate using the continuity equation or one of its forms. The cross-sectional area is determined by head rise and the wetted perimeter. The relational velocity is utilized as a representative velocity. That velocity is then used in the continuity equation as the representative rate of movement seen passing through the cross section.

Let us consider how the boundary conditions come into play in various flow-measuring technologies. In the case of the electromagnetic point meter, the boundary condition is the electromagnetic field projected directly on top of the sensor. In the case of transit time devices, the boundary condition is the actual location of the acoustic couplers, as they are fixed on the walls of the conduit or channel. In each case, the boundary condition is static. Doppler has a changing boundary condition, which is the surface of the water itself. Both head rise and the volume of water passing over the sensor have an effect on the measurement. Therefore, the integration of the frequencies as they shift with depth is imperative for an accurate measurement. That the boundary condition changes with every reading or number of readings is a consideration that must be addressed and noted by equipment manufacturers. Preamplification or preprocessing of the signal at the sensor itself is one way of ensuring that extraneous or otherwise erroneous readings induced by the changing boundary conditions are eliminated.

This argument about the changing boundary conditions may appear academic. However, if you consider the logistics of the measurement itself, the boundary condition is no longer trivial. A Doppler device measures a velocity in a volume of water. It projects sound pulses forward into an oncoming flow. Many layers are traversed as the pulses echo off the material being carried. The various shifts in frequency result from the traversing

14. American Digital Services, Huntsville Alabama, is the developer and patent holder of this technology.

Figure 2-11 Zero deadband level sensor with velocity measurement device. Proper tie wrapping of the sensor cables reduces exposure to solids deposition. (Courtesy of American Sigma. Used with permission.)

of these many layers, providing for a representative average. The boundary of the measurement is important because it is one of the elements defining the representative sample, which necessarily takes account of the current conditions within the conduit. Therefore, accurately known boundary conditions are essential if the measurement is to be of any value to the user.

The physical boundary has two parts: the surface of the water and the channel shape. The channel shape is programmed at setup as a function of the software. The device should be able to accommodate irregular conveyance conditions as part of the physical boundary. Known grit depositions and other unmovable features within the piping should be able to be accommodated or eliminated from the boundary zone. The user should be aware of this zone, whether it is sketched at the time of installation or designed prior to installation. This physical zone, with all the parts of the boundary, is critical to maximizing understanding of the signal's characteristic after its travel within the flow.

The electronic boundary is completely confined to the measuring instrument. It includes the minimum and maximum limits of the sound pulse, transformation of the signal to the manufacturer's specifications (Fourier analysis, signal propagation and timing, etc.), head rise determination, and the data field. There are limits that the signals can be processed for: signal rejection, accommodations for signal loss, and lost-of-parameter requirements. These boundary conditions place confines on the range and limits of the electronics so that which is reasonable can be distinguished from that which is unreasonable.

Identifying and allowing for the boundary conditions in Doppler technology is essential because the devices rely on data from the entire sound pulse path rather than from one point or a select few points. For this reason, averaging Doppler devices appear to be more representative of the actual flow profile than point velocity or point Doppler devices.

A frequency spectrum is generated from the echo shifts and resolved with any number of techniques based on manufacturer's preferences. The resultant value is converted into velocity and calculated into the cross section. One should understand the frequency domain as opposed to the time domain. In the case of time domain, time increases or decreases in reference to the measurement from a reference point. In the case of frequency domain, the frequency, not time, increases or decreases from a reference point. From this reference, velocities are generated based on the shift away from the base frequency. The base frequency, or reference point against which changes are measured, is determined by each manufacturer. Shifts from this base have been calculated and verified via a traceable and repeatable laboratory procedure to represent the velocity as desired. The phase shifts are then developed into an algorithm by the manufacturer.[15]

Until fairly recently, most Doppler devices required about 70 mg/l solids content for operation, which made them unsuitable for clear-water studies. Therefore, they were not applicable for infiltration and inflow (I/I) programs or river/stream influence programs. This has radically changed with improvements in electronics and signal processing. Now many manufacturers of area velocity devices have their independent laboratory use potable water for calibration and verification. These more sensitive devices may be readily installed for watershed studies, where installation conditions permit, in areas such as spillways, conduit channels, spring, and creek or

15. T. Day, *Demystifying Doppler Flowmeters in Sewage Flow Measurement,* Keystone Water Quality Manager, April 1996.

river feeds, and any other application where flow is required. Their ease of installation and maintenance makes them desirable for both temporary or permanent applications. Some manufacturers provide an insertion tool so the sensor and the mounting bracket can be installed from the top of the manhole, eliminating the confined space entry.

Area velocity meters work under silting and other debris conditions, which makes them applicable in difficult situations. Except for use in very large conduits and channels, area velocity meters are probably the best buy for managing the sewer infrastructure system, assuming that the model chosen can be mounted as a permanent fixture.

The advances in area velocity devices have fast made them the device of choice because of their relatively low cost, high quality, and large array of product offerings. As advances in electronics and software continue, these devices should respond well to market pressures and continue to adapt, thus ensuring their popularity for many years to come.

2.15 FLOW MEASUREMENT: FULL-PIPE PRIMARY METERS

Most sewers do not readily lend themselves to the installation of full-pipe meters. They are designed for applications in process, plant, or other environments where the pipe is exposed. However, they are found in specific locations, including plant influent entry, custody transfer, or plant effluent, either by design or by convenience. This excludes their application within the plant for process measurement. A brief description of these process meters is provided because they do, in fact, appear in the infrastructure. A complete discussion of these devices can be found in any reliable work on process instrumentation.

Full-pipe primary meters are used when need for accuracy is high. They are inserted usually after it has been determined that no other device appears capable of providing proper measurement for that application. In days gone by, this was the only accurate method of determining these flows. The attempt was to push the open-channel sewer into a full-pipe flow when such critical measurements were required. Modern technology for open-channel measurement now allows for a variety of reliable products to be installed without the massive construction and excavation efforts of backing up the wastewater in order to obtain a full-pipe condition.

Magnetic flowmeters are the most popular devices for full-pipe appli-

Figure 2-12 Magnetic flowmeters used in sewage transport applications. The devices are extremely accurate but impossible to install in most established sewer infrastructure applications. (Author photo.)

cations. Their straight-through design permits solids to travel freely without obstruction. They are designed to fit as a straight piece of pipe within the conduit, which is why the meter section is referred to as a *spool piece.* The meter itself is divided into two parts: the spool piece, which contains the magnets and electrodes, and the instrument package, which contains the signal converter and excitation circuitry. Magmeters are based on Faraday's law of electromagnetic induction, which says that if a conductor is placed in motion through a magnetic field, then an electrical potential is impressed across the conductor. The speed of the conductor's motion through the magnetic field is almost linearly proportional to the reference potential generated. This is directly referenced to velocity. In this case, the motion of the water through the spool piece (where the magnets are located) causes a potential to be impressed on the water that is directly proportional to the velocity of the water. The electronics package provides for output and display if so equipped.

A special calibrator reference is required to perform maintenance on the magmeters. Note that there is no need to determine level, as the pipe is always full; therefore, the cross section is known. The average magnetic flowmeter is accurate to 0.25%.

Differential pressure (DP) cells are the next most popular device. A full pipe of known size is sent through constriction, generating a reference pressure. A ratio of the differences of pressures between the throat and the intake is directly proportional to the flow rate. These devices are simple to operate and do not need the electronics that magnetic flowmeters require. However, they pose several problems with wastewater flows that magnetic flowmeters do not. The first is that—although the gradient tends to keep debris moving—larger debris does accumulate at or near the device's throat, leading to inaccurate readings and requiring frequent backwashing, sometimes weekly. An advantage is that the cell does not require any special device for calibration or maintenance. Pressure tubes or similar devices are all that is required, although manufacturers provide special calibration kits whose use is recommended. The average DP cell is generally accurate to 0.5%. It is of interest to note that the Parshall flume was originally called the *partial differential pressure device.*

These type of meters may appear in the infrastructure in cases where accuracy is critical. They are used in areas where the pipes are always full, such as siphons, wet wells, and lift stations. Both devices require external power sources to operate, so they can be difficult to install in remote locations. Therefore, the chances of them being widespread within most infrastructure systems is probably slim under normal requirements. However, in areas of billing or other critical flow measurement areas, one finds them installed, even at great expense.

2.16 VELOCITY MEASUREMENT: HANDHELD PROBES

A critical requirement in the installation of open-channel flowmeters is the ability to determine instantaneous velocities within a flow profile. Consider the cross section for a moment; points within each layer of velocity can be read and an average velocity calculated. Measurement of these points may be done directly using a point velocity measuring device. This is usually performed by an insertion probe that registers the point velocity at or around the sensor head of the instrument. This provides an insight concerning the flow profile at that particular moment. Several types of devices can perform this task.

The earliest device was a propeller meter. This is a device with a prop located on the end of a long probe, calibrated to spin based on the velocity of the water. Resistance from the effects of the propeller blades was a problem for these devices. Also, debris in wastewater tended to hang on the blades, slowing the motion and yielding unreliable results. Recent technological advances have largely eclipsed propeller meters.

Point velocity electromagnetic devices became popular because of their ability to shed fouling and other undesirables during the exploration of the flow profile. A small electromagnetic field is projected in front of a probe and, by Faraday's principle, a reference potential is generated. This reference potential is directly proportional to velocity of the water. Such a device is identical in structure and design to its meter double, but the electronics package is configured for carrying rather than mounting. These devices are very accurate and can provide long-term service. Because they are inserted and removed by the operator, the electrodes can be readily inspected, removed, and cleaned if there is a loss of signal at the instrument package. These devices continue to be popular.

Modern electronics has refined the point velocity device for using Doppler technology in its structure. The ability to interface Doppler technology in a point velocity device resolves the electrode fouling problem noted in the electromagnetic device. This evolution of the point velocity meter combines the durability of the Doppler device with the convenience of the handheld probe. As the Doppler can operate under clear-water conditions, it is a suitable device for river and stream use as well for watershed management and checking sewer outfall discharge points. Doppler technology here has taken an evolutionary step in handheld probes for velocity measurement.

Handheld probes are critical devices for installations, studies, and site conditions of the infrastructure. They should be used to verify accuracy of flowmeters, determine blockages, and establish baseline conditions of sections of the system. They should also be employed in the installation of any monitoring device to verify a proper installation. Every system operator should have at least one available to make these on-site determinations quickly and readily.

2.17 WATER QUALITY MEASUREMENT: REFERENCE SAMPLERS

A sampler is a device that can draw up a stream of water from a desired flow at some set interval and store it until removed by an operator or technician.

The samples are usually taken to a lab for analysis. The idea is to sample the waste stream at a specific time or at set intervals so generalizations can be made. The temperature of the sample must be maintained so the biological activity in it can be monitored.

Temporary or portable samplers are useful for compliance, regulatory, and remote sampling purposes. They are generally battery-powered, with icing to keep the samples chilled. Manufacturers provide algorithms for drawing and storing the samples, but several generic features remain. The first is the 24-hour individual bottle and the second is the 24-hour composite bottle. Between these two options, many combinations can be constructed, including flow-paced sampling during storms (with the addition of rain gauges or flow probes), triggers based on water quality (pH, DO, ORP, etc.), and triggers based on level or other external conditions. These are great features, as they mean that operators no longer need to be on the spot to take measurements during storms or incedents of illegal industrial discharge. One grade of temporary samplers is classified NEMA 4X for hazardous environments. The battery power allows these devices to be placed at any location, provided it is secure. Be certain that the device purges after the samples are taken to keep the sample tube free of debris.[16]

Permanent or stationary samplers are generally found at the entrance to a treatment works or in other areas where utility power and security are available. They possess the same options as their portable cousins but generally have refrigerators for keeping the samples at a constant temperature. As the devices are not as restricted by space or enclosure as the portable devices, they usually provide more options and are quite versatile, being capable of taking samples at specified intervals or when certain conditions exist. They also tend to provide a greater variety of bottle arrangements for storing samples. Usually, these samplers are enclosed in a fiberglass shelter, although some models are produced that do not require these shelters.

Samplers do little analysis of their own other than keeping track of the interval between samplings and monitoring specific conditions. The goal of the sampler is to provide a uniform and accurate sample for laboratory analysis at a controlled temperature. These instruments are imperative for both plant process and compliance because they provide a grab of the actual water quality. The time and event of these samples have long-term effects on regulatory, policy, and other matters involving the operations

16. American Sigma, Medina, New York, product literature.

and maintenance of sewage treatment works. They are a critical part of the infrastructure system and need to be strategically placed and handled.

2.18 SUMMARY

Proper selection, placement, and interfacing of measuring instruments are crucial for developing a complete view into the infrastructure. Local system knowledge and site inspections are vital to determining what conditions exist and the best instrumentation for monitoring them. Subsequent chapters discuss more detailed uses of the basic equipment and provide more specific and specialized product offerings that enhance system knowledge and performance.

No one technology is applicable to every flow or level measurement scheme. The applicability of products varies with the unique features of the infrastructure of interest. Therefore, this chapter gives a basic overview of the products available and a general understanding of the technology they use: listing the advantages and disadvantages of each type of device to make a preliminary determination of the best instrument choices for their own system, based on system type, application, or information requirements.

CHAPTER

3

System Layout: Choosing the Points of Interest

3.1 THE CRITICAL PATH

It is important to determine the optimal physical locations to employ the monitoring technology discussed in the previous chapter. This chapter describes a method for determining points of interest within your infrastructure. Because every infrastructure is different, the needs vary based on not only the sewers under the communities but the hydraulic effects of the control points within the system and the process effects of the sewage treatment works.

This chapter provides an example of mapping the system in a crude or rough form. Any number of other methods, whether internally or externally developed, can accomplish the same goal. However, the best learning comes from comparison. Readers should compare the method outlined here to their own methods, concepts, and ideas to create the best possible blend for their particular system.

It is important to know both what types of instrumentation are available and where to place the devices. Placing equipment in locations where significant events do not occur or sites of limited interest defeats the purpose of infrastructure monitoring systems. One must identify two vital features of an infrastructure system: the sites of interest and the territory in which these sites is contained. This can be referred to as the *critical path.* Choosing the critical path is as important as choosing the equipment and com-

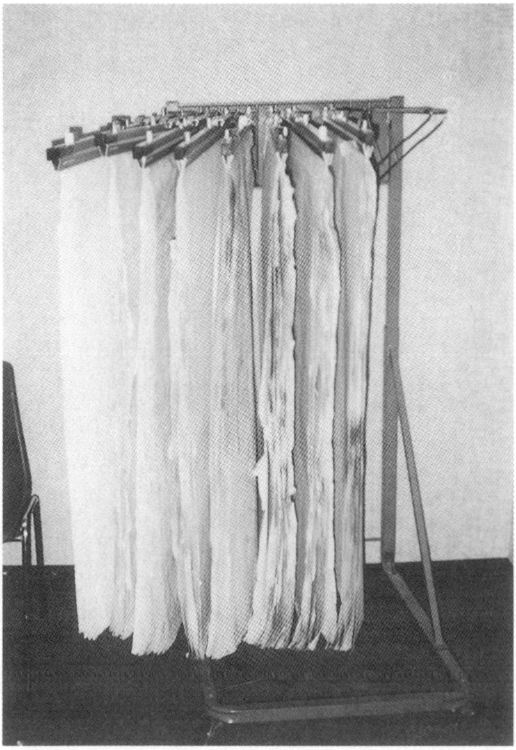

Figure 3-1 The best source to use in laying out a system is the standard sewer plats. The identification of all points of interest in the beginning, regardless of their likelihood of ever receiving a monitor, is the best first step in designing the monitoring system. (Author photo.)

puter system. If the wrong critical path is chosen, an inaccurate picture of the infrastructure system is generated.

The selection of critical path monitoring system points depends on the general system response and potential reaction to hydraulic control points within the infrastructure. The critical path displays the locations that offer particular interest based on their hydraulic reaction, whether conducive or adverse to the system at large. Critical paths cross through territories, where sites are grouped by the system operators according to geography, convenience, or political/governmental boundaries. Grouping sites into territories is an organizational technique. The organizing principle can cover responsibility areas, jurisdictions, maintenance, or any other housekeeping function. Critical paths are necessary for understanding the infrastructure system response based on the information received from relative locations along a hydraulic conduit of interest. The critical path exists regardless of territorial boundary; that is why it takes priority in the site choice.

A site is a location where instrumentation is or will be located to produce beneficial results for the users. A territory here is defined as an area containing more than one site of interest. Territories, therefore, contain sites that have potential interest for the system operators and planners. The sites are located along one or several critical paths that have a hydraulic effect that is of interest to the ultimate users. Critical paths can cross areas, towns, districts, cities, and, sometimes, countries. Sites thus can be described two ways: by territory and by critical path. The critical path is the effect that the system operators are monitoring. The territory is a grouping for organization. Therefore, the critical path always supersedes the territory for placement purposes.

Critical paths generally are already known to the infrastructure management and operations organizations within a utility or agency. They tend to be the major interceptors, collectors, or other conveyances that carry the bulk of the wastewater to the sewage treatment works. With the proper selection of monitoring sites along these paths, a full infrastructure monitoring system can be constructed.

3.2 INTUITIVE SYSTEM KNOWLEDGE

Definitions can be generated to delineate areas of potential interest within a system, but the bulk of information used in determining the critical paths comes from those who already operate the infrastructure system. No one has greater knowledge of a sewer system than those who work within it daily and respond to its problems. But because the concept of infrastructure management is relatively new, system operators may be uncomfortable looking inward for information that intuitively they would look outward for. The goal of this section is to show how the intuitive system knowledge that the operating agencies use so well can be organized and developed into a useful monitoring system.

Field forces, system records, and gut experience are the three factors that hold the system knowledge and keep the sewers running before monitors are installed. The field forces continually work the system, answering complaints, making repairs, and performing maintenance that keeps the wastewater flowing. System records can be a combination of completed complaints, inspections, and other methods of tabulating activities within the infrastructure. Gut experience is the body of historical information that every supervisor builds over time from answering complaints and otherwise maintaining the system. It makes sense that these factors can contribute to the information base and the determination of the initial monitoring points within the system. Desktop analysis may point to interesting locations, but many of the desktop systems are not yet mature enough to effectively pick every point of interest in the infrastructure. Use actual information on actual events as the baseline for laying out your system.

3.3 OBTAINING INTUITIVE INFORMATION

Stating that intuitive system information is an important component and an excellent starting point is all well and good; the problem is eliciting the information. Written records can be analyzed and tabulated. This is easy, although often tedious. The real challenge is getting the information out of the heads of the people who do the work day after day. The only way this can be done is to get the field forces and the first-line supervisors involved and make them comfortable with the concepts and goals of infrastructure management.

A lot of blue-collar workers and first-line supervisors are not in favor of any system that involves computers, electronics, or other remote unmanned sensing equipment. They believe that such maintenance systems eliminate jobs. As one who is setting up a system intended to increase production and response, you must explain that loss of jobs will not be a by-product of an infrastructure management system, ever. Until this unfounded fear is removed, your infrastructure system can never be a success. Field forces must accept and work with the information generated from it. Explain that these systems perform several functions.

The first function is to increase the general safety of workers in the field by providing knowledge of the system conditions before a work crew makes an entry. Properly placed equipment indicates where the problem is and its relative effect as well. The field crew can then determine safe entry locations. Without this information, entry are made haphazardly and, often, blindly. Use sample data to illustrate this point. Provide handouts showing the potential cases and how field forces can be protected with a little insight to the problem at hand. This approach alone generates a movement toward system acceptance.

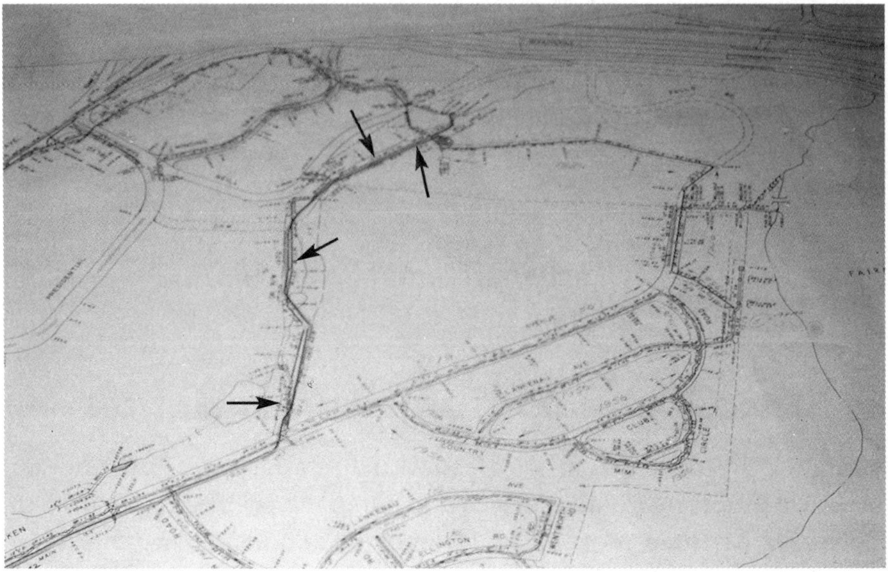

Figure 3-2 Charting the critical path. By highlighting it, the points of interest become apparent. (Author photo.)

The second function of the new system is to increase worker productivity in the eyes of the public. Hourly workers are rarely motivated beyond minimum effort without financial reward. However, remember the public scrutiny every field crew faces during the usual workday. The usual cry of "one guy working with five guys watching" is heard at every local council, utility management, or city agency meeting regardless of the effectiveness of the individuals or units performing the work. Why are five guys watching a job? Usually it is because a problem is being diagnosed and their particular skill is not yet needed. All field crews know that. The public does not. Infrastructure management systems provide advanced information on the problem so that crews can go to the problem site and start working immediately. Explain to the field forces that the public thus sees a more organized and efficient workforce and may harass them less. This can be a source of relief for many maintenance workers.

A third function is to proactively protect the environment. Workers are already aware that the faster they can respond to an event, the less damage there will be to the environment. Explain that a field complaint will not be the only method of knowing where a problem is once the system is in place. Also, some problems can be resolved before they affect the general public

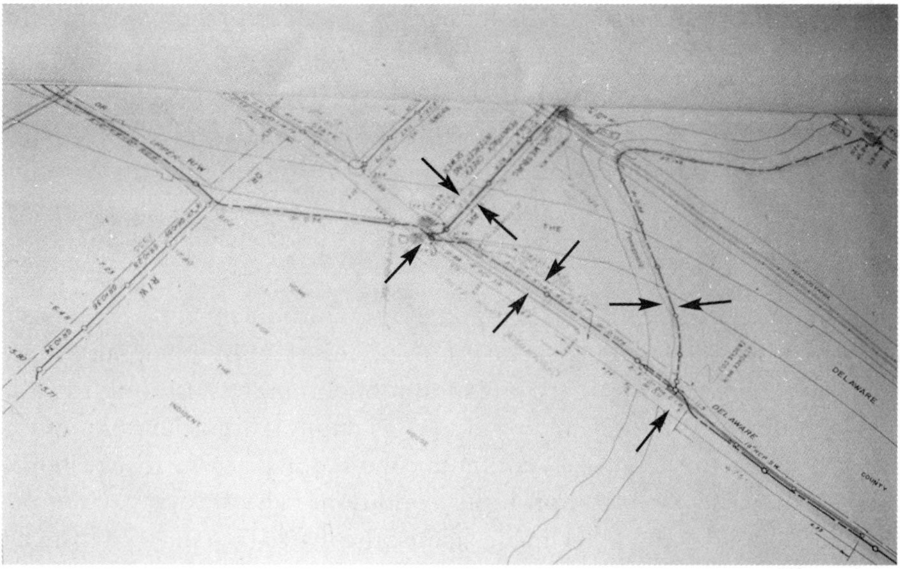

Figure 3-3 Merging of critical paths. Note where two paths merge into another larger critical path. (Author photo.)

at large. Most field workers live around or near the areas they work in. Showing protection for their own area certainly has appeal.

Gain the trust of the field forces by truthfully showing them the benefits of these systems and assuring them that their jobs will not be eliminated by automation. The information they can provide will make or break the system. Keep in mind that these are the people who will ultimately be performing the work. They need to accept the technology and the system to follow up with repairs. If they do not use the system to its benefit, it will fail. In the beginning, get them to accept it and the new system will succeed.

3.4 THE NEEDS OF THE UTILITY ENGINEERING DIVISION

Every operating agency contains an engineering unit. It may be a contract agency, a fully in-house design/planning unit, or a combination that suits the needs of the operating entity. Unit jobs can range from minor contract operations and preliminary studies to full bidding specifications and detailed equipment and systems designs. In the past, monitoring systems provided information sporadically for these units, with little other involvement. Another issue is that the engineering unit traditionally specifies the system for the field units, generally with no influence from the field personnel. These factors are likely to doom the system to failure. But with the recent influx of computerized hydraulic sewer models, engineering units need to maintain a channel to live data in order to calibrate and maintain the model. Therefore, the engineering units now have many considerations, including their own needs, in the layout and selection of the points of interest.

The subject of computerized modeling is introduced at this point as it is potentially volatile and can cause a lot of noise within the organization. Any computerized hydraulic simulation model is useless without live data to calibrate it. Although many references support this claim, common sense stands as the best one. A computer simulation run states the response of the infrastructure based on input conditions. The accuracy of any hydraulic model should be verifiable against the live data generated from actual hydraulic response in the system.[1] Therefore, the only way to verify the accuracy of the computer model is to compare it against good data. Vendors with integrity encourage you to verify the results of their computerized systems. Usually, the utility's engineering unit has the responsibility for such verification and thus must identify points of interest for data collection; therefore, the points of interest for calibration of the model and verification of its accuracy need to be provided by the keeper of the model, the engineering group. Simulation models cost a lot of money not only to purchase and set up but also to maintain. The model keeper must be aware of the locations from which information can verify and maintain the integrity of the simulation package.

Specifying the needs of the operating divisions (sewage treatment works, sewer maintenance division) and administration (legal, management) is also a challenge. The engineering unit must balance the needs of operations and administration with their own needs. This is a relatively easy task if all needs are thoroughly understood and considered. For example, points where overflow tends to occur in the computer model are probably the same ones that the operations and maintenance divisions chase regularly. If they are not, either something is wrong with the simulation or the sewer maintenance division is not catching all of the problems. As another ex-

1. E. Burgess, T. Day, J. Smullen, L. Roesner, *CSO Model Development and Verification Study,* (paper presented at the North American Environmental Conference, Anaheim, California, June 1996).

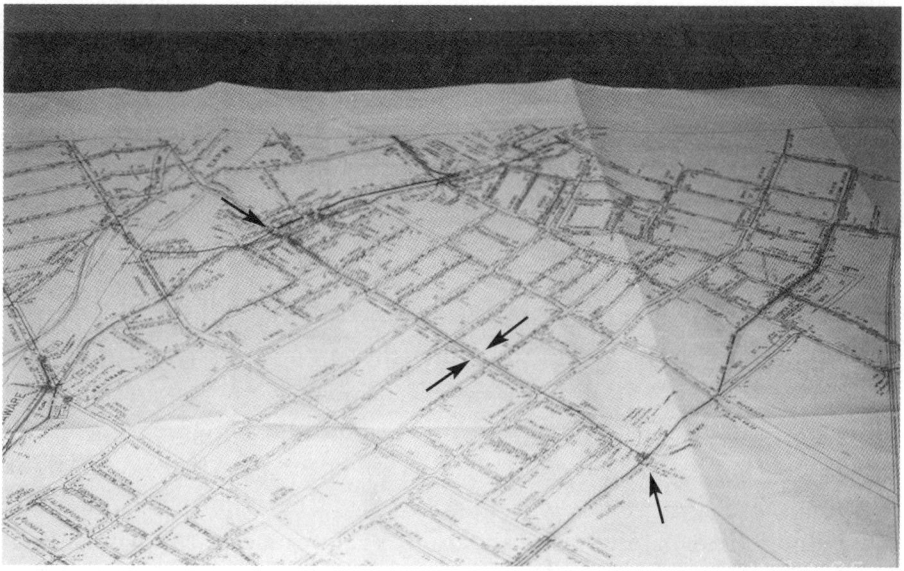

Figure 3-4 Junction of critical paths. The point at which the two paths connect is of tremendous interest to the sewer management systems team. (Author photo.)

ample, plant personnel need to know the effects of the plant on the system because it is the biggest hydraulic control point. Therefore, the simulation should be able to determine treatment effects and compare them with actual information. Finally, administration wants a return on its investment, and evidence of that can directly reflect on the engineering unit's success or failure in procuring a workable system. It therefore behooves the engineering group to use data from each location in ways that serve the needs of multiple units.

The engineering groups should be aware of an additional requirement. Sizing sewers and designs for repair and replacements is best performed with historical information based on usage of the pipes.[2] Live data provides information on customer usage, I/I, and storm capacities. This information is critical to cost-effective design. Selection of points along the critical path should include considerations for observing conditions for these parts of the systems within the territories serviced by the infrastructure. This calls

2. G. Szunko, *A Pollution-Focused Strategy for Sewerage Investment in the North West of England,* North West Water, Ltd. (paper presented at the Reducing CSOs Conference, Louisville, Kentucky, July 1994).

for a complete knowledge and consideration of the goals and objectives of the engineering groups and their planned projects over an extended period.

The engineering group thus has a substantial charge beyond system specification and design. In modern infrastructure management, engineering has an active need for the same site-specific information as the other potential users of the system. Therefore, it is paramount that engineering incorporate its own needs, the needs of other units, and potential uses for the system in its specification and design. The critical paths chosen should represent all of the issues for the engineering units.

Depending on the size of the utility, the engineering department may have heavy involvement in the areas of compliance, customer complaints, district boundaries, capital improvements, and other engineering-related functions. Many of these are discussed as separate topics later in this book. The goal here is to review the requirements of the engineering unit, which many times are separated from the utility's requirements as a whole. Engineering work grows separately from and its needs are not part of the mainstream of operations. Engineering work directly affects operations, but in subtle ways. The unit's plans and goals lead to programs that have long-term effects on the operations division, as they address capacity complaints, planning and connection issues, and a host of details with respect to which capital improvements are introduced. It must be understood that the decisions made in engineering units eventually affect operating units, so they need to work in concert.

3.5 THE NEEDS OF THE OPERATIONS AND MAINTENANCE UNITS

The operating units should use the infrastructure system information proactively. This is a different application than that of the engineering unit, which uses the information as needed for planning and projects. Operations can use the information for optimization of process, delivery, and maintenance activities. The way to optimize any of these targets is proper point selection along the critical path, with enough locations within any given territory to properly represent the system response. The needs of the operating units are discussed throughout this book; they are addressed here with respect to system layout.

The plants require information and alarms concerning system hydraulic changes, chemical content, and weather information. In the case of

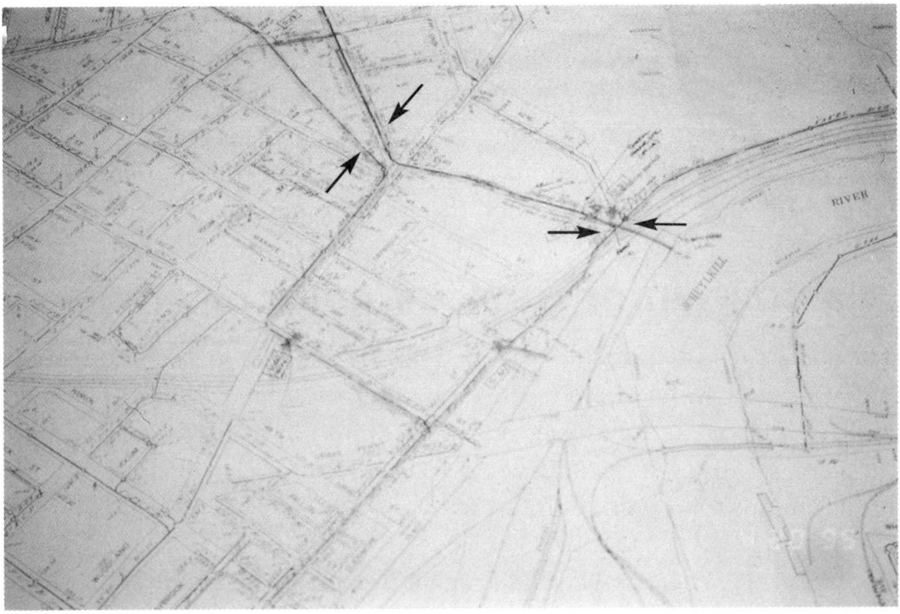

Figure 3-5 A complicated hydraulic pattern with mergings and junctions. Note the points of interest marked. Note also that the junction of the two critical paths is not the closest point. (Author photo.)

systems with combined sewers or known sanitary sewer overflow points, sensitive discharge point information is needed so the plant can make every attempt to keep these locations from discharging in excess of prescribed tolerances. Also, data from points along the critical path can help the influent operator or process control manager to maintain plant influent levels within specified parameters, thus permitting best possible use of the infrastructure.

A sewer maintenance unit has myriad uses that are discussed in full detail. The more points available to this unit, the more information can be utilized by the system operators. The problem is overloading the system operators with too much information overall or too much within one territory. Care should be taken in establishing proper points for monitoring along the critical paths to provide optimal information for this particular unit.

The industrial pretreatment unit also needs to monitor large industrial bases, either in part or whole. This effort includes a balance of chemical analysis along with rain and flow so personnel can determine the industrial contribution to the sewer infrastructure. The needs follow the critical path with points of interest distinct from all other users. This unit has heavy concentration in territories, as its interests generally coincide in industrial areas. It is not unusual to find imbalances in territories for this reason.[3]

In general, operating units require proactive information to properly maintain and plan for operations and maintenance activities. All of these needs should be assessed and tabulated so that placement of appropriate monitoring site can be done when the system is completely laid out.

3.6 CONTENT OF A SYSTEM POINT OF INTEREST

The above sections listed the units that have an interest in system layout and the input they should have to the discussion. Now let's look at what needs to be monitored at the various points of interest in order to provide a useful picture of the system. Following is a checklist for classifying equipment requirements for a point of interest.

1. **Level:** Where a hydraulic head provides significant information about the location. Points of interest may be manhole surcharging, storm relief chambers, and very small pipes. Interest groups: All.
2. **Flow:** Where level and velocity are required for a location. Velocity components provide true pictures of the system hydraulic response, as they actually detect liquid motion. These connections are also necessary for billing locations where inter-township billing is an issue. Interest groups: All.
3. **Quality:** Where water quality parameters provide useful information to the system operators. Discharges to watersheds, industrial pretreatment, permit compliance, I/I influences (including tidal), and POTW influent are potential candidates. Interest groups: operations (and engineering, to a limited extent).
4. **Rain:** The parameter that determines the beginning of an event. Spatial rain information is critical for runoff, monitoring, and other activities specific to the optimal operation of the infrastructure. Interest groups: All.
5. **Sample:** A site where water samples must be taken for further analysis. It can also be a location where the device is set to act on a trigger to draw only as part of enforcement. Interest groups: Operations.

3. T. Day, *Electronic Surveillence of Sewer Systems Using Monitoring Data to Protect the Environment,* Understanding Industrial Pre-Treatment, Indianapolis, Indiana, July 1996.

The system infrastructure can be laid out on a map and marked with numbers to note the potential equipment.[4]

3.7 SYSTEM POINT OF INTEREST: A SITE

Once it is determined which units need information and what type of instrumentation can provide the necessary data, the next consideration is where the equipment should go.

A site is a place where equipment is or will be deployed to provide specific information about that part of the infrastructure along a critical path. Sites populate territories; care should be taken to balance the number of sites within a territory. This is not to say there should be some mathematical average to populate a territory, based on drainage area or other usage factors. Balance is strictly a matter of ensuring the best picture of that critical path and the hydraulic response it makes by itself and in relation to the rest of the infrastructure.

Nomenclature for sites is already established within the management of the infrastructure operators. It is usually successful, as the names come from the intuitive or local knowledge of the site and match some logic within the operating organization. However, bear in mind that it may not be necessary to always refer to a site as, say, "Main and Elm Sanitary Overflow Site" in cases where "SSO-R15" will suffice.

It is strongly recommended, though, that sites be classified. Here is a brief outline of one way of denoting sites:

- **HC:** hydraulic control. This is a point that has specific hydraulic effects on the system. Included are junctions to interceptors, siphons, and other fixtures not motor- or otherwise electromechanically driven.
- **PS:** pump station. This site is classified as distinct from the hydraulic control noted above due to its variable nature.
- **SR:** A storm relief sewer or chamber that temporarily retains storm water.
- **CS:** combined sewer. This conveys both sanitary and storm water to a POTW. The overflow point usually designed into such systems is a **CSO.**

4. These definitions and those in section 3.7 are for illustrative purposes only. Individual utilities are encouraged to derive definitions to serve their own needs.

- **SS:** sanitary sewer. This is separate and not designed to convey storm water to the POTW. A point of unauthorized discharge is an **SSO.**
- **STW** or **POTW:** The sewage treatment works (STW) or the publicly owned treatment works (POTW) is a unique and generally the largest hydraulic contributor to the infrastructure system. Therefore, in this text, it is considered a separate function.
- **CT:** A custody transfer site or junction where a township's sewer joins another and a bill is generated for the processing of the sewage.

This classification covers all of the components within the infrastructure and enables sites and their contents to be determined. It is offered strictly for purposes of this discussion; most operating agencies and utilities already have successful nomenclature that should not be abandoned during the implementation process.

3.8 BRINGING THE SYSTEM TOGETHER

The task of satisfying all of the parties requiring information begins with picking site points along the critical paths and denoting the contents of each site chosen. This is really much simpler than it sounds. The following procedure is straightforward and produces the best results.

Lay out a map of the entire infrastructure system. Be certain that details of each section are available if the system is really big. Begin to map the critical paths and the junctions to the critical paths, which are the first round of hydraulic control sites. Critical path junction points are vital because the contribution of each upstream pipe contributes uniquely to the downstream conveyance. Note how the critical paths traverse politically or geographically bounded units that are of interest to the system operators, or are territories.

Now go to the input that was provided by the contributors. Sort it by site so all of the sites are laid out with the interests of each unit noted. Go territory by territory along each critical path from the farthest upstream point to the sewage treatment works. (This is why a numerical system is suggested; one need only place numbers on the site and make off-map references to locations later.) As each critical path is completed, the proper placement of each type and kind of device will become apparent. You will begin to see balances and distributions along each critical path and other points of interest that none of the groups considered. Complete each territory with the site information.

The task now is to filter the sites. Review each site along the critical path to determine if points of little interest or locations yielding duplicate information have been selected. It may be surprising how many sites can be eliminated due to duplication of effort. This filtering should not be performed for the sake of making the system smaller; its only purpose is to delete redundant points of interest. If you feel that resolution of the system will be reduced by eliminating a location, do not eliminate it, at least at this stage. For example, say that all of the HC sites are marked for flow on the first critical path pass. As additional sites are marked, CT sites are also marked for flow 1,500–2,000 feet above the HC. It is reasonable to eliminate the HC flow unless there is some specific reason why flow is required at that point.

At this point, calculate the cost of the infrastructure management system without any computer interface. Does sufficient funding exist to complete the endeavor? If the funds are available, you need only add in the computer and software solution and procure the system. If total funding is not available, consult potential vendors to be certain of the cost of equipment and your particular requirements.

Another filtering can be carefully performed at this point with available funding in mind. Perhaps plans can be made to add sites with additional fiscal-year funding programs. Consider these options as progress is made in the endeavor.

The picture of the new infrastructure management system should now be apparent. Addition or removal of locations under specific circumstances should become obvious through the critical paths and the territories that contribute to them. If the system has been correctly laid out, keen operators can almost picture mass balances strictly by the points that are monitored.

3.9 SITE VISITS

A critical part of any system layout is the need to physically describe what is there and what the structure looks like. It is not necessary to know the exact hydraulic response of the structure at this point in the development. That information is what the monitoring equipment will provide after data gathering. But the desktop layout of a system loses its relationship to the overall project if no actual street knowledge substantiates it. The view from the street coupled with the design information forge a powerful combination of what a specific point in the system was designed for and what it actually does. This is because many changes come during construction that substantially shift the focus of certain structures. The results from upstream modifications after final construction of the point under consideration also change the focus and function of the structure. These issues must be seen in the field to view their current usefulness.

3.10 SUPPORT INFORMATION

Other background information can assist the system designer in the layout and choice of points of interest. Geological survey and topographical maps can provide information on aquifer, water table, underground terrain, elevations of the surface, and other features that assist in system layout. The depth of a water table is a critical issue in studies of infiltration and inflow and has great significance in the placement of sensors to monitor these phenomena.

Much of this information is available at low or no cost in government libraries and geological services. Satellite imagery and precision computerized mapping provide cost-effective and reliable charts. These contain a wealth of information that expands local knowledge and increases the effectiveness of choices for points of interest. It is advisable to incorporate this layer of information when developing and outfitting a point of interest.

3.11 SUMMARY

The layout of an infrastructure system is a process that is more tedious and logistical than it is complex. Proper planning with the necessary input can reduce the stress of making the system appealing to everyone who stands to benefit from its use. The application of critical paths and territories makes the logistical task of choosing points much easier. Filtering of redundant points over several phases is also necessary for the most efficient use and application of equipment at a site.

Those individuals with intuitive knowledge, theoretical understanding, and firsthand information must contribute in making site selections and the quantities to be measured for. By satisfying all the groups, the system should serve the ultimate users with all the resolution necessary to provide a full picture of system response and activities.

It is also necessary to add as many layers of information as possible to the choice of a point of interest. The more layers presented, the clearer the picture of the site from the desktop view. Subsequent site inspections are necessary to bring the full picture of the point of interest into focus. Once this is examined, the structure can be designated a point of interest.

All points of interest follow critical paths. The critical path is defined here as the alignment of points of interest in a sewer infrastructure as transport functions to a treatment works. The path is labeled critical as it becomes necessary to transport the wastewater and its subsequent materials to the treatment works for processing as a function of safety, health, and protection. The critical weight of the path is strictly a function of the characteristics of the sewage and the importance to the surrounding communities of the need to make the connection to the sewage treatment works.[5]

5. Many schemes exist for producing this inventory. One suiting the needs of the utility should be selected and used consistently.

CHAPTER

4

System Definition: Components of a Sewer Infrastructure

4.1 INTRODUCTION AND METHODOLOGY

This chapter provides a basic, concise set of definitions useful in describing the fundamental physical features and elementary theory of a sewer infrastructure system as they are applicable to sewer system management. The discussion here is limited to open-channel systems—those where the conveyance is designed to operate partially full. Mention of full or pressure systems is reserved for the section on pumped flows and process measurements.

The descriptions of the sewer infrastracture components also are meant to be concrete illustrations of working units, with minimal theory. Readers can put them to use immediately based on individual understandings and experiences, regardless of the individual knowledge base. The definitions also provide some information about the origins of specific terms.

The topics begin with the subject of flow and its components because the main purpose of the sewer infrastructure is to transport sewage to the treatment facility in a manner that protects the health, safety, and property of residents and businesses. Following that is a practical discussion of basic infrastructure components.

4.2 FLOW

Flow—a volume of conveyed water, measured over time, transported from one location to another—is the business of a sewer system. As mentioned earlier, this discussion is limited to open-channel flow; other forms are presented elsewhere in the book as the subject appears. Experienced individuals will note that the terminology used in this section sometimes diverges from what is traditionally used. This has been done in order to provide the clearest explanation to the widest possible audience. The traditional discussion and modern applications as presented will be brought together at the end of the section.

Flow is a quantity composed of a volume (in this case, of water or wastewater) transported within a conveyance and the mathematical weighted average of all the layers of velocity within that given volume for a time interval. To say that flow is a quantity means that it has parts, each of which significantly affects the whole. The parts of flow, which are volume, velocity, and time, have subcomponents of their own.

The volume of a conveyed liquid is generally identified through its cross-sectional area. This cross section is determined by the geometric slice of the profile the flow volume makes at a given point along the conveyance. The cross section becomes a boundary condition by which the volume of water must pass through in its course. For purposes of discussion, the cross section has no width, only area on its face. The cross section takes the shape of the conduit from the bottom of the pipe to the water's surface and along the perimeter of the flow volume. The geometric cross section changes as the water level rises and falls. Therefore, the governing parameter for determining the overall cross section at any given point along the conveyance against the pipe geometry at any point is the level reading. Therefore, it can be stated that the level of the water generates a cross-sectional area.

Looking into the cross-sectional area, the various theoretical layers of velocities become apparent. Within the volume are interacting layers or waves. Seen at the cross section, this is the velocity profile, which consists of layers of similar waves of water interacting within the volume. Each wave contributes to the velocity layer. The layer consists of waves of water moving at similar speeds. Each wave within the profile contributes to the rate at which the water passes through the area of reference, which is the cross-sectional area. The boundary among these layers is not well defined but is discernible under laboratory conditions. The type and shape of these layers are affected by the upstream and downstream conditions of the transporting medium, including bends, breaks, obstructions, and sediment. Flow profiles tend to be uniform and consistent for smooth, straight, and unobstructed pipes. The more numerous and the greater the obstructions and changes upstream and downstream from the cross-sectional measuring point, the more irregular the flow profile.

The rate over time at which the volume makes the passage through the cross section is the flow under consideration. Consider the boundary condition of the cross-sectional area. Use this as a marker or reference. Now pass a volume through this reference. Note that the volume continues to increase the longer time is considered. Note also that the volume takes the shape of the cross section. This passage of volume that takes the shape of the cross section is called the *flow rate.* The actual volume that passes is called the *flow.* Flow rates are considered as a quantity, as cubic meters per second or million of gallons per day. Flow is a final quantity, as in cubic meters or millions of gallons.

To determine flow within any given conveyance, first measure the level within the conduit. From that point, the geometric boundary of the shape of the channel, as marked by the water, is formed. This is known as the *wetted perimeter* because it is the area within the channel that is wet due to the water's passage. This thin slice outlined by the wetted perimeter is known as the *cross-sectional area* and becomes a boundary condition for the measurement. In the cross section are an infinite number of points correlating to the varying velocities within the volume. These point velocities impress themselves at the boundary referred to as the cross section. These points tend to organize themselves into fairly defined layers. A generalized weighted average can be calculated for the point velocities the layers contain. A weighted average of the layers yields an average velocity for the profile.

This becomes the measurement zone. As the volume of water passes through the measurement zone, there are two considerations. The first consideration is the time or time increment at which the passage of the volume is marked. The second is the speed or velocity at which it passes. For purposes of this discussion, assume a volume that is passing through the boundary. Determine a time frame that suits the needs of the measurement. If the wetted perimeter is projected into a volume down the pipe for that time slice at the average velocity from all the contributing layers, this is a flow rate. The flow rate is the amount of water passing through the measurement zone in a given time frame. If the total time of the measurement is included in the calculation, this becomes the flow.

A flow rate is an open-ended measurement. It has no initial or terminal point. Flow is a definitive, fixed quantity. It is also the final quantity in determining system requirements. It is important to realize this in making reference to measurements, as these terms tend to be mixed at times. Other terms, such as *flow level,* that generally have little applicability to formal discussions of the infrastructure are also misapplied. Such complications should be avoided to keep the discussion of system as clear as possible.

The next sections break down each of the characteristics as described here and provide more extensive discussion. The characteristics of each component are identified and noted accordingly.

4.3 VELOCITY: AN ATTRIBUTE OF FLOW

The evaluation of the average velocity within any cross section of an open-channel conduit can be a tedious process unless certain principles are understood. The grasp of these concepts permits a greater understanding of this part of the flow profile.

Velocity, as used here, is the speed at which a volume is moving along a conveyance. Within the conveyance, every molecule contributes to the total velocity profile. Many molecules contribute to like areas, known as *velocity layers*. These are boundaries within the flow volume that contribute to the movement of the volume.[1] The combination of these boundaries contributes to the entire movement of the volume.

The layers continually slip and move over each other as the volume passes through the conveyance. Some characteristics particular to open-channel flow patterns are of note. The velocities tend to be highest at points of least resistance and lower at greater resistances. So the velocities can be expected, under ideal circumstances, to be greatest near the center and toward the surface of a volume, and least as the wetted perimeters are approached. Ideal conditions would show velocity layers fanning outward from the maximum velocity layer toward the walls of the conveyance.[2]

The ideal conditions would show uniform velocity patterns radiating outward from a point in the center and just under the surface of the flow volume. These ideal conditions consist of smooth, unobstructed conveyances with no bends or other factors that would impede or obstruct the movement of the water volume. However, component velocities are directly affected by both upstream and downstream conditions that tend to perturb the profile and the associated velocity layers. It is important, at this point, to properly ascertain the conditions of the upstream and downstream piping at least 20 pipe diameters from the measurement point. Bends and

1. G. Tchobanoglous, *Wastewater Engineering,* 2nd ed. (New York: McGraw-Hill, 1979), Tables 2-19 and 2-20, 50–51.

2. R. Giles, *Fluid Mechanics and Hydraulics,* 2nd Ed. (New York: McGraw-Hill, 1962), 160–161.

faults in the pipe also perturb the profile and may make it nonuniform. Faulty bricks, grit deposition, and other debris act as a dam and also perturb the profile. It is important to note any condition that would be considered a deviation from the ideal conditions, as it may have a direct impact on the distribution of velocities within the volume.

The average velocity is derived from the components of the velocity layers that are components of the flow volume. Each velocity layer contributes to the total movement of the volume within the conduit. It can be considered that the movement of the volume has some average speed or average velocity at which it moves. This average is the weighted contribution of each of the layers within it. Finding the average velocity practically and with available technology is addressed later in the material.

4.4 LEVEL

Level is the measurement that determines the height of the water volume within the pipe as it makes its passage. It is probably the most arbitrarily used term within infrastructure management and, potentially, the most misused. The concept is again reviewed in various chapters as it pertains to instrumentation and the methods commonly used in making the measurement. However, an understanding of the basic concept of level is required, separate from instrumentation, to fully comprehend the application of any measuring device.

The discussion of level revolves around two perspectives. The first is the level of the volume within the conduit, which is the depth of the water taken from the water's surface to the lowest part of the pipe. The second is the level over or otherwise attributed to the sensor, which is known as the *height of the water column*. This may appear to be the same measurement, but subtle differences do exist. Proper discussion is necessary to align the terminology. Manufacturers will use one or the other based on their technology.

For the height within the conveyance from the water's surface to the maximum depth, the point from which to measure is easy to identify in uniform piping arrangements: it is simply the bottom of the pipe at the maximum point of depth. In nonuniform piping arrangements, the point is more difficult to identify. However, nonuniform piping arrangements are generally are not encountered in sewer infrastructures. If they are, the maximum depth within the pipe is to be decided upon by the system operators, engineers, or other knowledgeable party in the course of determining points of interest.

The height of the water column over a particular sensor is theoretical because the water column is assumed to have no width, only height. It also is generally not a direct measurement of the maximum depth of the water but some relational operator through which the desired depth can be obtained. The water column measurement is useful when extensive obstructions or other occlusions prevent direct measurement of the lowest point in the infrastructure. Use of the height of the water column involves profiling and other extensive knowledge of the system at the point where instrumentation is being installed, otherwise the height being determined has little meaning.

Profiling is the surveying—either through checking various level points, or direct observation—of the pipe or structure of interest within the infrastructure. A profile should be performed before any equipment is introduced into the system so the characteristics can be observed and unexpected results can be related properly. Many unknown hydraulic characteristics of a point of interest become visible with the use of intrusive instrumentation. Without proper profiling, these now-visible characteristics may potentially be dismissed as equipment malfunctions and valuable information about the system may be wasted or lost.

The generation of a proper level measurement produces the basis for determining the boundary condition of cross-sectional area. The better the level measurement, the more reliably the cross-sectional area can be determined. Although the accuracy of the physical profile ultimately determines the regularity of the cross section being measured, proper translation of the measurement into the boundary condition is required. Any inaccurate measurement by the level sensor renders the best physical profiling useless. For that reason, the integrity and applicability of the level measurement is critical in the translation of any boundary condition.

4.5 CROSS-SECTIONAL AREA

The generation of the cross sectional area is the formation of the boundary condition of flow. By generating the boundary condition, limits can be placed on the rate and division of measurement and required by the system operator. Although a discussion of the cross section boundary condition preceded this section, a more formal presentation is given here to complete the reader's understanding of the basic concepts.

As discussed previously, if it were possible to cleave the flow volume, the end would be the cross section. Also being cleaved would be the velocity layers that run throughout the flow volume. This is the boundary condition or a start/stop point for measurement. The cross section is one of the essential components of generation of a flow or flow rate, which makes its own individual components important as well.

Taking the height of the volume as noted above, tracing the wetted perimeter, and calculating the area that it borders yields the construction of a cross section. The wetted perimeter is the portion of the flow that finds its boundary against the walls of the structure or pipe of interest. The area generated from this procedure becomes the boundary for determining flow. The shape of the wetted perimeter is the geometric physical features of the conduit. The level that the volume produces within the conduit determines the portion of the geometry of the conduit or structure that forms the wetted perimeter. Therefore, the wetted perimeter is the edge of the boundary from which the cross section is formed.

The assumption, when developing a cross-sectional profile, is that the volume being considered is uniform in its level or depth to the cross section. That is, the height does not vary over the volume being measured. This is a correct generalization if the boundary condition is one slice, as noted earlier. It can remain that one slice if the flow volume is considered passing through that stationary and fixed boundary condition. Within this text, unless otherwise noted, this is the condition referenced for the boundary in any discussion of flow. Variations are noted as they arise. This establishes basic parameters as a starting point in the material for application of devices to properly survey the infrastructure and increase its overall effectiveness to modern management techniques.

4.6 THE CONTINUITY EQUATION

The discussion above introduces the concept of flow and its calculation. All parts of flow have been evaluated and its characteristics described. The next logical step is to bring all the pieces together and develop expressions that can accurately describe the conveyance of wastewater within the infrastructure. Also, this provides the basis for deploying monitoring, control, and communications equipment to better manage these assets.

Flow is designated Q in most engineering applications unless otherwise noted. The units of flow determine if the quantity being sought is flow or flow rate. A quantity is a measurable whole, a numerical amount of a desired result. Flow or flow rate (designated by Q) is a reasonable result of the Continuity Equation.

Another component of the continuity equation is the average velocity, as seen from the cross-sectional viewpoint. The average velocity, generally referred to as *V*, is one of two components required to generate the quantity. The velocity, or speed at which the water is moving, has direct and proportional impact on the *Q* being calculated. Velocity is the rate of movement of the volume within the pipe. Its units are in length per time. The continuity equation requires the unit length per time interval for proper result generation. The usual units related to sewer infrastructures are feet per second or meters per second.

The second component is the cross-sectional area, or *A*. This is a two-dimensional view as one would look into the cleaved portion of the flow, as made by the wetted perimeter. The units are in area using the wetted perimeter for the boundary. Two boundaries must be considered here. The first boundary is the face of the cross section. The second is the contour as made by the wetted perimeter. Together, this makes a boundary that the volume will pass through and be measured. The usual and ordinary units related to sewer infrastructures are square feet or square meters.

The continuity equation is:

$$Q = VA$$

where *Q* is the quantity of flow, *V* is the average velocity, and *A* is the cross-sectional area of the flow. Velocity, or the speed at which the water is moving through the conduit, is expressed in units of distance over time, usually feet or meters per second. *A*, the cross-sectional area of the water in the conduit, is expressed in square feet or square meters.

The continuity equation is the basis for all values of flow,[3] including all the complex weir equations and free-flow matrices that generate potential *Q*s.* All more complex equations basically seek to generate a flow *(Q)* from the rate over time at which a volume is moving *(V)* through the area of some fixed boundary condition *(A)*. This is an important concept to understand, as one can easily be misled by the complex numerical analysis required to generate *Q* from some elaborate hydraulic network.

4.7 STRUCTURES

Up to this point, all basic definitions revolved around the movement of water and its expression for transport to some destination. Now let us turn to the methods of conveyance. Many good texts are available about the design and construction of structures contained within a operating utility's venue

3. R. Giles, *Fluid Mechanics and Hydraulics*, 2nd ed. (New York: McGraw-Hill, 1962), 160.

of responsibility, the description here is limited to a basic presentation of the more common structures from the point of view of management.

A structure is defined, for the purpose of this text, as any fixed, unmoveable physical construction within the infrastructure that conveys wastewater to some terminal point. The terminal point may be be another structure, a receiving body of water, or a treatment works; it makes little difference in this context. Structures are generally buried or otherwise removed from the general public. A structure can divert, temporarily store, relieve, or directly convey water to a specific area as designated by the system designers.

4.8 RECEIVING AREA

All structures within a drainage district have a point of termination—that is, a place where the water is targeted to go after being collected. This is called the *receiving area.* The receiving area is the delivery point at which, treated or untreated, the wastewater is suppose to arrive from the particular drainage district or pumping system. A drainage district is a designated area or territory that an infrastructure serves. Generally, the drainage district is delineated by the infrastructure piping system. However, political and social influences also affect the boundaries of such districts. Each drainage district is targeted to a receiving area. The receiving area can be a drainage pond, river, lake, stream, ocean, or treatment facility. It may include structures that regulate the delivery of wastewater.

There are two principal considerations when dealing with receiving bodies of water (rivers, lakes, ocean, etc.): the water being delivered and the effect of the receiving waters coming back into the delivery system. Keep in mind that, particularly during a storm event, both the level of the infrastructure delivery system and that of the receiving waters potentially rise with runoff. This is also true in tidally affected areas. It is therefore important to realize that design conditions must be met in order to keep the receiving waters out of the infrastructure and the delivery water flowing as well as possible. Most engineering studies done prior to design take these matters in consideration and the most common or economically feasible obstacles are accounted for in the implementation. Many times, compromises are made between the cost of specific construction and the residual effects. All of these factors are particular to the individual location and the regulations surrounding it.[4]

4. See *Effects of Polluting Discharges on the Thames Estuary,* Water Pollution Research Technical Paper #11, for an excellent example of the vast detail that can be achieved in such investigations.

Receiving basins, such as the ones seen near shopping centers and housing developments, retain the water so it can be reabsorbed into below-ground aquifers and underground streams. Here too there are two considerations: the capacity of the basin to retain the water and the clearing of debris based on the frequency of buildup. These are site-specific details and probably not known until after the system is fully operational and long-term observation is performed.

A treatment plant is considered a receiving area as well. It is also the most critical hydraulic control point along any given collector system. The tanks and pumps in a treatment plant are factors that influence the flow of wastewater. It is necessary to be familiar with the operating strategies of any facility that influences a collector system of any type.

Wet wells and pump stations are also receiving areas; however, they are of minor importance compared to the treatment facility. The newer wet-weather treatment facilities currently being constructed also will influence the systems, depending on their connection point and design functions. These stations are design specific and have slightly less impact on the collector system in its receiving areas than receiving waters or a treatment facility have. Nonetheless, they need to be noted and their function understood for the entire picture of the collection network to be clear, especially in view of how the receiving area functions within it.

In sum, the receiving area is a target for the delivery of wastewater. It may contain structures or include structures that regulate the interface of delivery and receiving. However, all structures terminate at the receiving area for purposes of wastewater delivery.

4.9 STORM SEWER SYSTEM

A storm sewer system is a drainage system that redirects water runoff resulting from a weather event away from areas of public or private use where its residual effects would be undesirable. Uncontrolled stormwater runoff can be a mere public nuisance or inconvenience, or it may have serious adverse effects on public health, safety, and property. Uncontrolled stormwater runoff is classified, in many areas, as a non-point pollutant source. Non-point pollution is a large and critical contributor to watershed pollution.[5]

5. "Urban Pollution Prevention and Control Planning," U.S. Environmental Protection Agency, EPA/625/R-93/004, 2.

The sizing of such systems is strictly based on the requirements of the drainage area they serve and the amount of potential runoff expected for that drainage area. The sizing begins with the determination of an average weather event, known as a *design storm.* The design storm is derived from meteorological information and information about relevant local hydrological and geological factors (such as large bodies of water, elevations, peripheral mountain ranges or deserts, and prevailing winds). Most states keep records, as does the National Oceanic and Atmospheric Administration (NOAA), of the timing and duration of weather events. From these records, the type of storm that occurs most frequently can be determined. Also, the largest expected storm(s), or 100-year storm(s), as they are called, can be examined along with frequency of appearance. Also, regulatory groups influence the factors of a design storm. There are few set formulas for coming to this determination, but good judgment, experience, and accepted engineering practices make the design storm a useful tool.

Once the design storm is established, it can be used as the basis for determining the size, design, and location of the storm sewer system. Usually, the storm runoff is channeled to either a receiving body of water or to drainage ponds from which the runoff can be reabsorbed by the groundwater system. Storm systems that flow to receiving waters generally have length and channeling considerations. Storm systems that flow to drainage ponds generally have sizing considerations. In all cases, their movement through populated areas poses issues of location, depth in the ground, easements or structure placement rights, and, potentially, the disturbance of established groundwater locations. Political issues and the influence of regulatory and interest groups also play heavy roles in the systems' eventual placement and design. Therefore, when looking at existing systems, consider that what appears to be strange engineering should be closely examined for the sometimes hidden factors that motivated the system's ultimate design and construction.

The surface feeds to these stormwater systems generally are the catchment basins seen on the sides of curbs, gratings on a street, or feeds from road drainage.[6] Most are fairly simple openings that allow access to the underground piping system to provide some path for the runoff and keep property from flooding. Although such inlets can be made elaborate with

6. "Planning Phase: Preliminary Assessment of Wet Weather Problems," UPM FR/CL 0002, November 1994.

traps and screening systems, the basic idea is to get the stormwater off the surface and into the underground system, thereby allowing the runoff to be channeled to some more desirable destination.

A storm sewer system contains structures of its own. The simplest system contains only the piping that delivers the runoff to a receiving water or detention area. Some systems join other as a relief to ones that are potentially overtaxed. These connections are strictly pipes joining pipes. The more elaborate arrangements feature large underground holding tanks, called *relief chambers,* that temporarily store runoff so as not to overtax a system during storm events. These holding structures are built right into the line and are allowed to fill and drain based on the capacity of the upstream and downstream conduits. They allow the system to absorb more runoff per unit length by storing the water until the target destination is better able to receive it. These structures generate hydraulic conditions particular to their location and receiving/discharge capacities. It is important to note if there are any storage structures along the traverse of a storm sewer. These structures, additions from building improvements, and other features can make the predevelopment rates vary greatly from the postdevelopment runoff rates.

Within the structure of the storm sewer are other structures that assist in the movement of the flow. Some of these are called *pump stations* or *lift stations.* These are pumps of a specific size and array that lift stormwater into a higher portion of the sewer so it may travel by gravity to the receiving area. (Pump and lift stations have other uses in other types of sewers.) These stations run intermittently based on the depth of a wet well or the upstream capacity being received by the station. Pump and lift stations are installed when their cost is lower than that of constructing gravity sewers, or when hydraulic constraints require other unique solutions. For this reason, careful economic analysis is important in the evaluation of the pumping option.

Pump stations, as related to storm sewer systems, are generally found in low-lying areas and are meant to keep stormwater out of public and private structures, such as roadbeds, stadiums, residential developments, and other areas prone to flooding. Each scheme for their operation is unique, which limits specific discussion. The important concept to bear in mind is that a pump station drains areas where stormwater could otherwise cause property damage. Sometimes storm pumps are used to minimize groundwater levels at the same location. In many design cases, a wet well is constructed and the pump drains the well based on a set level known as the *operating point.* Again, the array and operating protocol, along with station design, is unique to each system and location. Pump stations should be understood in relation to the system at large.

A lift station, as related to storm sewers, takes stormwater from a low-level gravity-driven storm sewer and raises it to a sewer with higher potential gravity. Lift stations are common in flatlands where the elevations do not change much over distance. A lift station is also the preferred alternative where construction of a full gravity sewer would be too costly or would disrupt specific aspects of a targeted section of the drainage district.

The concept of flushing is critical to understanding the operation of the storm sewer system. The first flush of a storm sewer usually carries a lot of debris that accumulated during the dry period. The first flush is defined as the first wave of water carried through the storm sewer after a prolonged dry period, generally a period of weeks to months. The longer the dry period, the more debris accumulates and the more is washed to the receiving area. Subsequent flushings carry less debris.

It is possible to characterize the amount of debris carried by a particular system through observation and screening. Screening is the process of placing a screen at the end of the storm sewer and inventorying the accumulated debris at the end of the storm event. This is a tedious process and very labor-intensive, but one that provides a reasonable characterization of the mass carried during particular events by that part of the system. The important concept here is that the content of stormwater depends on the interval of rainfall and the volume of stormwater it carries.

The monitoring system has the potential to assist the storm system in monitoring for planned oil dumping programs, bypassing endeavors, and other abuses of the storm system. Very small quantities of material may be difficult to detect, but repeat offenses and trouble locations can be patterned. Therefore, the characteristics of a particular storm system are important to understand so its classification as a point of interest along a critical path can be ascertained.

Stormwater systems are of all sizes and shapes, so they cannot be generally described except by usage. They are constructed of numerous materials, ranging from small corrugated steel tubes to large and fancy brick and stone designs. They can serve small areas or entire cities. Size is no longer a classification for storm sewers. Construction sites of all types are now required to have stormwater systems in place right from groundbreaking. New stormwater regulations also require the operator to characterize this runoff to the regulatory agency. It is imperative to realize that storm sewers can be of any size, shape, and design; all are specific to their designated applications. Each storm drainage system within a given area should be understood for its purposes and all of its construction particulars categorized from the extremely small to the extremely large.

4.10 SANITARY SEWER SYSTEMS

The most commonly recognized sewer type is the sanitary sewer, which is designed to carry domestic sewage and processed industrial wastes to a treatment facility. Domestic sewage is that wastewater generated by residential households. Processed industrial waste, for this discussion, is wastewater that has been certified for processing by a biological treatment process in a given drainage district. These are the sewers in which most chemical reactions and compound mixings takes place in the sewage along its route to the treatment works. The sanitary sewer system is relied on to provide communities with safe and reliable transportation of sewage from its point of origin to a treatment point. Systems where the sanitary and storm components are separate entities are generally referred to as *separate sewer systems*. Separate systems are preferred in modern sewer design.

Sanitary sewers are sized to the drainage district they serve based on the number of households, residents per household, businesses, and the potential usage of the business. Generally, a community assumes a certain consumption rate of potable water per household; that is the daily discharge rate used in the calculation of sewer size. This is an important consideration in zoning and licensing. Communities may deny building permits if the development will exceed the rated capacity of the sewer system. Cities where zoning was loose or nonexistent experienced difficulties in development when high-rise apartment buildings were placed in areas serviced by inadequate sanitary systems. It is difficult to simply dig up a sewer and replace it as the need arises. Therefore, a balance in long-term construction planning, zoning, and development is required to keep properly sized sanitary sewers available for land improvements.

The concept of sanitary sewers embodies the idea that only planned wastewater will be conveyed by the piping—that is, wastewater that is knowingly introduced into the system for purposes of conveyance. In practical considerations, infiltration and inflow are always issues in system maintenance. The design storm is taken into account, along with potential groundwater sources. This is because storms influence groundwater levels, which, in turn, have the potential for infiltration and inflow to the sanitary system. The integration of these areas is at the discretion of the design engineers, as they establish the system based on local geological and hydrological influences. These influences are unique to the area and each area has to be classified accordingly.

The feeds to a sanitary sewer include but are not limited to the following connection points. From a private home, the cleanout vent and roof leader service outside are connected to the infrastructure. Most new construction and revisions of local regulations call for disconnection of the roof leader, but in many communities, the roof leader remains connected. Within the home, the toilet, sinks, washer, and other appliances requiring water drains are connected to the sanitary service. In some instances, local drains in front of a home also connect to the sanitary system using a connecting pipe known as a *lateral.* From industrial users, roof leaders and outside drains are connected, as are lavatory services. The various processes for which the industry requires drainage are also connected. If the regulations require processing of the wastewater prior to discharge, the principal connection point is the pretreatment facility within the industrial site. Otherwise, the discharge is direct to the sanitary system. Also included are external drains designed into the properties per local ordinances.

Lateral connections recently have become the subject of investigations in many communities. House laterals are intended to join the sanitary sewer system. However, for any number of reasons, from honest mistakes to deliberate deceit, many laterals are connected to the storm system, which rarely discharges to a treatment works. The result is known as a *cross connection* within the sewer systems. This makes the storm system a source of pollution, as it constantly introduces untreated sewage to an area that expected only storm runoff. Laterals and their connections are thus of tremendous interest, especially in old communities where regulations may not have existed at time of construction, and in newer construction where the storm sewer location was preferred to the sanitary sewer. Raw sewage running within a storm system is usually a sign of lateral cross connections. It is a direct hazard to health and carries the potential of property damage, depending on the severity and conditions of the cross connections.

Domestic sewage is fairly uniform in strength and content, and the output of the average residential customer is usually predictable: fecal material, ground food residuals, (as from garbage disposals), wastewater from washers and like appliances, and runoff where connections are established. Industrial waste, in contrast, tends to vary in complexity depending upon the type of industry. Also, there is the requirement that much of the industrial waste must be pretreated prior to discharge into the municipal system. This variety of materials, constantly mixing and being agitated on its way to the treatment works, can produce fairly nasty compounds that are highly toxic in nature and sometimes produces loadings undesirable to the treatment facility. The treatment facility can generally handle these loadings as long as they remain within the parameters of the design of the plant. The point is that the sanitary sewer is, in itself, a bioreactor. The sewage product entering the system is substantially changed by the time it gets to the treatment works.

The sanitary system contains structures. One of the structures causing hydraulic impacts is the entry manhole. Because many manholes are deep and, at times, bigger than the pipes they service, they may act as a series of overflow chambers or relief chambers if the system becomes overtaxed. Assume, for example, a 12-inch sanitary sewer being serviced by a 30-inch manhole at a depth of 15 feet. Locate a manhole every 200 feet in the system and say there is about 5,000 feet (or 25 manholes) between the beginning of the run and the treatment facility. Given these parameters, there is the potential for about 118 cubic feet of static storage for each manhole, or about 2,900 cubic feet of static storage along the entire line. Of course, these are rough calculations to be used for discussion purposes only; the important point is that the manholes have the potential to act as relief structures. Therefore, the location and placement of each manhole within the sanitary system is an important to understanding the hydraulic action of the sewer network.

Other structures within the sanitary system include relief or bypass chambers, bypass sewers, and pump stations. Relief or bypass chambers are known in some areas as *off-* or *on-line storage tanks.* They provide underground storage for the sewage during high flow situations and keep the material from entering private property, public conveyances, and commercial locations. They are placed at the time of system design or added later in response to unanticipated or changed surface conditions. Bypass sewers rechannel wastewater to relieve the flow volume within a targeted system and provide additional conveyance capacity to the treatment plant. These structures are common in areas in which unexpected development occurred and the loads had to be accommodated.

Among the more common structures within the sanitary system are pump and lift stations. Their function is similar to that of storm sewer stations, but with notable exceptions. Lift stations move sewage from a lower-elevation sewer to a higher-elevation sewer to maintain gravity flow. They also are used where it is not economically feasible to construct full-flow gravity sewers. The lift station generally services a wet well, which fills and drains based on a set level. Any number of variations of this design make it adaptable to the particular circumstance and need of the infrastructure system.

Pump stations also maintain the flow of sewage from the sanitary system to the treatment plant. They maintain service where it would be unworkable or economically prohibitive to construct a full gravity system. However, the sanitary pump station is expected to operate continuously, as opposed to a storm station, which operates only during weather events. Once a pump station is put into service, it is difficult, if not impossible, to stop its operation. It is therefore critical to the proper operation of the line that it services. Many designs for these stations have redundant pumps so that one can be removed for repairs or maintenance while keeping the site operational. Pump stations also may act as junctions connecting several smaller feeds to a larger sewer service. This increases the delivery capacity of all lines involved, so the loadings can be greatly increased beyond general gravity calculations. A pump station within a line is a vital component of the sanitary sewer it services.

Infiltration and inflow are critical to the sanitary sewer in relation to flushing. If the system experiences heavy infiltration and inflow, the flow rises and falls based on groundwater levels and storm events. This may cause scouring when the level drops too low within the system or flushing when the levels get too high. Scouring strips the grit base and causes particulate matter to arrive at the head of the plant. Flushing may cause trash and other floatables to arrive at the head of the plant. It also dilutes the waste stream, thereby essentially reducing the percentage of solids (a biological treatment works requires a certain amount of solids to operate properly). If these conditions are unexpected by the treatment works, their presence may cause unwanted negative effects. It is therefore advisable to remove as much infiltration and inflow to the system as possible to reduce the delivery of undesirables to the head of the plant.

The types of sanitary sewers range from cast-in-place poured concrete to brick double-lined to stone to PVC systems; some older brick sewers in major cities have elaborate patterns and structures and are like works of art. Sanitary sewers are engineered to the needs of a specific service area. In relation to storm sewers, they are an order of magnitude smaller and are expected to flow continuously. Their shape can range from circular to egg-shaped to rectangular to any other configuration the system designer thought would properly service the drainage district. It is important to catalog and inventory the construction material, shape, and potential flows within a sanitary system because the interfacing of all the unique structures affects the hydraulic conditions they generate. Keep in mind that the original design criteria for a system can be radically different from its current use because of changing surface conditions over time. For this reason, it is important to truly understand the makeup of the sanitary infrastructure for purposes of zoning and development, rehabilitation, and intrusion to the sanitary system at large.

4.11 COMBINED SEWER SYSTEMS

The sewer infrastructure system that has received the most attention in recent years is the combined sewer system, a hybrid of storm and sanitary systems. Usually found in older urban areas, combined systems channel sewage to a treatment works along with stormwater during storm events. When the hydraulic capacity of the combined sewer is achieved, the mixed sewage and stormwater is expelled through control structures and into receiving waters. Until recent years, this was an accepted engineering practice for channeling wastewater to a treatment works, but research over the last 30 years or so has shown combined sewers to pose health and other environmental hazards to watercourses and receiving watersheds, hazards that were unknown at the time of their acceptance. These are now the target of scrutiny and massive regulation efforts.

The concept of a combined sewer system is interesting and, when the system was installed correctly, a tribute to engineering practices. The intention of civil engineers and municipal managers was to channel as much water from as many areas as possible to a treatment facility for processing before returning it to some watershed or water body. Studies were performed based on the mean elevations of the receiving waters, tidal influences where applicable, terrain and geological influences, hydrology (including the design

Figure 4-1 This is a typical type of difficult installation in which the conduit has a difficult fit and cabling in places must be exposed. Note debris buildup around the conduit in the lower part of the picture.

storm), and expected growth of the service community. Agreements were reached on reasonable hydraulic capacities, then the combined sewer was designed and a regulating structure introduced at the end of the sewer run.

The theory was that the regulating structure would retain as much water within the sewer as possible until the plant had the ability to process it. The regulating structure is usually equipped with an overflow weir or regulating orifice. This weir is nothing more than an opening designed in or around the top of the regulating structure that allows sewage exceeding the capacity of the system to be discharged back to the receiving waters. The size of the overflow weir is calibrated to the upstream capacity so the maximum water is retained in the system with minimal discharge to the environment. The orifice is a metal plate that can be either fixed or adjusted; it controls the flow of wastewater back through the regulator to the treatment works. Adjustable orifices can determine the amount of upstream storage possible in the combined system. In many cases, the regulating structures have feed-forward and feed-backward systems that assist in the regulating of sewage flow and in storage schemes.

The operating sequence is that all dry-weather flow is delivered to a sewage treatment works and that the first flush of a storm is also delivered for processing as well. Any discharge over the weir is dilute and has little detrimental effect on the surrounding watershed. From a design viewpoint, this was achievable based on the information available at the time and the design and modeling tools of the day.[7]

The combined systems are extensive through older sections of Europe, the British Isles, and cities in 16 U.S. states. They were installed in a period when municipal managers and environmentalists were becoming more conscious of the necessity to treat wastewater prior to its entry back to a receiving body of water. Although details are particular to each system, the systems are designed to treat most of the combined sewage for most of the predictable events in a given drainage district. Thus the intent was to get as much of the water to the plant as possible. The residual that was discharged was assumed to be inconsequential to the environment—that is, the environment would be able to absorb it with no ill effects.

The only problem was that although the system planners and engineers anticipated future needs as best they could, they could not know exactly how communities would develop. Over the years, changing regulations, building codes, budgets, external political pressures, changing topology and hydrology, and the unexpected caused the combined sewers fall short. At the time of this writing, the second phase of a massive regulation program is being introduced in the United States to combat the effects of combined sewers on the watersheds to which they discharge. Regulators and mitigation programs have existed in the United Kingdom for many years and have continued even after utility privatization in 1989.

Combined sewer systems traditionally are sized for two to three times the expected dry-weather flow.[8] This particular baseline for system design is

7. Introduction, U.S. Environmental Protection Agency, 382-B-95–002, 1. This is the standard U.S. guidance document for CSO LTCP development and implementation.

8. ASCE Engineering Practice #77, WEF MOP #FD-20, *Design and Construction of Urban Stormwater Management Systems,* 1992, ch. 10, 403.

modified by each drainage district, taking into account the factors described previously. A discussion of the actual sizing techniques is no longer needed, as in the United States combined systems are no longer constructed. The problem with existing combined sewers usually is related to the growth of the community. As the population expanded, standalone sanitary systems with separate storm systems were added in the newer areas; however, the new sanitary system was connected to the old combined system. Therefore, one of the biggest potential feeds to a combined system is not the established communities it serves but the expanded communities that require wastewater drainage services. Usually, this was not anticipated in the original design, and the systems became overtaxed very quickly. This appears to be a major reason why combined systems have failed in many areas. It is not necessarily because they are ill designed. It appears to be—and this needs to be assessed on a case-by-case basis—that the intended use and the actual use differ greatly because of unpredicted community expansion. This point is discussed here because the history of the system expansion can explain a lot of the use, success, or potential failure of the infrastructure.

A common combined sewer system includes the combined sewer, a regulating structure, a collecting sewer, and an intercepting sewer. The combined sewer is the pipe that collects the wastewater from its various feeds. The collected water is delivered to a regulating structure, which is a device or fixture that causes the storage or retention of wastewater within the combined sewer. It also accepts the overflow, which is discharged to a receiving water. The collecting sewer is the pipe that receives the regulated water from the combined sewer for delivery to the treatment works. The collecting sewer joins the regulating structure and combined sewer to the intercepting sewer, which delivers the wastewater to the treatment facility.

Each of these parts is integral to the proper operation of the combined system. Some variations of these elements exist, such as a slot regulator. In this case, the collecting sewer is omitted because the regulator is cut directly into the intercepting sewer. Everything else is unchanged. Such variations are common in the customizing of these systems for individual drainage districts.

Combined sewers are fed in two ways: from the same feeds as a sanitary system, and from the same feeds as a storm system. The various wastewaters mix and produce different strengths depending upon whether the weather has been dry or wet. They have the same infiltration and inflow problems as storm and sanitary systems. This makes the flow, if these conditions exist, vary with groundwater level. In extensive systems, large changes in seasonal flows are possible for potentially faulty lengths of combined sewers.

The combined sewers feed what is known as an *intercepting sewer,* which moves the regulated flow from the combined sewers through the regulating structures and brings it to the treatment facility. An intercepting sewer services many lengths of combined sewers. The regulating structures assist in keeping the intercepting sewer from being surcharged during weather events. If the combined sewer becomes overtaxed, the excess wastewater is returned to the receiving waters rather than overloading the intercepting sewer or flooding the streets and property above it.

Many structures are attached to a combined system. They fall into two classes: regulating and storage. The regulating chamber, which regulates the flow to the interceptor and stores what it can within the limits of the combined system, was already described. A detailed discussion of each type of regulating structure is beyond the scope of the text. For the subject matter at hand, it suffices to say that there are many varieties of regulating structures and they should be noted during system inventory. The uniqueness of each drainage district almost demands that each structure be custom-constructed to its current and projected requirements. For example, areas affected by tides have devices that prevent the inflow of tidewaters while allowing discharge of sewage in overload conditions. Some of the structures are completely unique to the geology and hydrology of a particular drainage area and have no reference other than the community's documentation.

Until recent decades, most off-line combined sewer overflow (CSO) storage facilities were in the United Kingdom and Europe.[9] More recently, these storage facilities are being installed in sensitive watershed areas in the United States to protect these waters from CSO spills. Detention basins, storage tanks, swirl concentrators, and like devices are being installed or have been installed in many combined sewer systems. They serve two distinct purposes. The first is to provide off-line relief from overtaxed combined sewers. The second is to remove solids and debris before returning the wa-

9. The UPM offers extensive information concerning the use and placement of on- and off-line CSO systems based on implementation experience.

ter to the receiving waters or to the treatment facility, when capacity permits. A swirl concentrator, as an example, does not provide off-line storage so much as it removes solids and disinfects, to some degree, prior to the waters' return to the system. An off-line storage tank provides more water storage than solids removal during a storm event with essentially no disinfection potential. Therefore, a distinction can be drawn between the passive and active storage of wastewater. The choice, of course, depends on the needs of the community served.

Pump and lift stations also service combined sewers. Generally, however, pumping is performed within the intercepting sewer portion and not in the combined sewer itself unless specific needs arise. Pump stations combine several intercepting sewers into one larger sewer for final transport to a treatment facility. They also service the combined portion in the case of low-lying structures and where groundwater may cause damage to property. The waters are fed into a wet well and pumped into the combined system at a given setpoint level. This is common at the base of roadways and buildings, where a combined sewer is readily accessible. Stations are located in areas where full gravity stretches of piping are prohibitively expensive or would cause too much disturbance to other existing underground utilities.

The excess wastewater is returned to a receiving water, generally referred to as the *watershed*. In years gone by, these overflows to the watershed essentially went unnoticed. However, as noted previously, they are the subject of massive regulations and remediation programs to minimize or, preferably, eliminate their existence and effects on the watersheds. The repair and remediation of CSOs is a project of massive expenditure for any community. This is mainly due to the construction costs for underground utilities and the need to cast in place many of the fixtures necessary to prevent CSOs. Again, most of the solutions are custom-designed and require extensive engineering. The best way to minimize the cost of CSO remediation is a complete system inventory, system study, and accurate remediation plan based on actual work to be performed. Proper understanding of the entire system and its hydraulic operation during a variety of storm events allows commonsense solutions financially palatable to all concerned. There is nothing inherently complex about the remediation of a CSO. The logistics of implementation is where the dollars become difficult.

Combined sewers and their overflows are interesting subjects of hydraulic ingenuity. Their source of pollution is generally due to unplanned expansion and capacities for which they were not designed. Their uniqueness to the communities they serve and the individual character of the regulating structures can pose fascinating engineering challenges. CSOs became of interest once their potential pollution impacts were more clearly understood; these were not apparent at the time of system design. The correction of these overflows will be of great interest for many years to come.

4.12 HYDRAULIC CONTROL POINTS

Flow moves through the infrastructure through a series of systems. Within these systems are points that affect its transition. Hydraulic control occurs where some action is required during the transit of the waste stream. Up to now, our discussion covered flow, its components, and its conduits. Now variables that shape and form its movement within the infrastructure are introduced. The hydraulic control point is not a structure or station; it is anything in the system that imposes control on the flow of wastewater between its origin and its destination.

Water can be controlled with something as complex as a multi-pump station or as simple as a small dam in the system. The important concept is that anything that changes the aggregate, uninhibited movement of water through the system is a control point for purposes of this discussion. Hydraulic control exists to efficiently move water through the infrastructure and to overcome obstacles that would otherwise impede its efficient delivery.

The largest hydraulic control point along any sewer infrastructure is the treatment plant. All the sanitary and combined systems end there. If the plant lowers its gates and stops the flow to it without proper planning, the results could be very undesirable. This action is the key for determining hydraulic control points. The treatment plant is the biggest entity, but anything else that affects the wastewater flow becomes a point of interest along the conveyance. These points include but are not limited to pumping and lift stations, regulating chambers, weirs and dams, off-line detention and storage tanks, junction chambers, and relief sewers.

Previous sections have described many of the major hydraulic control points. Some of the lesser hydraulic control points need to be outlined as well. The term *lesser* here is not in any way a measure of importance; it is only a classification for a control that fits into or is part of a major control

Figure 4-2 A slot regulator. The simple design and effectiveness of the device provides great difficulty in either level or flow measurement. The interceptor is below the slot, which makes the cable runs difficult.

point. The types are not actually classified officially, as their use and sizing, again, are particular to the individual system they service, making each device unique.

4.13 DAMS

A dam is a structure placed within a service system to prevent, split, divert, or restrict the flow of wastewater. It can be as simple as stacks of bricks located within a pipe to redirect water flow or as complex as structures that channel flow among systems as the level rises in the sewer. Dams provide hydraulic jumps to keep solids in suspension and can also cause turbulence that, at times, may be useful for proper transport of the waste stream. A dam can be part of the original construction or added as hydraulic characteristics dictate during system growth.

Generally, dams prevent flow in a given area or direct the flow of water in intended directions. On the plus side, they effectively control water flow

Figure 4-3 Complicated hydraulic patterns make the monitoring process difficult. Here, the free flow during dry weather makes the measurement less difficult. Depending of the loading and lessening of free flow will determine the accuracy of the measurement by any means.

without any maintenance other than structural repairs due to wear. Their downside is that their total effectiveness is not often visible on the surface because, like any internal device, it cannot be directly observed without proper instrumentation. Therefore, the best judgment of the system designer and engineer is required both in the placement of dams and when retrofitting of the system is performed.

The location, design, purpose, and structural content of dams is of critical importance to the infrastructure. Dams, depending on their application, retain quantities of flow and release them based on the upstream level. Undocumented dams may make the system appear to be functioning abnormally. However when their presence is known, their effect on flow behavior is predictable.

Dams are meant to keep out as much as they are to redirect. In the case of combined sewers near watersheds, dams keep the water from the watershed out of the sewer system as much as keep in the sewage. In tidal areas, the size of the dam is estimated from the highest known high tide and a localized algorithm is employed to determine tide height. The opening over the dam is the relief for the combined system; it is from here that the overflow, or CSO, emanates.

The arrangement of dams can provide storage within systems if implemented with foresight. If the dam is constructed with relief systems built into the bottom, it will provide a baffling effect that retains water for a given time frame. Long series of these baffles can retain water for periods as the system requires. Another example of hydraulic uses for dams is the protection of tide gates. In cases where surges are known to occur in a system, a dam can be erected so the tide gate does not open unless the flow during low-tide conditions exceeds the damming conditions. This not only generates storage within the system but provides an extra measure of protection to the watershed it services. Another use for a dam is to develop head pressure to push water through a regulator or other structure, depending on the engineers' requirements. Such configurations also provide in-line storage as well as harness water power for hydraulic functions. The uses of a dam are limited only by the imagination of the system engineer, designer, or operator.

Measurements of flow can be estimated by calculating the amount of free fall over a dam, and in many case, a dam is installed for just that purpose (though profiling the velocity for given levels is a preferred and more scientific and repeatable method of generating the flow and flow rates). If a dam is used for measurement, there must be free flow over the dam, unrestricted flow downstream, and laminar flow upstream. Standard rectangular weir equations are acceptable for generating estimated flows under these conditions. If the velocity for a given head rise is known, the cross-sectional area becomes fixed based on the structure; the continuity equation then can be employed, assuming the velocity is repeatable for a given set of levels, which under free-flow conditions is a valid generalization.

4.14 WEIRS

The weir is a lesser hydraulic control point that is the subject of intense debate regardless of where and how it is installed. A weir is more a measurement tool than a restrictive device; however, the flow must be restricted if the weir is to be effectively utilized in this capacity.

Weirs are simple to employ and cost-effective to implement.[10] They are increasingly being equipped with electronic instrumentation, because they must be constantly watched and recorded unless equipped with an electronic level device.

Generally installed over a dam or within a pipe, a weir produces tested velocities for a given level, thereby producing a theoretical flow or flow rate. Weirs come in various shapes and sizes, depending on the application. Before the days of instrumentation, the weir was a primary measuring device for most temporary and many permanent flow-monitoring applications. The geometric shape of the restriction determines the type and classification of the weir. The crest of the weir is the surface of the water that passes over and through the device. The nappe is the stream of water leaving the weir. Free flow, sometimes called *critical flow,* is the unrestricted pouring of water through and over the device into its next point of passage. Submerged flow, sometimes referred to in texts as *subcritical,* is where the passage of water over and through the weir is restricted by the downstream level. Submerged flow measurement using a weir tends to be inaccurate for any number of reasons. It can approach accuracy if compensated for using compensation tables based on upstream and downstream level measurements. This is cumbersome and not necessarily desirable. Weirs are best suited to areas of free flow and proper downstream drainage.

The airspace under the free flow of a weir is important to the accuracy of the measurement. Space for air passage should be sufficient to prevent restriction of the free fall of water. If airspace is insufficient between the nappe and the wall or structure, a vacuum may develop and impede the free flow of sewage. This generally occurs under two conditions, both of which may produce less than free-fall conditions and perturb the surface of the nappe. The first is improper sizing and installation of the weir; too large a weir does not provide proper upstream restriction and, therefore not enough free flow. The second condition is insufficient height from the free flow to the receiving pool.

A standard flat-topped dam may be used as a weir. One type of weir consists of a depression in the dam that is less than the width of the sewer. This type is useful in low flow conditions. The depression increases the head pressure behind the weir and provides the necessary environment to generate

10. M. Grant and B. Dawson, *Isco Open-Channel Flow Mesaurement Handbook,* 26.

a free-flowing nappe. The second type is the suppressed rectangular weir, where the width of the channel or sewer is the same as the width of the dam. In both cases, the minimum height from the crest to the bottom of the receiving area should be twice the minimum head estimated to flow in the channel.

Weirs come in many types, schemes, and geometric forms. Reference books contain complete printouts for every head rise over every potential incremental flow or flow rate. Electronic level measuring devices allow weirs to be used for remote or long-term measurement by translating the free flow level into a flow rate and storing the results over long time periods. Even as more complex and elaborate instrumentation is developed, it appears that the weir will remain a mainstay in measurement for a long time to come.

4.15 FLUMES

Flumes are measuring devices that work in a manner similar to weirs. They are, for the most part, permanent structures, unlike weirs, which are often temporary. The older materials used in flumes are poured concrete or iron fabricated in place. There are quickly being replaced with plastics and other composite materials. More detailed description of the measurement properties of a flume are described in subsequent chapters.

Broadly, a flume is the shaping of a conduit to restrict velocity to a known rate. When this velocity is correlated to height, a flow or flow rate can be determined mathematically. The various configurations and designs of flumes cause various restrictions within the upstream water volume because they are an integral part of the piping that it services. The range of flume configurations is wide, and there are numerous books on the style, size, configuration, installation requirements, and flow rate of flumes.

The structure of a flume generally depends on the buildup of solids within the restriction region. For this reason, the design of most flumes allows for self-cleaning; that is, the flume helps keep solids in suspension within the wastewater flow as it passes through its hydraulic restriction for measurement. This is important, as solids deposition within a sewer is always an issue to be considered; solids deposition, depending on its placement, produces a false level and yields an inaccurate measurement. The point here is that the better-designed flumes keep solids in suspension and passing through the restriction with little or no deposition to the bottom of the conduit.

A characteristic of older flumes is the fabrication in place within the pip-

Figure 4-4 Level monitor installed at a combined sewer regulator outlet. Note the potential for inaccurate readings if the devices are installed incorrectly because of the change in conduit shape.

Figure 4-5 An in-place flume measuring flow. The free flow is needed for proper flume measurement when level is used to convert to the volume quantity.

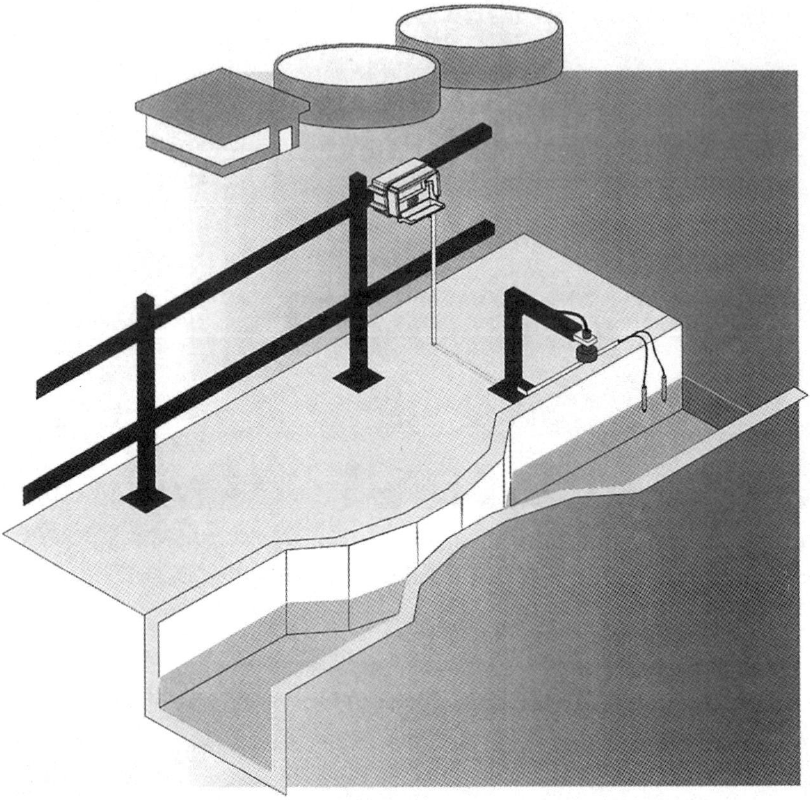

Figure 4-6 The measurement of levels in a Parshall flume. This configuration shows pressure and ultrasonic level measurement.

ing. This makes them integral parts of the infrastructure, more definitively a structure all on their own. Regardless of their characteristics for measurement, flumes are hydraulic restrictions; all such devices have the potential to generate solids deposition and to restrict the passage of the flow volume. The flume should be recognized as a structure within the sewer system and treated as any other fixed object that affects the flow to the treatment works.

4.16 TANKS OR STORAGE

Water presented to the infrastructure system can sometimes be diverted to storage in structures referred to as *tanks,* which can be in-line or off-line, depending on their use and application. Tanks retain wastewater for some predetermined period to keep the system from becoming overtaxed and to maintain consistent delivery to the treatment works. The discussion here is meant to identify tanks and their basic function with respect to the needs of the sewers. Reference was made to these earlier in discussions about relief chambers.

A surface tank can be a basin or pond where storm runoff is held and metered slowly into the infrastructure. This storage is useful during intense weather events because it gathers and retains sudden flows until the sewers can accept them. Surface tanks can be as simple as basin catchments and as complex as concrete structures fed by pump stations. Their basic characteristic is that they are visible from the surface, unlike most infrastructure applications, which reside below ground or otherwise out of sight.

In a sense, sewer tanks attenuate flows within a system. The constant filling and discharging of the tank based on upstream and downstream conditions is calculated in the design process and then observed and modified as necessary when the system comes on line. Sewer tanks are used also in areas of development where straight runs of sewers are economically infeasible and the predicted quantities of sewage would ordinarily overtax the design capacity of the line. Sewer tanks are usually concrete poured-in-place structures directly in line with the sewer system. Because of the retentive nature of these tanks, they are susceptible to grit deposition and should be monitored and receive maintenance regularly.

A CSO tank takes the excess of a combined sewer to its designed capacity in order to reduce or otherwise prevent the occurrence of CSOs. This type of tank can be on- or off-line, depending on the design function. Generally, a CSO tank is designed to store during periods of high flows until the treatment works can effectively process the water. The function of the tank is not only to prevent CSO discharges but also to prevent surface flooding where applicable. These tanks are gaining interest through the various worldwide CSO remediation programs. Although most of these structures are underground, some designs may be classified as surface tanks. Because they are attached to a combined sewer and, usually, part of a remediation program, the label *CSO tank* remains with them regardless of location.

A storm tank retains excess storm runoff. Storm tanks are generally made of poured-in-place concrete. Like the CSO tank, it operates during times of heavy flow, such as during storm events. The storm tank removes a volume of water and retains it until the storm sewer can receive it to reduce or otherwise prevent surface flooding. The prevention of surface flooding

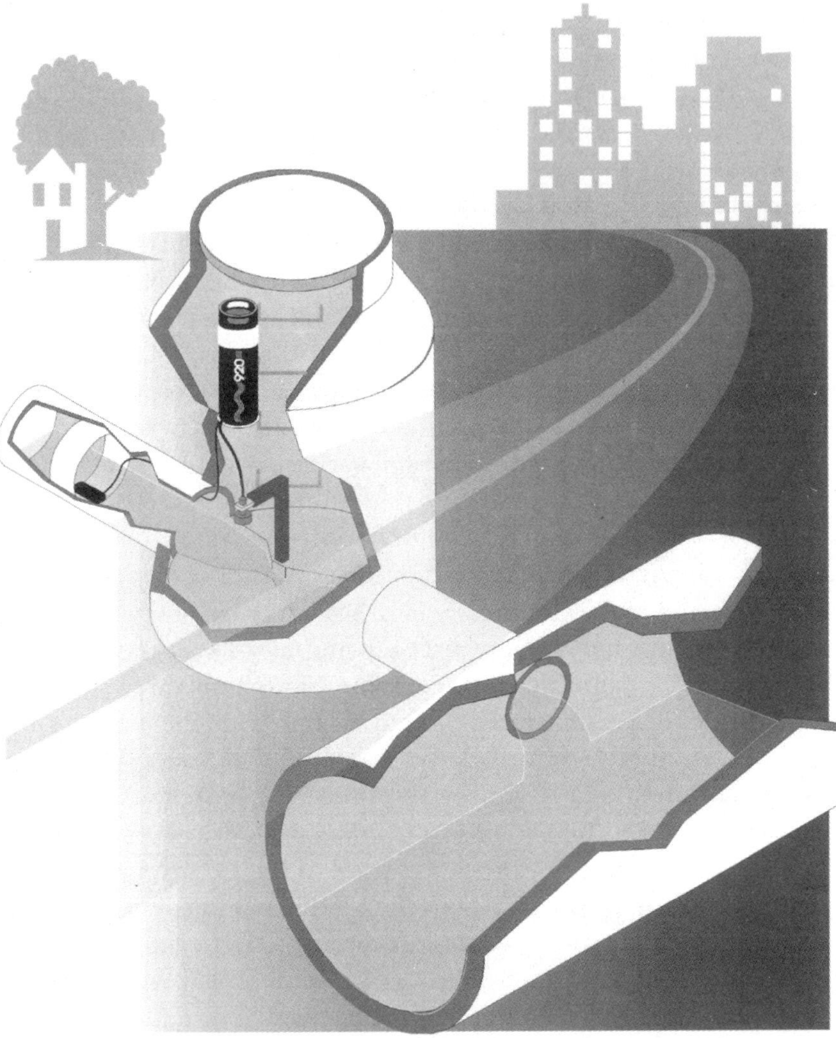

Figure 4-7 A method of verifying a Parshall flume under less than amicable mounting conditions. A band with a sensor is inserted upstream of the Parshall to verify its accuracy.

protects property from damage and minimizes the potential for pollution. These tanks can be off- or on-line, depending on the design criteria. In some applications, they are referred to as *relief chambers*. Generally, solids deposition is not a major issue. Floatable debris, such as trash and other surface litter, does become an issue, especially during the first flush after a particularly long dry spell. The storm tank assists in reducing the potential for surface flooding by offloading water when the storm sewer begins to surcharge.

Like tanks of all types, storm tanks are expensive to install because they involve excavation costs and custom concrete pours. Designed correctly and implemented properly, they do protect property, health, and safety. Tanks have hydraulic characteristics that make most flows nonlinear, an effect that needs to be taken into account whenever flows are considered in that drainage district.

4.17 DEEP TUNNELS

A very expensive alternative to tanks is the deep tunnel. These structures require specific geology, are always pumped-flow at some point, and are difficult to maintain. These large tunnels or holding tanks are deep in the earth and retain wastewater until the treatment facility can process it.

Although not common, they are mentioned here because where these major structures do exist, they may be the only reasonable alternative for cities and countries where population concentrations restrict the use of other technologies and wastewater storage strategies. A deep tunnel may well be a solution for the future as technologies mature and other alternatives become outdated.[11]

Deep tunnels are cut directly into a bedrock base. Pump stations introduce and extract the flow through a predetermined control scheme. Deep tunnels are difficult to place due to their surface pumping and venting requirements. Also, unlike more near-surface storage strategies, large air separation chambers are required to manage air exchange. This necessitates a complex design and dedicated operation and maintenance strategies. Therefore, the tunnel is not simply a cavern into which wastewater is pumped and stored but a compete facility like any treatment plant or pump station.

The depths at which the tunnels are built require specific designs for the entry and egress of water from the system. Wastewater presented to the tunnel system is first screened for floatable debris and other trash. Dropshafts deliver the water to the tunnel system. The water is delivered to

11. "Combined Sewer Overflow Control," U.S. Environmental Protection Agency, EPA/625/k-94/003. The great interest expressed several years ago in this technology has waned due to its cost.

the dropshaft by either surface drainage or pumping. Because the shafts go so deep into the earth, air becomes entrained in the water as it goes from the surface region to the tunnel storage. Before the water enters the tunnel itself, it must be deaerated. An air exchange chamber is required for this function. At the point that the treatment facility can handle the flow, the stored wastewater is pumped out and conveyed to the treatment facility.

Deep tunnels may find their way into developing countries where the addition of underground tanks and additional sewer lines is not physically possible or economically feasible. They are appropriate for areas of dense population and overtaxed underground utility structures. They can be dug below known archeological sites so these areas are not disturbed. They can also be dug well below near-surface structures so as not to impede current services. With greater ability to dig with remote boring equipment, deep tunnels may become more economical to construct. The tunnels can be sized to accommodate huge amounts of water, theoretically in excess of any surface storage scheme. Therefore, they have potential where near-surface storage is not practical.

4.18 PRETREATMENT FACILITIES

The ability to remove difficult or unplanned loads from the treatment facility is the job of a pretreatment facility. This is a step up from tanks and tunnels, as some treatment is actually performed. Many pretreatment functions are being or have been added to tunnel systems, as it simply makes sense to do so. Therefore, the difference between deep tunnels and pretreatment facilities may be hazy at times.

A pretreatment facility takes wastewater from a system, performs predetermined treatment on it, and returns it either to the waste stream or to a watershed. These facilities have become popular for stormwater pretreatment and CSO compliance efforts. They vary hydraulically in their characteristics depending on geology, elevation, and expected use. Pretreatment facilities generally screen wastewater for floatables and trash and then perform some disinfection with sodium hypochloride or another disinfection agent. This permits otherwise untreated or unscreened water to be returned to a watershed with less confidence about damage to the environment. Examples of pretreatment facilities include swirl and vortex separators, screening facilities, pump stations equipped with screening and disinfection, and

CSO treatment stations. All these require dedicated maintenance and monitoring for proper operations.

4.19 SUMMARY

The major parts and components of infrastructure systems were introduced in this chapter. There are potentially many more, depending upon the needs of a particular utility. The components were presented to allow a developing understanding, so as the monitoring devices are included, the knowledge base can be proportionally increased.

Many of the structures not common to sewer systems, such as siphons and drop manholes, are not discussed because they are difficult to monitor.

In such cases where the point of interest appears difficult, a substitute point should always be considered for monitoring. This is why intimate system knowledge is so important. The local, intimate knowledge of the system can indicate a substitute point that can yield the same information.

CHAPTER

5

Hydrology and Climatic Effects on Infrastructures and Systems

5.1 INTRODUCTION

The basic role of management systems in the infrastructure is to detect and report conditions that are not considered steady state. The infrastructure systems can monitor for other circumstances, but those conditions that deviate from a utility's accepted normal operations are the most pressing. These conditions place stress on the drainage basin and the structures that transport the water to the treatment works or the receiving waters. They also have the potential to cause health and safety problems as well as property damage. Protection of the watershed, preservation of health and safety, and the care of property come with understanding the hydrologic cycle and its effects on the infrastructure. This chapter describes aspects of the hydrologic system that directly affect the infrastructure, producing changes to the usual and ordinary states of accepted operations. The word *accepted* is used here for the operating parameter as what one community would see as normal. For this reason, only the conditions and effects are described. The definition of normal is left to the individual utility to decide.

In terms of sewer infrastructure management, hydrology is the study of water and its effects on the surrounding drainage districts. It covers aspects of rain, runoff, tides, and groundwater as direct contributors to the sewer infrastructure. Hydrology theory is supplemented with a practical description of the effects that water has on the sewer systems and why it does. The purpose is to thoroughly discuss actions and reactions with respect to their effects on the infrastructure so they may be clearly identified at the time electronic systems are installed.

Climate is described as those effects of weather that directly affect the infrastructure by virtue of their actions on the drainage district. By keeping the focus narrow, only those aspects that relate to sewer responses within the drainage district are addressed. This allows concentration on the subject matter at hand and focus on resolving the issues within the infrastructure.

The discussions here conclude with the lead-in to proper placement of electronic instrumentation and the methods of obtaining the best readings for conditions as changes occur within the drainage catchment.

5.2 THE HYDROLOGIC CYCLE

It is proper to introduce the subject of water's causes and effects with a description of the hydrologic cycle. This is the cycle in which water—the most versatile compound at the disposal of mankind—circulates throughout the earth.

The hydrologic cycle begins with the action of the sun on the earth. Evaporation from oceans, lakes, and rivers in the form of water vapor is collected in weather patterns and moves over land masses.[1] The movement over land results in the collision of air of different temperature gradients, causing precipitation. This becomes runoff and finds its way to a treatment plant, watershed, or other body recharging the groundwater tables. This cycle continues all over the world and provides the planet with constantly moving water in various forms. People have learned to harvest water for all of their needs.

The local climate, location to water, sunlight exposure, structures, and the geology of the region are of great significance to the manner in which the weather forms and affects the area of interest. Few generalizations can be made concerning the weather in any given zone or province, as climatic conditions can be affected simply by the placement of high-rise buildings. It is therefore important to understand the cycle of hydrology specific to

1. Viessman, Knapp, Lewis, and Harbaugh, *Introduction to Hydrology,* 2nd ed. (New York: Harper and Row, 1977), ch. 3.

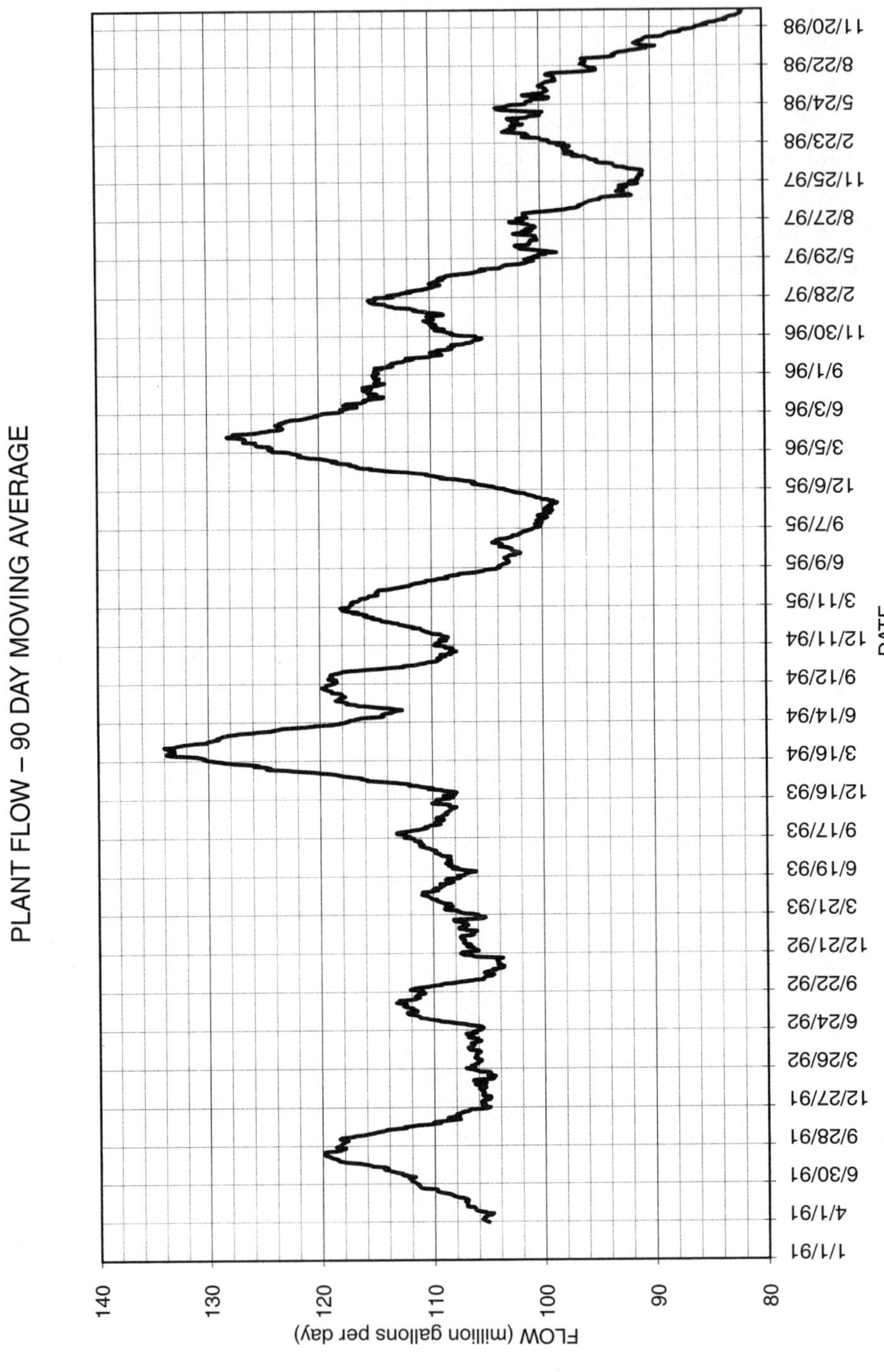

Figure 5-1 Graph for introduction to the localized hydrologic cycle. Note that there is a peak each September even though the intensity is different

any given area. Because the hydrologic cycle is so site specific, a complete understanding of weather, climate cycles, seasons, and other particulars of an area is essential. It makes sense to understand all the natural and artificial factors that have an effect on the cycle.

5.3 UNDERSTANDING THE LOCALIZED HYDROLOGIC CYCLE

The theoretical hydrologic cycle is of little interest to utility operators dealing with the day-to-day functions of a particular infrastructure system. They need to know what *will* happen and why, not the theory of what should or might happen.

It is highly important to understand the climatic and hydrologic history of an area. The past trends that have been recorded serve as the basis for the initial understanding of an area's cycle. Most communities support agencies that develop weather information. Satellite imagery has made the recording of weather and weather patterns over this planet a much simpler task than it once was. This makes the initial step gathering weather data from as many potential sources as possible. It may be a tedious task, but the ability to determine long-term weather patterns removes many questions about what appear to be disruptions or radical changes in localized rainfall and temperatures.

Generating a usable database requires rainfall information, temperature highs and lows, groundwater data, treatment plant flows, and any other data that relates to the use and reuse of water in the area. Rainfall totals are generally maintained by the local weather service. Time and duration of events is valuable information and is discussed below. For this purpose, the totals over 24 hours are sufficient as long as the data cover long periods of time. It is assumed, for this discussion, that there is little or no instrumentation in the infrastructure and that the information gathered will determine its placement. Lay out the data, correlating 24-hour plant flows, 24-hour rainfall totals, temperature highs and lows, groundwater levels, if available, and any other available data relating to long-term records of climate and weather. This can be a big job and can take a long time. Cover as broad a time frame as possible. Arrange the data historically, concentrating first on plant flows and rainfall. Specific trends in rainfall that correlate with plant flows and groundwater levels (where available) will immediately become apparent. In order to make the data more manageable for trending, make monthly summaries; with these, long-term trends can still be seen, and areas of interest can be investigated more closely with the 24-hour information. Bar graphs of the monthly data can provide a pictorial view better suited for trending. This is the first part of mapping the localized hydrologic cycle.

The addition of temperature information segregates the rain events into seasonal changes. Seasonal variations in regard to plant flows, rainfall, and temperature are important to an understanding of the localized hydrologic cycle. Typical patterns emerge in the long-term studies, as do variances. As an example, an El Niño year is a variance of a local pattern and the effects can be readily identified simply by rainfall, temperature, and plant flow correlations. However, long-term data must be available for these patterns to manifest.

The localized hydrologic cycle is now apparent in the compilation of the data. Graphical representation of the information allows trends to be seen and variances identified. The localized hydrologic cycle can tell the system operators much about the loads their systems must carry and thus provide a tool for developing system enhancements.

5.4 STORM EVENTS

The hydrologic cycle is driven, from the practical viewpoint, by the rainfall or other storm event events occurring in a given province. These events are based on the climatic conditions particular to an area. In order to properly understand a storm event, it is necessary to become familiar with its components.

A storm event, for this discussion, is a weather occurrence that produces precipitation over a given drainage district. Storms are composed of cells; each cell is composed of varying amounts of precipitation potentially capable of falling over the area of interest. As these cells go over areas of high humidity or water bodies, they tend to pick up moisture for redeposit. Over drier ground, the potential is to release the precipitation. It is important to understand that precipitation can evaporate prior to hitting the earth. Depending on the mix of air masses, precipitation seen falling by the satellites may never reach the ground. The distribution of precipitation in any given storm or storm cell is not predictable, even with a string of supercomputers. However, predictions can be made concerning general trends based on historical information and known climate variations. Beyond identifying trends, little can be done except tracking the weather system and the cell patterns within it.

A storm event is described using these three components: time, duration, and intensity. Time is described as the total elapsed time of the event from start to finish. It is probably more accurate to refer to this interval as *elapsed time.* The time in between events is subjectively assigned by individual weather agencies and is a parameter that should be known and agreed to by the utility agency. A rule of thumb is a six-hour delay from the ending of recorded precipitation from one event until the beginning of another event. Contained within the time parameter is duration. Duration is also a time measurement, but it is the slice of time for each occurrence within the time interval—that is, for each individual storm cell passing through. Each storm cell produces an intensity. If the storm is well defined, then the precipitation is recorded as a continuous event and no individual durations appear. If the storm has many defined cells within it, the contribution made by each cell is the duration in which each produced precipitation. In many cases, the time and duration of the event are the same.

Intensity is the rate at which precipitation falls. The varying intensities and the duration of each event as contributed by the individual storm cells are of particular interest to the system operators. The intensity generally has little bearing on the total amount of precipitation that actually falls. This is because the intensity of a large, multicelled storm varies greatly. Hence, only the cumulative sum of the total rainfall is accurate for the time and duration of all contributions to the total event. Intensity has some valuable recording functions as well as tracking properties. It is generally expressed in a rate, such as millimeters or inches per hour, and the rate can vary per duration within the time slice. Intensity is a factor in classifying the severity of storm events, and thus in determining return intervals.

A return interval is the likelihood of a storm of a particular intensity and duration returning to a region or drainage district. Most meteorological offices worldwide have developed localized chartings of estimates for such return intervals. When determining a return interval, the time is converted to the duration of the recorded intensity. The interpretation of the return interval is recorded by measuring the amount of precipitation that fell for the duration and matching it to the estimated likelihood of return on the local meteorological table. A large database of such information is the most practical tool for generating the design storm used in planning and engineering programs.

In the case of liquid precipitation, or rain, these parameters are readily measured. In the case of frozen precipitation, they are not as readily measurable and their impacts are delayed until the precipitation melts. Frozen precipitation events are difficult to gauge because most instrumentation requires that the snow or sleet be converted to water. A water equivalent measurement is utilized, but it is not very accurate. This measurement is performed by taking representative core samples of the final frozen precipitation (such as snow), measuring it, then melting it and determining how much water it yielded. For frozen precipitation, duration and intensity information is not possible to obtain. Time, the only parameter available, is recorded as the time the event began and the time it ended.

These methods for classifying storm events constitute a basic step in documenting the localized hydrologic cycle. Because rainfall and its runoff govern the loadings to the sewer system, the storm that generates the runoff needs to be properly classified. This practical approach provides usable, long-term data for planning engineering and maintenance programs as the infrastructure systems mature and develop.

5.5 RECORDING WEATHER EVENTS: LOCATING RAIN GAUGES

The most reasonable and accurate method for recording storm events begins with determining which events cannot be accurately recorded—for example, frozen precipitation. Once this is decided, the deployment of a rain gauge network is probably the most economical and effective method for recording storm events. The tipping bucket rain gauge is a low-maintenance, highly effective device for obtaining this information. Use of more complex and sensitive devices should be deferred until a practical experience base is generated with the tipping buckets, unless such expertise already exists within the utility.

Many engineering theories pertain to placement of these devices. However, the elusive nature of precipitation and its variability across a drainage district makes many of these theories impossible to verify. It is therefore more reasonable and practical to implement simple and verifiable techniques. These programs may lack flair and glitter, but they produce tangible results usable immediately by the system operators. This discussion involves practical approaches to the placement of these devices and commonsense analysis of the information obtained. At all times, common sense

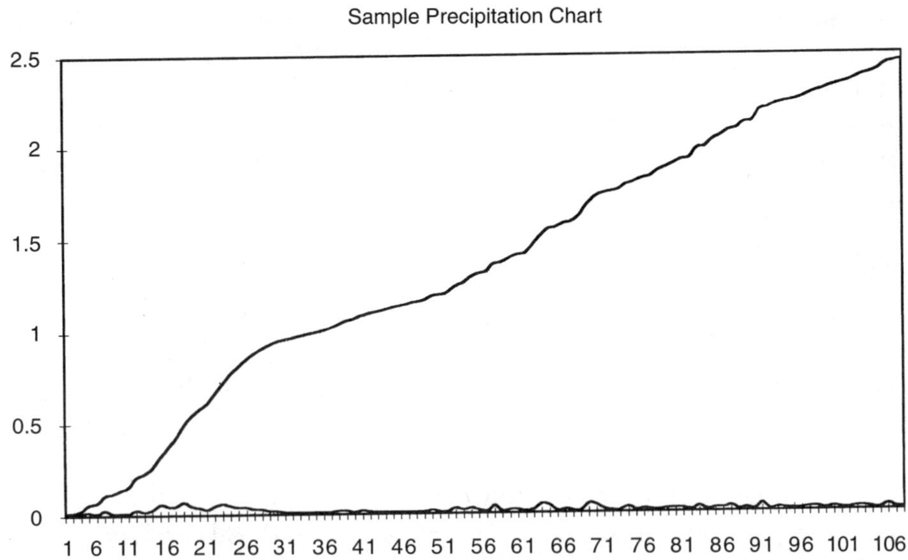

Figure 5-2 A sample precipitation chart. The *x*-axis is in minutes. The increasing line is the cumulative total. The bottom line is the five-minute total. (Data courtesy of American Sigma. Used by permission.)

should supersede adventurous theories and postulates, which do not keep neighborhoods from flooding and make poor explanations during civil trials. An adventurous theory is one that would be nice to try but may compromise the potential practical result. Precipitation cannot be controlled; it can only be studied and recorded. It is wise to stick with practical and proven techniques until the experience base permits the luxury of further investigations.

The design and installation of a rain gauge system is reviewed and elaborated below. The subject is mentioned here because this is a good time in an investigation of the hydrology cycle to gather rainfall data. The deployment of rain gauges can only assist in long-term planning and subsequently increase the experience and database of the system operators. Therefore, at this juncture, rain gauge deployment in any part of the system for practice and development is advised. Added system knowledge from any accurate and reliable source is encouraged.

Few texts agree on appropriate numbers and zones of rain gauges. Those agreements that do exist are more conceptual than practical. This lack of guidelines means that each utility must adopt and support a scheme that will work for it. This is based on climate, geology, varying elevations, and location of water sources, to name just a few sources. However, it is possible to establish rules of thumb for placement of rain gauges. Because nothing here is hard and fast, the reason for the decisions should always be documented and periodically reviewed to be sure that it serves the best interests of the utility.

Rain gauges come with installation instructions from the manufacturer. These instructions must be taken into account during the decision making, because some of the best schemes may not be adaptable to the manufacturer's requirements, which renders the readings questionable at best. It is always a given that, due to the sporadic nature of rainfall, the more gauges deployed, the more accurate the reading. The question becomes how many is enough and how many is too much.

One potential scheme is to follow all the watercourses in a drainage district at equal elevations. Because watercourses tend to represent the low points in a given drainage district, this approach leads to consistency in installation. Placing the gauges at equal elevations leads to consistency in when the precipitation is sampled. However, this approach is not workable for communities in which watercourses do not adequately cover enough territory to provide representative samplings. Also, utility and other services may be at a premium in these locations.

A more versatile scheme is to divide the city, town, or province into equal areas, assign each a weight, and place a gauge in as many area centers as possible. In the event one gauge fails, the adjoining area gauges can be used as a substitute reading. This allows for storm tracking and required maintenance. Such weighted areas are known as Theissan areas. The problem here is to determine the number of weighted areas required for good data resolution. This is a question of budgeting and logistics. It is desirable to place as many gauges in an area as possible and in as many diverse areas of geographic significance as possible. Therefore, once budgets are ascertained for the program, an agreement to, as an example, one gauge every half square mile can be made to properly represent the class and frequency of storms that pass over an area.

Many determining factors contribute to this decision. In areas of diverse and frequent precipitation, varying climate cycles, population concentrations, and large geology and structures, the resolution should be close. An example of close is a gauge every one quarter to one half square mile, a concentration typical of such areas as New England and the Great

Lakes region in the United States and the British Isles as a whole. In areas of large expanse and drier and less diverse climates, a less refined resolution is acceptable. An example of larger resolution is one gauge per square mile, as used in the southwestern portions of the United States. It may be necessary to vary the resolution—for example, fewer gauges in peripheral areas and a greater number in more populated or otherwise interesting areas.

The question now is where to put the gauges. Putting them at equal elevations generates a false plane and probably will not yield a completely representative picture of actual weather conditions. This is due to winds and evaporation of precipitation prior to groundfall. It does make sense to place gauges at rooftop level, unobstructed by overhangs, as there is little difference between the top and bottom of an unobstructed building in relation to groundfall. Also, the contour of the terrain will be more consistent with the heights of the gauges and runoff can be more properly represented. It is probably impossible to find structures at the exact geometric center of the zone agreed on, but something should be available for a rain gauge attachment, such as a telephone pole. In major urban areas, the ability to find sites close to utility services is quite good, and height is the issue that needs to be considered. As the area goes from suburban to rural, utility services and structure attachments become the issue. In rural areas, height is generally not an issue unless the point is within a valley. All treatment plants should have a rain gauge.

The gauges should be provided with utility services and telemetry. The system operators find that once they are installed, information is required more frequently. Suitable telemetry that can operate in storm events should be chosen. In most cases, utility electric service is adequate to power the equipment. Solar cells to charge batteries can be used in place of electrical services; however, a model should be developed to determine the true cost of utility service and maintenance and replacement, as solar cells and batteries do decay over time, even with recent refinements to the technology. Therefore, it is advised to perform a complete study comparing utility service to any innovative or alternative technology.

Once the system is in place, a list should be made of each gauge and the gauges surrounding it. The gauge of interest is called the *target gauge* and the surrounding devices are called the *relational gauges*. These can be used as substitutes in the event the target gauge is out of service; data from one relational gauge or an average of several may be used as a replacement

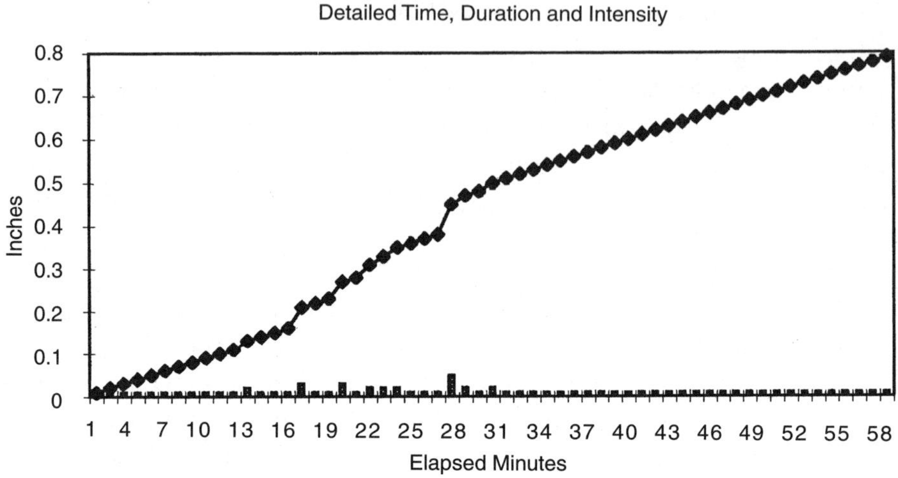

Figure 5-3 Further detail of time, intensity, and duration. Intensity detail runs at the bottom of the chart. (Data courtesy of American Sigma. Used by permission.)

valve. It is suggested that, before any substitute system is agreed on, weather patterns be followed for a while to determine if one of the gauges experiences wetter weather than another. More than likely, differences will exist among gauges in nearby zones. These can be weighted accordingly to provide reasonable estimates for those devices out of service. Estimated data not only appear better but lend more credibility to a report than if the device were taken out due to equipment failure.

The gauges themselves require correction for storm events of specific intensity. This stems from the physical limitation of the tipping devices, where the arms cannot keep up with the pace of heavy precipitation. This can readily be compensated for by two factors. First is the unobstructed location of devices, so false introduction of water from overhangs, gutters, and tree limbs is prevented. The second is the inclusion of the manufacturers' compensation algorithm for the intensity and duration of storms in excess of the physical abilities of the device.

It is now possible to see that much understanding of the local terrain is required to make commonsense decisions for equipment deployment. However, the framework presented here allows for structured approaches that, combined with good, sound engineering, produce the best potential for precipitation recording.

5.6 TIDES

All areas are governed by the storm events that pass through their regions and provinces. Not all areas are governed by tides. Those utilities affected by tides will attest to how they can affect the system in sometimes unexpected ways. In the many areas unaffected by tide, much of this material is not applicable. However, large populations are in areas where tides do affect them.

The gravitational force of the moon is the driving force for tides. As the moon orbits the Earth, the waters of the planet ebb and flow in accordance with the direction and relative distance from the Earth to the moon. If the position of the moon within its orbit were always predictable, so would the height and occurrence of tides. However, only a range of distance and orbits can be calculated with currently available technology, hence the variations in estimated tidal heights. Also consider that the Earth is in orbit around the sun, which has its own variation that translate into variations in the lunar orbit. We can, however, estimate tides well enough for planning purposes.

Tide is the rise and fall of a water body in response to the orbit of the moon and its relationship to the orbit of the Earth around the sun. While there are several other factors that influence tides, we can estimate them well enough fo planning purposes. Seasonal variations in tides are charted and predictable for many rivers as well as coastal areas. However, other factors affect tides as well. The moon is always the largest influence on the tide cycle; this must be stressed. Everything else discussed here merely adds to the effects of the moon on tides.

The first major influence we need to be concerned with is storm water runoff. Depending on the body of water and the watershed that contains it, the effects of runoff can markly increase or restrain the effects of the tide. If the runoff occurs during peak cycles, it makes the tide that much higher. If it occurs between high tides, no tide change is apparent. Combinations of effects are based on the severity of the storm event and the cycle of tide and its aggregate affect on the water body. The greatest effects of storm runoff are seen where a river is fed by many tributaries. When the tributaries increase in flow because of runoff, the tidal river also is affected. Oceans tend to be little affected because of their vastness, but bays and backwaters are usually heavily influenced by tide-enhanced runoff. It is imperative that catchment areas influenced by tides be monitored for potential influences of runoff from storms.

The second major influence is wind. Though it does not happen often, windstorms can essentially stop the tidal progression in rivers and streams. Before tides were examined with electronic monitoring equipment, mariners knew of the condition, but it was not common enough to spur further study, at least in the general sense. It is good policy to anticipate windstorms and their potential disruptions to the tidal cycles.

Eclipses of the sun and moon can affect the aggregate height of a tide. Depending on the alignment and relative position of the eclipse, the tides can be much higher or much lower than normal. There is no hard and fixed formula for this condition, as it is completely dependent on the position and location of the eclipse in the sky. The important consideration is that these astronomical events can generate mild to severe effects on tides.

The tide and its extremes strongly affect utilities with combined- and storm sewers in tidally affected areas. The combined sewer overflow relief, the discharge point over the dam, or the storm outlet can act as an entry point for unanticipated tidewater. This entry dilutes the sewage going to a treatment plant, which in turn can disrupt the treatment works, as the plant needs a particular loading in order to continue the biological treatment normally. The dilution causes a lower percentage of solids concentration per flow to reach the process. Also, in communities with percent removal requirements, this may dramatically change the percent removal value, giving the impression that a treatment facility that is, in reality, very effective is not meeting effluent standards. A flap over the CSO point generally is enough to stop the inflow of tide back into the combined system, though it requires maintenance. It is understood that many storm sewer systems do not empty into any treatment facility. In communities with separate sewer systems, pretreatment for stormwater is becoming more common, and so even in these systems, tides should be considered in the overall appraisal.

In low-lying and coastal areas, tides can cause inflow and infiltration to marginal sewer systems because they can introduce water where none is expected. In the case of storm sewers, extreme high tides can back up into an unprotected system and cause street flooding and property damage. Duck bills and flappers can protect these systems from such events, but these too need maintenance.

It is important that the effects of tides, extreme tide events, and that which disrupts their cycle be understood and added to the localized hydrologic cycle. Because many people live near tidally affected areas, the protection of the systems from inflow is almost mandatory. Tides are generated by a force not yet understood by physics. However, it is a force to be monitored in order to protect the plant and infrastructure from damage.

5.7 GROUNDWATER CONSIDERATIONS

Hidden components can cause the most difficulty. Groundwater, in the form of springs, small creeks, and underground porous rock, among others, is one of those components. Many texts discuss the subject at length, including the theory of recharge and three-dimensional volume movement. The purpose here is to discuss how groundwater affects the infrastructure, regardless of its source.[2]

Older sewer systems have a characteristic that appears strange today but 100 years ago was perfectly acceptable. When communities were first constructed, groundwater was often channeled, by design, into the sewer system. This was done to prepare a more suitable foundation for construction—in other words, to move the path of the water to accommodate settlement. Today, environmentalists consider this to be far from the best approach. The problem in correcting this very common situation is that many of the original design documents either are lost or never existed. This means that the extent of the rerouting problem is unknown, both within communities and across the country. Frequently it is undiagnosed before a full system evaluation is performed.

This issue is brought up because infiltration and inflow (I/I) are major contributors to the sewer system that need to be eliminated. Here is a way of determining if this condition exists in a given drainage district. Refer to the localized hydrologic cycle database. If there is a correlation between low plant flows and periods of no precipitation, then precipitation increases the potential for groundwater I/I. Note that this example only indicates a potential for groundwater I/I and is not definite evidence. However, these indicators can provide clues concerning a situation within the system that requires attention.

I/I reduction programs are now in the planning stages in many communities. The goal is to rechannel this natural water flow out of the sewers

2. Viessman, Knapp, Lewis, and Harbaugh, *Introduction to Hydrology,* 2nd ed. (New York: Harper and Row, 1977), ch. 8.

leading to treatment plants. The long-term dilution from this I/I is a source of concern for system operators, plant personnel, and regulatory bodies who have to decide what to do with the water once it leaves the sewers. If not properly dealt with, it will undermine the structures it was redirected from in the first place. Also, if it is not rechanneled correctly, the water will become a pollutant carrier, picking up undesirables from the sewers and residual areas and carrying them back to the watersheds. Therefore, it is important that groundwater issues be integrated with the localized hydrologic database.

One way is to perform a boring and provide a level monitor at a point or points over the local aquifer. By monitoring the level of the aquifer, the groundwater table can be known and when the elevation, where applicable, exceeds the elevation of the invert of the sewers, investigation for entry points can be performed. Once they are located, remedial action can be implemented.

Faulty sewer systems are another cause of I/I from groundwater. Leaks in joints and cracks in any portion of the system become a point of connection for groundwater sources. As the groundwater level rises and falls from aquifer recharge, the evidence should be visible at the head of the treatment plant. Small sources probably will not be noticed, and it may cost more to remove the I/I than to treat it. That is simply a matter of economics. However, large sources of inflow noticed at the plant should be removed. The faults within the sewer system then should be investigated and corrected.[3]

Another effect of groundwater on the capacity of the infrastructure is especially relevant if the system is combined. Infiltration and inflow from groundwater sources reduces the aggregate conveyance capacity of the sewer system as the water table rises during the recharge after a storm event. This can produce discharges through the CSO sooner than expected. In such cases the I/I should be removed, not necessarily as a cost-to-treatment issue, but as a health and safety issue.

These groundwater issues have been buried within the infrastructure systems for many years, long before the current system operators were appointed. Only now are these issues coming to light as research, study, and commonsense analysis detect system operation problems. Groundwaters

3. G. Tchobanoglous, *Wastewater Engineering*, 2nd ed. (New York: McGraw-Hill, 1979), 24.

Figure 5-4 The effect of tide on sensitive watershed areas. This particular scene terminates at an industrial facility, which complicates the issues. The combination of population concentrations and industrial influences interacts with the already significant effects of tides.

should be returned accordingly so the watersheds they service can be maintained for future generations. Many designs accommodated groundwater return, but through the years of change and growth of an infrastructure, much of the design integrity may have been lost. If possible, the design criteria for each infrastructure should be determined or obtained to identify accommodation for groundwater control.

5.8 THE WATERSHED AND THE LOCALIZED HYDROLOGIC CYCLE

The discussion, to this point, involved the description of patterns of hydrology and their effects on the infrastructure. It also described how to use identify cycles and patterns within the drainage district. These all have an effect on the total area surrounding the drainage district, known as the *watershed.* The definition of a watershed is constantly being changed and modified by regulators around the world. The climatic and hydrologic cycles all affect the watersheds. This is why the topic of watersheds is presented.

The ecosystem that contains a reserve of water that people utilize is called a watershed. Each part of the ecosystem is important to the purity of the water, the continuation of the resource, and the preservation of the life that it supports. The balance in the ecosystem determines if the watershed will survive or, eventually, fail. Failure of the ecosystem minimizes the quality of life in the region as a whole. Failures of many watersheds can be disastrous to any province or country. It is therefore important to be able to identify and boundary watershed ecosystems.

A critical force affecting the ecological balance is the climatic cycle. This in turn affects the localized hydrologic cycle. The ability to track cycles of intake and discharge of water to the watershed and document changes, whether they are seasonal, related to usage, or natural, is critical to the preservation of the watershed and the life it supports.

The development of public and private property within and surrounding a given watershed has varying effects and variances on the ecosystem. It is also an issue among the residents. Few fixed guidelines exist for protection of these areas, regardless of what part of the world the ecosystem is located in. A constant battle for rights to use of the watershed and to what use it should be put exists among the factions. Such is the case now made for monitoring and control of these areas, and understanding how infrastructure facilities may assist or detract from the ecosystem is an important issue for the system operator.

The goal with respect to public and private property usage along a watershed should be a balance between efficient use and protection. The development of protection measures begins with expanding understanding of the localized hydrologic cycle. A complete inventory of ground and surface water sources, currents and tidal movements, flora and forested areas, and ultimate use of the water supply should be documented and cataloged. Water supplies used for drinking, bathing, recreation, and other human contact require great care. Preventing the spread of waterborne diseases from pollution spills and other undesirable sources protects the public. This may not appear to be the task of the sewer infrastructure team, but disease prevention is one of the most important functions of such systems.

5.9 TEMPERATURE EFFECTS ON THE WATERSHED

The determination of climatic effects on the watershed is a starting point in understanding the behavior of the ecosystem. Because climatic effects are so localized, an understanding of the factors that make up localized climate and their parameters should be available to the localized hydraulic database.

The temperature of water is one of the more important parameters that affect the equilibrium of the watershed's ecosystem. It is also the easiest to determine. Seasonal variations and tributary contributions may significantly alter the water's temperature gradient. Most temperature changes in any watershed are seasonally dependent and gradual. Aquatic life adapts to these slow, gradual changes quite readily. Sudden cold or hot baths, on the other hand, tend to shock the systems they are presented to and potentially cause discomfort or damage.

Two forms of temperature shock can affect the watershed. The first is inflow and the second is discharge. The first case occurs when water of a different temperature gradient is introduced to the sewer infrastructure. This may have detrimental effects on the plant process. Unexpectedly warm or cold water, when introduced to the pipes as unwanted inflow, does little to upset the balance in the sewers, except possibly to reduce their capacity to carry sewage to the treatment works. It *can* upset the balance at the treatment works by shocking the biological process there if the plant is not prepared for it. Domestic and most industrial sewage, even in the winter, rarely drops below 50–55°F as it is heated in the homes or plants before being introduced to the sewers. This is what is expected at the head of the plant. Process shock can disrupt biological action until the plant's biological organisms become accustomed to the temperature change. Cold water can also dramatically increase pumping costs because water is more dense as it approaches 34°F than it is at, say, 70°F. This puts greater pressure on the cost of processing in a time when costs are already under great scrutiny. Tide inflow from a CSO point or sewer break can also produce the same alteration in temperature gradient.

A radical drop or rise in temperature is a significant indication of inflow to the sewers. Another clue that something is wrong is when watershed life shows up in the sewers—as, for example, when fish show up at the bar screens of a pumping station or treatment works. It can then be expected that algae and other food for aquatic life is in the inflow as well. The accompanying dilution factor is also undesirable.

The first line of defense is a temperature probe placed near sensitive watershed areas where inflow potentially could take place or at a reasonable monitoring point along the main sewer to the treatment works. Recorded radical drops in temperature are evidence of inflow to the system. If data is reviewed historically, the duration of the inflow may point to the source, whether it is a break in the structures or tide moving through a CSO point. Noting the point where the temperature change is recorded narrows the search area down considerably. Common sense and local knowledge can lead to reasonable deductions concerning location of the trouble spot.

The second form of temperature shock is a discharge. The best example is the discharge from hydroelectric stations to watersheds. Before the effects were completely known, heated water from power plants was introduced to watersheds without thought about the effects. However, when fish kills began to occur and changes to the foliage took place, power plant operators and environmentalists wondered why such things would happen. Evidence pointed to the sudden change in temperature. Sewer discharges can have the same effect, along with the introduction of pollutant loadings to the watershed, which also increase the exposure of the watershed to harm. Tempeature shock is thus an indicator of potential pollutant loadings.

Again, the best defense is temperature probes. However, they are difficult to introduce into watersheds because of the currents, depths, and size of the water bodies. This is why a complete understanding of the hydrology and localized hydrologic cycle is critical before the sensors are placed. The best locations are at or near known potential discharge points. Handheld probes and sensors are better than permanent placements for testing for pollutant plumes in water bodies unless appropriate locations are identified.

The temperature of the water is an excellent indication of the function or lack of function of the infrastructure system. Temperature shock is a serious condition that can significantly affect the quality of aquatic life and the efficiency of a biological treatment plant. It can also be an indication of pollutant loadings to the watersheds. Strategically placed temperature probes can assist system operators in quick and efficient pinpointing of potential failures in the infrastructure so repairs and permanent corrections can be made.

5.10 OTHER CLIMATIC FACTORS

Weather and its effects on the ecosystem, as noted in earlier sections, constitute a localized and individual issue with respect to each watershed. Operators should understand local climatic occurrences in terms of their relevance to watershed monitoring.

The localized hydrologic cycle table should include temperature and precipitation tracking, as noted above. One of the columns should include the 24-hour rainfall total. This total is an indication of degrees of wetness for a period. If precipitation occurs many days in a row for the same period, year after year, the system operators may make operating changes to accommodate these cycles of wet weather. Such system changes are specialized and require understanding of local weather patterns.

Prevailing winds also have effects where there are open bodies of water. Shifts in the winds and their intensity can shape and move water bodies. If tides are considered, the combination can produce the effects of weather without the event. Radical departures in predicted tide levels can cause unanticipated conditions within the infrastructure. In most cases, the recording of winds is not a requirement in the localized hydrologic database, but it may be of significance in maritime areas, which experience varying effects from tides and other factors affecting water body levels.

The areas that surround or are in known bands of climate or weather occurrences should be heavily documented. Those drainage districts within areas of hurricane activity, snow belts, theoretical frost lines, or other delineations of known extreme weather activity have particulars worth noting in the database. Many of these localized functions are significant contributors to regular precipitation. Though these climatic phenomena have so many variables that it is impossible to document them in one reference, available indicators should be tracked. It is important that such weather events are recognized and documented for their local significance, so the variances from the repeatable patterns can be recognized.

Consider a watershed that is near a major highway with a pump station—a common configuration in urbanized areas. In the winter, there is a potential for salt runoff from the highway from deicing activities and emergency discharge from failure of the pump station, neither of which may have any substantial effect on the watershed. The pump station infrastructure may never cause a discharge and, in normal winters, the salt runoff may be completely absorbed by the watershed without hazard. In an El Niño year, the level of the watershed may rise to the discharge level of the pump station due to runoff of heavy rains, introducing inflow dilution to the sewer system. Because this occurs only every seven years, roughly, it may be unidentified by the treatment works initially. The introduction of cold water would shock the plant and biological activity would be significantly affected. In a snowy or icy year, excessive salt may overload the watershed's ability to absorb the runoff effectively, causing damage to vegetation during the thaw. A simple remedial be a flap valve installed at the discharge, preventing inflow as a result of El Niño. Salt runoff cannot be prevented. The example is given because sometimes there are no potential solutions, only explanations.

Awareness of the trending and variations in climatic factors is important to planning, engineering, and long-term operations of the infrastructure, where the goal is to provide maximum service with the minimum disruption and exposure to the watershed.

5.11 POINT AND NON-POINT POLLUTANT LOADING SOURCES

The purpose of much of the equipment targeted for sewer infrastructure management systems is to protect the watersheds from point pollutant loadings. It is therefore important to understand the differences between point and non-point pollutant loading sources and how these sources can be identified. It is also important to note the degree to which loadings can be tolerated by the watershed, as some may, strange as it may seem at first, be natural to local geology and terrain.

A pollutant is that which will cause harm to the watershed.[4] Pollution takes many forms, from trash containing contaminants to silt with residual pesticides. It can be introduced to the watershed from the infrastructure through a fault in the system, deliberately discharged from the infrastructure, as in the case of a discharge from a pump station overflow point, or from runoff following the natural terrain as it slopes to a water body. Pollution can be nonhazardous or hazardous. The exact terms used to grade pollutants vary among jurisdictions and from country to country. Many countries developed lists of what they believe are hazardous and nonhazardous pollutants; these are available for general inspection.

It is important at this point in the discussion to categorize sources of pollution. A point source is one that the system operators can probably repair, minimize, or correct. Examples of point sources include CSO discharge locations, pump station overflows, and emergency bypasses at treatment plants. A non-point source is one the system operators probably have no control over. Examples include runoff from fields and farms and other general introductions to the watershed that have no distinct point of origin. Point sources have a reasonable chance of being identified and successfully remediated. Non-point sources are generally uncontrolled and unlikely to be fixed.

Natural runoff back to a watershed must be understood. In recent years, there has been great interest in remediating all pollution sources running back to a watershed district. Certainly attention should be paid to the point sources; techniques for how to monitor for them and what equipment to introduce are discussed later. However, care must be taken to allow that which is natural to return whence it came. Too often, concerned citizens forget that certain amounts of runoff are required by a watershed to survive. Though well-meaning, they pressure infrastructure system operators to prevent or limit runoff from reaching the watersheds. Sewer system operators must be aware of what needs to go to a treatment plant through the sewer conveyance and what needs to be returned to the watershed to continue the ecological balance. Having these tangible answers is necessary to assure regulators, local government officials, and interested

4. "Urban Runoff Pollution Prevention and Control Planning," U.S. Environmental Protection Agency, EPA/625/R-93/004, Table 1–1, 3.

parties concerned for the well-being of the watershed that their resources, both the infrastructure and the watershed, are being maintained.

Point pollution sources, in many cases, can be identified and characterized. The characterization has to do with the typical loadings that would be experienced by the watershed if pollutants were introduced at that point. These characterizations are extremely useful for building economic and cost models used in developing capital programs for rehabilitation and expansion of the infrastructure. If the point pollutant loading estimate suggests significant detrimental impacts on the watershed, it may make sense to go forward with a given project. If the characterization shows minimal impact other than neutral silt introduction, expenditures may be deferred at that point in favor of more sensitive and pressing programs. The identification and characterization of point pollution sources thus are important steps not only in the protection of the watershed but also in planning the installation of electronic monitoring equipment.

Studies now under way seem to show that non-point pollution sources contribute more hazardous loadings to a given watershed than point sources, by magnitudes of three to five. Although still under evaluation, this could change the thinking of regulators. If true, and it may well be, the entire approach of monitoring and characterization of a watershed could change. This is brought up here because most infrastructure monitoring and systems are geared to point source pollution. The introduction of non-point pollution sources will significantly affect the entry and delivery of sewage, runoff characteristics, and new ideas and models for infrastructure management. This is an issue to be considered in weighing the contribution of any point source to a watershed and its overall effect.[5]

Industrial point sources are also at issues. In most cases, industry either pretreats waste or dumps directly into the sewer system, depending on the industrial process. Water washes and cooling waters that do not carry any known harmful pollutants may, if local authorities permit, be returned to a watershed and bypass the sewer system. This is a point source, but one that generally has no effect on pollutant loadings, unless the manufacturing process fails. Monitoring of these nonpolluting point sources should be constant in order to provide assurances that there is no process failure or the introduction of any residual pollutants to the watershed. Generally, in-

5. Several presentations pointing in this direction are still in the publication process; they should be available for review before midyear 2000.

dustry is aware of the necessity to process waste and does so with little coaxing or regulation. It's a matter of keeping all information available and allowing the free flow of this data to authorized examining parties. Monitoring also provides a deterrent to businesses that might seek to avoid properly processing their wastewater in accordance with local clean water codes. The infrastructure managers need to be acutely aware of these noncomplaint businesses so industry failures, if they occur, can be traced and remedial action taken.

The identification and characterization of point sources can provide much-needed data for short- and long-term planning efforts that, when implemented correctly, can make the best use of ratepayers' funds and provide targeted remediaton for those points that pose the greatest risk to the watersheds. This targeted approach also reduces the exposure of the watershed to long-term pollutant loadings and provides objectives with reasonable expectations of the programs based on the information derived from the point source classification.

5.12 RIGHTS GROUPS

Groups of citizens from all walks of life, from many countries, and from all socioeconomic levels are bonding together in defense of the environment—here, the watershed. These *rights groups* seek to protect one or several very local watershed areas. Other groups are of international scope, with great political clout and influence. Some are well informed, with batteries of databases and information resources to assist their efforts. Others are poorly informed and self-centered, and ill-informed rights groups with selfish motives tend to cause more harm than good. It is proper, at this point in the discussion, to present a method of handling these political entities so the interests of all parties are served.

A rights group has a good and common goal: the protection of a watershed in which members have a generated or vested interest. These groups push lawmakers to take interest in environmental issues. It is through their efforts that lawmakers develop policies and appropriate funds that affect the way system operators must conduct the business of running a sewer infrastructure. It is therefore imperative and in the best interest of all parties that the goals of rights groups be understood. It is equally important that the methods by which they attain their goals be understood as well.

A well-informed, well-organized rights group can potentially be the best friend of a system operator. Its members can understand reasonable explanations of specific actions, both remedial and long-term. A loose band of uninformed fanatics can be disruptive and engender a lot of unneccessary work. It is sometimes difficult to distinguish the types, especially at the site of emergency situations when several groups are jockeying for attention and something is missed in the interaction with the system operators. Steps must be taken to ascertain that all facets of an issue, or as least as many as possible, are covered in these situations.

The guidelines are easy. Keep presentations concise and short. Make statements clear and keep them free of technical jargon. Do not try to pacify any particular group. Just state the facts as they happened. Where several rights groups are represented, do not address any one group individually; the others may feel slighted. Remember, regardless of politics, they all can assist the cause—that is, the preservation of the watershed. If the numbers related to a potential pollution issue are small and support your case, state them. If the numbers are large and damage control is required, state the numbers and the plan for remediation. Hide nothing. Do not provide information that can be twisted. As an example, if one dead fish is found at an outfall site, is this considered a fish kill? If a dead goose is found with the fish, is this a fish and wildlife kill? Uninformed fanatical rights groups exploit these situations just to give themselves credibility. Informed, reasonable groups see the absurdity of the complaint. This kind of situation should to be avoided, or at least minimized. The example illustrates how easy it is to twist something simple and make it serious, at least by hype. Learn to recognize these situations and deal with them swiftly.

One of the best weapons in dealing with rights groups is the proactive strike. This is not to strike against them but with them. Keep them informed about events and situations within the watershed district. Also, if they insist on becoming involved, give them work to do. Explain what their efforts will achieve and how they will be assisting in the preservation of the watershed district. Remember, the system operator is still considered the ultimate expert regardless of the influence of the rights group. Few rights groups will take the responsibility for carrying out remedial action. Most only report what is done incorrectly. By giving assignments and getting members physically involved, many of the frivolous complaints drop away. This hands-on approach is probably the best way to maintain order among interest groups. Always take the high road with rights groups. Their members are also the ratepayers as well.

5.13 INDUSTRIAL DISCHARGES IN WATERSHED AREAS

The case of industrial discharges in watershed areas drives home the basic need for instrumentation in monitoring for industrial accidents. Such events are usually the result of human error or equipment failures. These conditions cause the most public criticism and noticeable damage. Deviant industrial discharges are generally hidden and their results are not known for long periods. Both cases are examined in this section, which addresses a seldom examined player in watershed areas, the industrial user.[6]

A battery of governmental regulations pertains to the actual discharge of the industrial user. The shortcoming is that regulations generally do not consider the sewer infrastructure itself; the focus tends to be on the plant's biological process or the action on the watershed. This discussion, however, addresses the effects of industrial discharges on the infrastructure.

The industrial user is generally missed when regulations are implemented. Most regulations are geared to the utility operators, and those affecting the industrial areas are deferred to local authorities. This is not the best approach because it does not generate uniformity among industrial residuals. There is a large body of toxic discharge limits; the effects of exceeding these limits are well documented and available in many good reference sources. The problem is that regulators usually present violations to the sewer operators, who then must seek out the industrial violator and extract damages accordingly. In the case of specialty chemicals, this may not be difficult. When generic or common toxins are involved, the case against specific industries is a bit more problematic.

This situation is underscored by the rights groups. Many times these industrial discharges, regardless of their source of origin, are labeled by the rights groups as the sole responsibility of the sewer managers. This can make the system operators uneasy because the industries pay usually heavy taxes to the local communities along with water and sewer rates. This makes

6. T. Day, "Biomonitoring Equipment and Techniques: Modern Instrumentation Applications to Monitoring the Ecosystem," Understanding the Industrial Pre-Treatment Program, Indianapolis, Indiana, July 1996.

the businesses large-scale customers who deserve respect. Therefore, a balance between the allegations of the rights groups and the permitted discharge from the sewer system by the industrial users must be maintained by the sewer operators. This is a delicate balance and quite trying, at times.

The watersheds have the potential to absorb many of the toxins that are released inadvertently by industrial customers. This is evident in areas of shellfishing beds. Many shellfish beds are closed due to high levels of contaminants. After a time of purging, the creatures cleanse themselves and shellfishing can resume. The same is true for plant life and fish. They naturally purge themselves of minimal amounts of toxins. This is not to say that any limited industrial discharges are acceptable. It simply makes the case that some mistakes can be handled without significant damage to the watershed. This is a safety margin system operators should be aware of in the event of discharges from the system. It is also one that the rights groups can be informed of in the event of a crisis. The rights groups that are reasonable will understand this. Keep these groups well informed; communicate any such findings to them. It also helps to know the limits of the toxins' aquatic life within a particular watershed so these tested figures can be conveyed accordingly.

The point is that much of the responsibility for industrial pollutants becomes the task of the system operators. The reasons are various and have different weights. The system operators must be acutely aware of industries along the watershed, what their processes are, what their residuals are, and the toxic limits to which aquatic life responds and cleanses itself in the event of a mishap. Industrial accidents may ocur with no fault to anyone. The preparedness of the utility can make the difference in the success or failure of remedial action.

Industrial users are generally required to pretreat their effluent stream before its introduction to a sewer system. Pretreatment, depending on the industry and the effluent, is required to meet loading limits for toxins, heavy metals, biodegradable oxygen demand, total suspended solids, and other loading limits. These limits are imposed so that the biological process at the treatment works continues without chemical or other loading shock. Temperature shock is one of the parameters for which industrial users must compensate. Many industries use process steam. Retention tanks are used to return the reclaimed saturated steam to more or less ambient temperature. This not only protects the quality of sewage being delivered to the

Figure 5-5 Both point and nonpoint sources can contribute to the pollution within a watercourse. Note the oil slick on the water. Here, both sources contributed to cause the contamination. Sewer infrastructure monitoring assists in detailing potential sources of pollutants.

plant but also protects workers from scaldings when working within the infrastructure.

Here is another reason for pretreatment of industrial wastes. Many rights groups and all regulators constantly speak of the protection of the watersheds and of health and safety hazards to the public.[7] Seldom, if ever, is the health and safety of the workers within the infrastructure considered in regulations. The issue for workers involves protection from mixtures of industrial, residential, and medical (which is considered industrial for this discussion) inflows. If the sewer is a big bioreactor itself, then obviously the churning and mixing of wastes can produce compounds of virtually unknown structures that may constantly change in the transition to the treatment works. It is simply not known what resides in the bioreactor at the

7. T. Day, "Proactive Use of Sewer Monitoring Data to Protect the Watershed," American Water Resource Association, Ft. Lauderdale, Florida, September 1996.

time the workers are there. Therefore, the short- and long-term hazards are also not known. For this reason, every effort must be taken to protect the people who enters the system. Their greatest protection is knowledge. Meeting industrial effluent criteria can significantly reduce the risks to these people. This should really be a prime consideration, but it is often missed in the effort to protect treatment plant processes and the watersheds. The protection of workers through proper industrial effluent standards meets all these criteria, and serves all these areas of specific interest.

The aggregate effects of climate on the stray industrial discharge can vary based on the compounds released. Because of this, constant awareness of winds, temperature, tides, and precipitation cycles assist in determining the potential impacts of the release on the environment. Cooler water temperatures, such as those that approach freezing, may limit damages that would ordinarily be caused if the receiving body was more tepid. Remember that cooler water temperatures inhibit the BOD requirements in wastewater and make it more dense. Winds may blow a chemical plume in undesirable directions; the plume can be contained with booms to restrict its influence. In many cases, the effects of algae blooming in warmer temperatures may assist in remediation efforts, depending on the circumstance. This type of natural processing is good to track and assists in the efforts of protection and remediation. Remember that this type of remediation is not unlike the actual biological process found in many treatment facilities. Climate is a big player in the tracking and remediation of any industrial discharge.

At times, prohibited industrial discharges are knowingly made. This subject has been deferred until this point because the more usual scenario involves good-faith attempts by all parties to remediate genuine accidents. The accidental discharge of industrial pollutants is generally the result of a failure of human operation or a breakdown of equipment. Sometimes, however, discharges are not accidental. The deviant discharge of pollutants to bypass regulatory efforts is difficult to monitor because the effects are hidden and generally not traceable. It is these discharges that potentially can cause the most hazards over the short and long term. Rights groups generally do not see them because they occur away from known discharge points or are made to look like errors in sewer system management. Great care must be taken in monitoring and tracking such situations.

Illegal discharges of industrial pollutants are usually made in one of two places. The first is the sewer system, where the pollutant blends and mixes with the usual wastewater flow. The second is a hidden location where the actions of the polluters cannot be readily traced. Other sites exist, but these are most frequently used. The problem is that the damage is done before anyone knows that the discharge was made. If the pollutant is identified, then remedial action may be taken to neutralize the area. Many times, however, the problem becomes a long-term one, as most industrial discharges are not identified readily and the damage can continue for extended periods. This hidden effect can present long-term problems requiring expensive studies and novel solutions.

The first suspect in any industrial discharge is a sewer overflow point. This is because the overflow point is known, it usually is observed, and the effects are usually noticed quickly. Infrastructure managers are informed of such conditions through citizen- or rights group complaints or via some law enforcement condition requiring the reporting. This immediately puts the utility operators on the defensive, as it is difficult to explain that the sewers only convey the wastewater; they do not generate it. As simple a concept as this seems, it may not be readily apparent to the rights or law enforcement groups seeking to remediate the situation. Monitoring and reporting techniques are discussed in upcoming sections. It is important at this juncture to be able to identify such potential sites, the compounds that they may discharge, and the industrial customers who may be ultimately responsible. A complete and sound database of this information can dramatically reduce the time it takes to find the deviant discharge source and the parties responsible.

The protection of the watershed through proper industrial effluent monitoring and control assists in all health and safety issues. The deployment of monitoring equipment both at the industrial site and through the sewers can assist in protection and provide remediation information in the event of an industrial accident. It is important to characterize potential sources of discharge and to document the pollutants that may be introduced to the watershed. Although not necessarily part of the localized hydrologic cycle, this information can supplement such a database.

5.14 UPSTREAM AND DOWNSTREAM CONSIDERATIONS

It is probably more accurate to use the terms *upwater* and *downwater* with respect to pollutant loadings than the terms *upstream* and *downstream* to illustrate the effect of pollutant loadings. However, the purpose here is to show that pollution monitoring and control requires a combined effort by all communities living along a watershed area. All must take responsibility for the protection of the watersheds that serve their populations. Therefore, these up- and down the watershed considerations must be observed and understood to properly align the result of point and non-point pollution loadings.

Consider a river along which many cities are established. Imagine that only half of the communities implemented watershed protection measures. The other half did not and discharged freely into the watershed. It is probably impossible to tell who is the originator of an illegal discharge of any sort, industrial or not. This is the issue with adjoining communities on the watershed.

In many cases, communities both draw water from and discharge processed wastewater into the water body. There is a constant in and out of water through these towns, the next lower town receivng the effluent of the one above it. Pollutant loadings are carried not only into the town just below the discharge but, potentially, to all the communities below the discharge point. This is a serious consideration because the plume of pollution generally spreads outward from discharge; it usually does not collapse inward. Thus a heavy polluter upstream can have detrimental effects on all the communities living downstream. If serious and detailed investigative measures are not undertaken, the regulators will look at the community that reports finding the pollution in the first place. This is the difficulty with sharing a watershed.

Many communities around the world have already identified this situation and formed associations to develop uniform monitoring, reporting, and remediation actions as situations emerge. There are many receiving waters whose communities still have not formed successful associations for dealing with this situation. This causes the more careful cities to be on guard and expend unfair amounts of ratepayer dollars to defend against from the less regulated communities. When the balance cannot be struck, defensive measures must be taken to protect the citizens and their financial interests from unwarranted attacks from rights groups and regulators. This is an unfortunate situation but one that needs to be recognized in fairness to those industries and utility operators who operate their systems to the best of their abilities.

The best defense is an understanding of the upstream and downstream

Figure 5-6 Industrial customers are generally self-policing in matters of discharges to the environment, but this institutional practice can sometimes hide important problems. Hence the need for external monitoring and regulations.

characteristics that may affect the watershed. This requires expanding the potential pollutant database to include those points along the entire watershed, or at least those of significant concern, that may have a detrimental impact on the community. This makes the business of tracking and treating pollutant discharges less stressful to the system operator and puts the blame for carelessness on those to whom it belongs. By understanding the up- and downstream potentials for pollutant discharges, more of the characteristics of the watershed can be identified and protected.

5.15 SUMMARY

The people who work daily to maintain the infrastructure are experienced in the effects of weather on their workplace and the surrounding environment. Climate and other factors have a tremendous impact on the sewers in general. This is seen readily along the entire drainage district and the watershed it services. Climate and hydrology issues often produce pressure on the sewer systems to perform well above the parameters anticipated by their design criteria. The amount of flow transported above the design rating or below its capacities is a function of the weather as a part of climate and hydrology.

The localized hydrologic cycle is an attempt to completely understand the way that a drainage catchment is affected by the continuing cycle of weather. The goal is to document and pattern the effects of weather and climate on the area of study. This method allows a beginner to appreciate the weather effect that many engineers, meterologists, and scientists respect and study.

CHAPTER

6

Installation and Functions of the Monitoring System

6.1 INTRODUCTION

The infrastructure has been outlined and inventoried and basic equipment described. It is now possible to describe the installation of a monitoring system that conveys to system operators the data on system characteristics and hydraulic response that they need in order to meet the needs of the utility and its operation.

Note that the setup, planning, data presentation, and analysis of results are harder than the installation. If the system is not properly planned and its output correctly interpreted, it has little value. So much depends on the plan and following set procedures. Therefore, the plan must be well thought out, carefully reviewed, and in complete agreement with the philosophy and business endeavors of the utility that it serves. The actual introduction of equipment to the infrastructure is straightforward and methodical; the supporting plan has the greatest effect on the success or failure of the monitoring system.

The bringing together of a monitoring system requires that many facets fit perfectly. These include proper siting and local knowledge, selection of the best technology available for the task, awareness of the needs of the system, and a commitment to long-term support to maintain the transfer of data to the parties who may desire it. A broadly applicable way to meet all these requirements is presented and the information and revenue-generating uses of the system are discussed.[1]

The intent here is to provide a system-wide view of the process of introducing electronics into the infrastructure. It should not matter if the installation is temporary or permanent. It does matter whether the approach has been thought out logically and with consideration of all the facets of the organization. This rational plan should remain in place through changes in administration, personnel, and policies. The ability to transcend these transitions is the glue that makes the systems functional over long periods; it substantially benefits the ratepayers who ultimately funded the endeavor.

Well-placed and maintained instrumentation, along with proper targets and allocation of data streams, make monitoring system a solid and sound investment for the community it serves. People will begin to rely on the data generated as they find it conveys the ability to make informed decisions that are not possible without the eyes in the ground. The more a system is used, the more functions will be required of it and the more it will expand. The system expansion will be relatively easy if the basic principles generated at the time of the initial design are followed. The solid foundation laid by the system designers support the building blocks of a useful system.

The discussion here focuses on an approach to generating targets, placements, and methodologies geared to the community a system serves. It addresses temporary and permanent monitoring systems and the uses of both types. Another point is who should finally deploy the equipment and how. As reiterated throughout this book, local knowledge of an infrastructure is critical for the success of its monitoring system. The comprehensive view that each agency develops will provide the best return on the ratepayers' investment in both direct and indirect dividends.

6.2 DEVELOPMENT OF AN ANALYSIS MATRIX: NEEDS ASSESSMENT

The development of an analysis matrix to generate a long-term plan is an essential beginning to the design of any monitoring system. This is the same as developing a business plan for any corporate entity. The strategy deter-

1. T. Day et al., "Alternative Design of a Wastewater Monitoring Network," WEFTEC '95, Miami Beach, Florida, October 1995.

mines the purpose of the system, what will be monitored, why it is to be monitored, what equipment will be used in the system, and how the data will be distributed. Also included are plans for long-term maintenance and other support factors that keep the system operational, and a plan to expand the system as technologies change and the needs of the end user mature.

An analysis matrix can be used for new system analysis or for evaluation of existing systems. This discussion is meant to provide a stepwise approach to determining new, improved, or changed requirements of the monitoring network. It is therefore the logical progression of steps that is outlined here, and therefore emphasized. It is meant to assist even when regulatory mandates are pressing urgently for a monitoring and management endeavor.

The first and foremost action that should be taken is to determine what a utility wants in terms of system performance. It is impossible to enter reasonable negotiations with system houses or equipment manufacturers unless the utility can determine what it wants from introducing monitoring systems into the infrastructure. A target must drive the program and outline the expectations. At this juncture, the expectations need not be reasonable; the only requirement is to begin a sound thought process and recording method with respect to the needs of the operating utility. No contact should be made with any service or equipment provider unless many of the internal targets have been developed and reasonable expectations generated. Know what the utility wishes to achieve and why it should be achieved. A consulting engineer may assist in this endeavor, but caution must be taken that plans so generated reflect the wishes and desires of the utility agency and not an external influence.

A reasonable strategy is to list every possible way in which the system might be utilized, realistic or unrealistic. Everyone should be welcome to make suggestions. Many factions within local governments and utility operations desire input on such projects and programs. This approach gives everyone a chance to participate and contribute. Some ideas may be unworkable in themselves but contain a kernel of insight that provides great assistance in concept formulation. As the ideas appear on paper and are reviewed, more organized thoughts about the task will develop as a result of a deeper understanding of the wishes and desires of the utility as a whole. Ignore how realistic or ridiculous any function sounds or appears; simply make this comprehensive list. It will take a long time. Once the list is generated, an iterative process can begin. This is an iterative process. It means taking steps, and must be done at various phases of the program. Cuts are made as more facts become available. For this cut, review the list to see which entry is so far out that it makes no sense to pursue it. Realize that some of the seemingly far-fetched entries may in fact have some potential. Leave these in until technological limitations and financial restrictions are determined.

The next step in the development of a strategy is to review the system characterization developed earlier. From this, a series of drawings or prints is made that describes all of the sites that may have particular interest. Critical paths connect these points of interest (see chapter 3). Also incorporate data concerning the localized hydrological cycle (chapter 5). These two elements, along with the system characterization, form the basis of the initial system design. Thus planners now know what is in the system, what appears to be of hydraulic interest, and how precipitation falls in the drainage district. This data needs to be organized in a manner suited to the utility operators' policies and practices and to the current constraints of their staffing and other resource availability. Again, the extent of the tabulation and the number of points or paths chosen do not matter. The goal is to get all the potential points of interest and all the potential critical paths down in some format that is logical to the utility doing the design.

Two structures within the matrix now require review: the potential targets and goals and the the potential points and paths. Remember, these matrix structures represent *potential* sites. An iterative process of elimination now begins. The two major considerations that come from the second structure are resources currently committed and the ability to commit future resources. Therefore, the concentration is on the first structure, which constitutes targets and goals.

The elimination process can begin based on two criteria: targets/goals and points/paths. This is a first-cut elimination; therefore, do not fear of losing any aspects or abilities of the final product. The first elimination begins with examining the wish list and simply dropping, but noting, all the extraneous desires for the system. Many of these items just appear infeasible. If an item appears even slightly workable, leave it in. This is a judgment call and judgment must be trusted. Rework this list after the first cut and develop a smaller matrix (it may not appear to be much smaller). Now perform the first cut on the target/goal matrix. The potential commitment

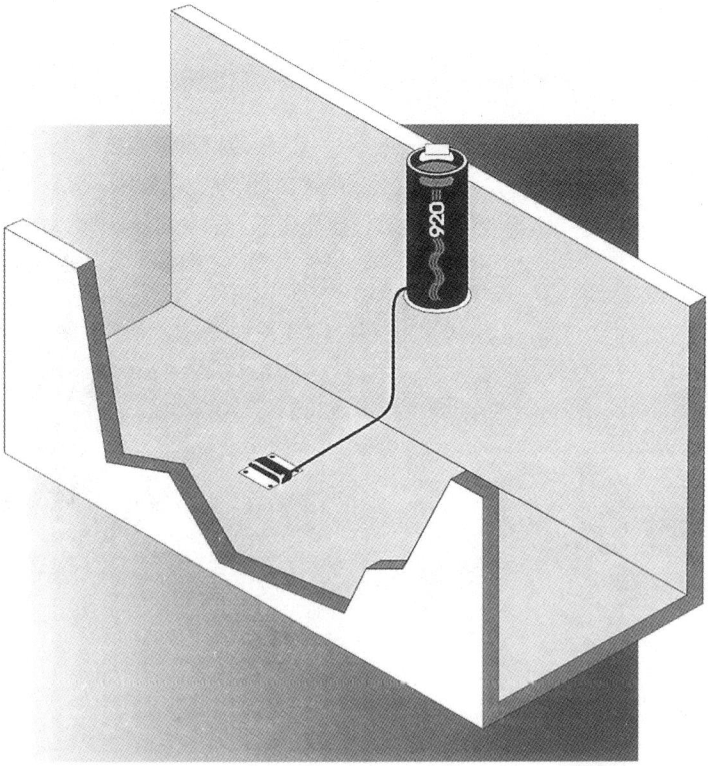

Figure 6-1 Mounting a sensor in a rectangular sewer where mounting bands are not applicable. The invert should be clear of grit and other debris.

of resources for the program must be determined, followed by the resources for a set period after construction. Five to ten years is a reasonable range to choose in performing this cut. If the program is technology driven, use five years. A technology-driven program is one that emphasizes equipment. If the program is manpower driven, use ten years. A manpower-driven program is one that utilizes manpower for normal operations and maintenance of the system. It cannot be emphasized enough that commitments of resources is required for both the initial steps and for years to come as part of a system commitment. Many systems are installed without the proper long-term commitment and their success tends to be marginal as a result. The commitment of future resources is as critical as the initial funding for the program.

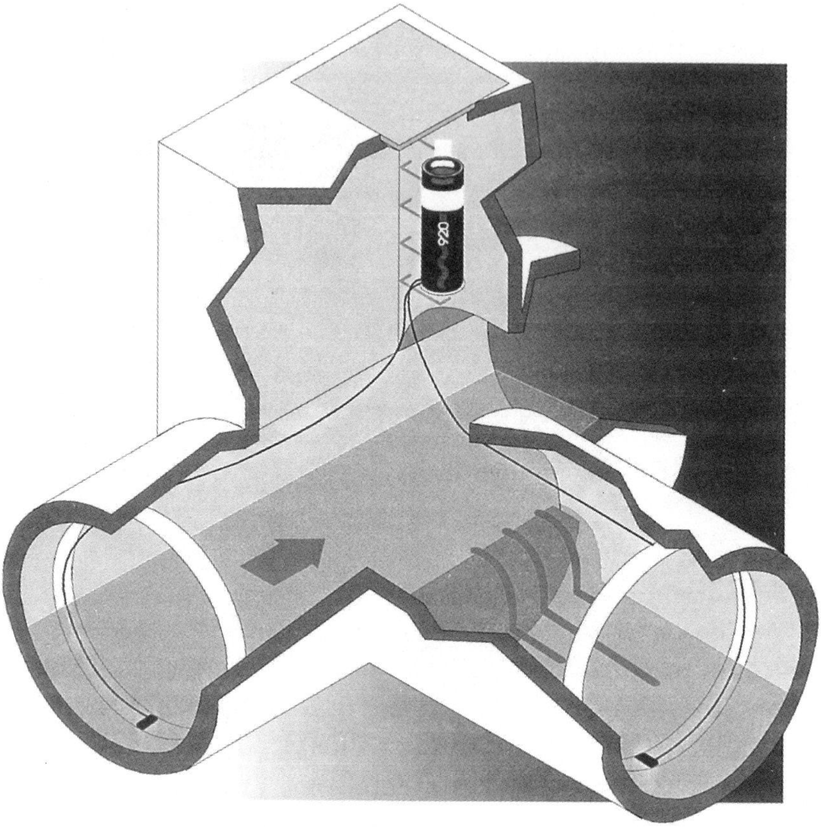

Figure 6-2 Ring banding on a complex flow measurement scheme. Note dam causing restriction. Such a configuration can record discharge as a result of excess flows.

The cut for both matrix structures now provides a working tool with which to start the system design. The wish list is more manageable and short- and long-term funding availability better understood. No text can outline methods for tabulating and editing such lists. Operators of municipally owned utilities already have relevant policies and procedures in place; the suggestions here simply enhance these diverse but usually effective practices. Operators of private utilities assemble for this task tend to have more leeway in compiling the matrix structures and may not require as many outside interfaces as a municipal operator. However, each approach, again, is unique. The iterative process of elimination from potentially large matrix structures does much to stimulate the thought process and bring to the surface many ideas which would not be apparent otherwise.

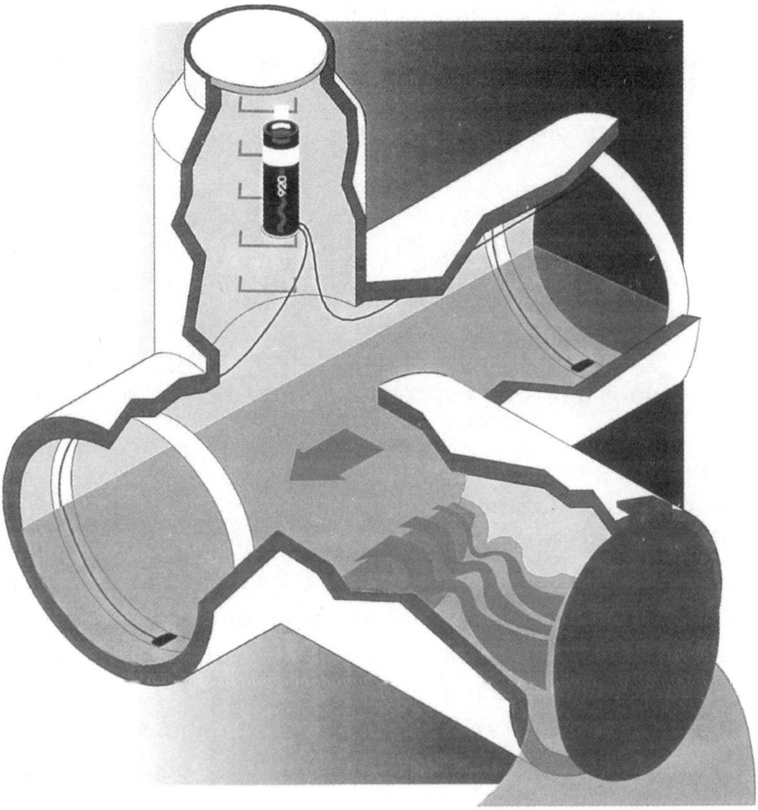

Figure 6-3 Method for determining discharge if the hydraulic configuration does not allow sensor placement. A subtraction is performed from one sensor to another, and the mass lost is the discharge.

The wish list items now can be categorized as either sites to be monitored or results to be obtained. A *result* here is a computerized data presentation. At this point, many items blend into data presentation concepts. The wish list shrinks and the distinction between field issues and computer room issues becomes obvious. Most of the time, the largest wish list issue is a report that few will care about. However, why that report is requested bears on how the project should proceed. The report structures themselves can be traced to a repeatable pattern of information requests that currently cannot be fulfilled. These unfulfilled requests, referred to here as *data holes,* become a convincing argument to use with the agency's authorities.

The identification of data holes through the iterative analysis of the wish list is critical at this juncture. Many more wishes will surface after the design is approved or after the system is installed and all the current data holes are plugged. It is therefore important to determine where all targeted key employees are positioned at the time the system is designed and at the time it is installed.

The result, after all the cuts and analysis, is a structured tool that can be used to guide the beginning steps of a system design. The needs assessment, referred to as the *wish list,* absorbs broad input and the iterative elimination condenses the list into a manageable working document. The financial analysis keeps to the forefront that the entire project is constrained by available funds and that expectations must be tailored to match the funding route.

The amount of time required to complete this analysis is a function of the size of the utility and the monitoring system desired for installation. It is difficult to assign a time of completion for this portion of the endeavor but it should nevertheless be estimated. It is important to allocate as much time as necessary without compromising any of the particular aspects of the matrix. The more complete the matrix, the more questions that can be answered in the planning and implementation phases.

At this point, contact can be initiated with equipment manufacturers, service providers, and system houses to generate a technology assessment of available products and services. Depending on the type of matrix generated from the needs of each situation, the assessment can be released to these vendors for their analysis. Sharing this information is a good way to get their feedback based on their experience and product offerings. The vendors can also readily tell if technology that meets the stated needs is currently available or under development, or if another approach should be considered.

The input of manufacturers and providers is critical in the system design. No group knows a manufacturer's equipment better than the manufacturer itself. This may sound logical, but it is missed many times. If the technology is sound and there is a match to the needs assessment, a manufacturer is best qualified to make that match. Reputable manufacturers will indicate if a request exceeds the state of technology or simply of their equipment. Input from many manufacturers, along with their pricing structures, provides the tools for an additional cut to the matrix. Service providers also can offer excellent feedback, as they generally use the products of many manufacturers side by side. Their strength is the ability to mix the best from all the manufacturers and customize a blend suited to the needs assessment. Service providers' feedback is critical in determining performance criteria and in verifying manufacturers' claims. The information received from all these parties is targeted to the needs assessment, making it more useful. Also, discussing the system points and equipment particulars with users of the equipment generates parameters that may have been overlooked in the original matrix generation.

The next cut eliminates anything identified as technologically infeasible or borderline by the manufacturers and service providers. The utility agency must decide how far to experiment with available technology. This is strictly an internal judgment call and one that should be considered carefully. Attempting to extend the boundaries of current technology is the nature of innovation, which should be encouraged at every juncture. However, it must appeal to the utility and definitely have commitment from the manufacturer. This cut leaves the system designers with a matrix that is feasible, financially sound, and versatile. So far, the only factors considered have been from a desktop. Now, field information must be integrated. This will be the final step in generating the data for a proper system design.

6.3 LOCAL KNOWLEDGE

The gathering of local information is the last step in system design. Local knowledge is critical in fully understanding and utilizing any infrastructure system. Most field forces know when water is flooding the street from a manhole or a discharge based on overcapacity in a relief point. The next step is to determine why these things occur and the best approach to surveillance.

The reliable method for generating local knowledge is the temporary monitoring. Permanent monitoring generally follows based on the data set generated from temporary installations. Refer to chapter 3, where the critical path was discussed. The critical path includes points of interest that draw the attention of the system operators for reasons ranging from the engineering staff's attempts at ascertaining conditions within the infrastructure for purposes of planning and modification to system operators' need to respond to complaints. The reasons for this interest are varied and include pure engineering curiosity and flooding or other complaints. These reasons are usually known to the system operators, who deal with the complaints every day, or to an engineering staff attempting to ascertain conditions within the infrastructure for planning and modification purposes. Because it is impossible to obtain firsthand information during a storm event from within the infrastructure by direct observation, the only reasonable method is temporary monitoring.

The commitment of resources for any monitoring program should incorporate provisions for site surveys that include temporary monitoring of specific locations. It does not matter how the information is generated or in what format. It only matters that initial site information is obtained so an understanding of the system's hydraulics and its response to storm events can be ascertained. As points along the critical path are monitored, specific determination can be made concerning characteristics of that path and the points that possess those characteristics. It will become apparent that not every point along the path requires monitoring. It will also become obvious if any spot is overlooked or missed. The temporary monitors can note similarities of hydraulic characteristics of sites on the same critical path and those along different paths.

A program of temporary monitoring is also an excellent addition to any sewer maintenance program. It enhances the existing sewer programs while providing site-specific data required to support or deny claims as registered by public and private programs. Whether it is executed by contract or by in-house resources, such a program can provide insights concerning those sections generating particular interest within the infrastructure. Many individuals, rights groups, consulting engineers, and regulatory agencies require specific data from such sites regardless of whether a permanent system is in place or not. As noted, public and private complaints and other inquiries can be readily addressed when the ability to deploy monitoring equipment on an as-needed basis is available. Therefore, the ability to quickly install and generate site-specific information is a definite asset to any infrastructure program. It also enhances any permanent monitoring program by providing verification information, calibration data, and, in the case of new sites, hydraulic overview responses that assist in equipment selection and deployment.

Reference is now made to the deployment of a rain gauge network, as discussed in in chapter 5. This system was described and, for purposes now discussed, allowed to be loosely installed just to gain general insight for the localized hydrologic cycle. From this point on, a more structured approach is taken to place, record, and maintain this equipment. It was necessary just to get something out into the street and develop some experience in it. If the utility already had such a loose system, a study of the existing data stream is in order to fully understand the localized hydrologic cycle. In either situation, the precipitation characteristics in a given area are critical to understanding the loadings due to runoff into a sewer infrastructure. Therefore, it is necessary to review this data and make associations with complaints based on these conditions.

This paves the way for the deployment of the temporary monitors. Temporary monitoring is done for two reasons: to generate information based on hydraulic characteristics and to investigate complaints or comments. Two studies commencing at once makes good use of the equipment. Armed with the precipitation data, the complaint information, and the points of interest along the critical path, sites can be chosen for the installation of equipment.

It has been noted that complaints are a source of local knowledge. Complaints in a sewer maintenance setting are the driving force for many repairs and remediation programs. This is brought to light here because one of the many sources of local knowledge is workers in the field. Day in and day out, these people handle complaints from a variety of sources in the utility environment. They know where the trouble spots are and are experienced in remediating the situation. They know the difficult locations, sites with recurring problems, and those places that have particular impact on the community the infrastructure serves. Caution must be taken, however, in obtaining information from this source. Much of this data is used to the advantage of the maintenance workers. As efficient a job as they do, maintenance workers sometimes closely guard data in a way that works to their advantage; it probably should be subject to careful interpretation.

Local knowledge of the system, regardless of its size, is a critical element in determining the starting point of, and its application to, the localized hydrologic cycle. Despite any caveats, though, the ability to develop local knowledge and the understanding of the hydraulic response through temporary monitoring is a critical beginning in the use of electronic equipment not only for short-term information but also the long-term value such infrastructure management systems can provide to the system operators and the community.

6.4 UNDERSTANDING THE GRIT BASE

The grit base is the deposition of solids within the sewer system. The characteristics of grit deposition are probably the aspect of the system understood least by design engineers and managers and best by the people who enter and work within the infrastructure. The grit base is given a separate section here because of the complexities that it can add to any infrastructure monitoring program.

Grit is a combination of the organic portions of waste material and inorganic matter entering as washdown or runoff sediment. The living organisms within a grit base are necessary for the operation of a biological treatment process necessary to maintain the parasites that destroy *Escherichia coli* bacteria, a carrier of disease organisms in warm-blooded mammals. Killing the *E. coli* kills the diseases being carried by it. By the same token, the longer the grit sits as sediment, the more microorganisms it can breed both the harmful and helpful kinds. Obviously, grit must be conveyed to a treatment plant.

The base, or false bottom, that grit generates as a result of deposition is formed from solids that could not be suspended long enough to be transported to the treatment works. A velocity of approximately 3 ft/s or 1 m/s is required to keep most fine grit particles in suspension suitable for transportation. This is also a design criterion for most sewer systems, and if it cannot be met, a pump station is considered. Most primary treatment tanks in a biological treatment process slow the incoming flow (or velocity of the flow) enough to settle out the solids. The appearance of a grit base is generally the result of the velocity of a transport volume dropping below this 3 ft/s or 1 m/s parameter. The appearance of grit around structures is a direct result of the velocity of the transport layer dropping below these parameters. The structure provides the slowing and the solids settle out.

How the solids settle determines the extent and shape of the grit base. It is usually irregular, with varying depths and layers, depending on its composition. It is the nature of a grit base to be irregular because of the particular settling of solids. Grit settles in sloping layers away from a structure or in patches throughout the sewers when it does occur. This non-uniform base presents its own set of challenges when introducing equipment to the system. Because the bottom of the system is now not uniform, nor is it as designed and constructed, standard prints and drawings are useless in proper equipment deployment.

Figure 6-4 Grit base as seen within a primary measuring device. The clear water allows for viewing the layers. The various layers have colors, which makes the depth of the grit visible. Grit artificially raises the level within a conveyance and provides false readings.

The ability to identify and locate the grit base is of crucial importance when installing equipment in the sewer systems. This is a process called *grit profiling*, and it should cover the entire area that the equipment is intended to monitor. Grit profiling is determining the depth and extent of the grit base when it exists. Most field crews have well-developed methods for determining the existence of a grit base within a sewer system; this can be done fairly effectively with a rod pressed into the grit base. The distance the rod penetrates into the grit base is the depth of the false bottom. Grit profiling takes the next step and determines the distribution of grit in a larger area. It is just as important to realize that the compactness of the base relates to the ability to utilize it as a bottom for determining level or flow. A compact grit base can be solid enough to be used as a base as long as it is level enough.

Understanding the grit base is important because grit can reduce hydraulic capacity, cause maintenance issues, and produce unpredictable conditions within the sewers. The structure and formation of the grit base is important also from the monitoring viewpoint, as extensive grit bases limit the deployment of sensors and inhibit their accuracy. Further discussion of grit appears throughout the text as it applies to each component of the system.

The geometry of the base as it relates to the shape of the profile being generated now needs to be determined. If a grit base is formed in a rectangular sewer, it is a simple matter of raising the base using the top of the grit as the new bottom. It is therefore reasonable to measure the height of the sewer using the fairly uniform grit base as the bottom and experiencing a reduced design capacity. The solution is not as simple in sewers of spherical or other curved shape. The grit base fills in the curve and forms a flat bottom, unlike the design shape. If the base is to be utilized and not removed, the new geometric shape must be taken into account. A profile must be generated of the cross section including the grit base and the reduced capacity included in the new calculations. A grit base should be included in any drawings made of the installation or of the point for this reason.

In many cases, there are uniform spots within the base where it can be considered the bottom of the sewer and the actual depth of flow taken from the upper surface of the grit base. It follows that some grit bases are fairly uniform on their own because of the action of the flow and the scouring it performs. Care must be exercised during the profiling to maintain a uniform base if one actually appears. If one does not appear, then a determination must be made if the sewer should be cleaned or simply disturbed to get the grit back into suspension. Either way, it is imperative to assess a grit profile properly and act to prepare the site for equipment installation. If the grit base is uniform, it can be considered the bottom of the pipe or structure and installation can proceed accordingly. If the grit base is irregular, it must be removed or leveled. If the grit base is spotty, judgment must be utilized to determine if the spotty depositions will significantly perturb the flow profile and thereby skew the measurement.

Grit is easily disturbed. When it is, so are the microorganisms contained within it, including disease-carrying bacteria. The undesirable washdowns of industry and farming are also disturbed. Care must always be taken in handling this material. Therefore, when profiling, extensive grit bases should be noted and consideration given to cleaning the system not only for the accuracy of the program but for health and safety issues.

The profile that the grit makes at a point along the critical path within the sewers can have significant impact on the final accuracy of the level or flow being measured. An accurate profile and an understanding of its extent is critical before the introduction of any electronics or other measuring devices can be properly staged.

6.5 LIGHTNING AND ITS MITIGATION

Lightning is a serious issue with any field installation. For temporary monitoring, little can be done unless the vendor includes lightning arresters in the instrument package. However, for a permanent installation, this can be remedied by using lightning suppression in the cabling. Lightning suppression should be between a ground point and the instrument package and between the sensors and the electronics package, depending on the manufacturer's design.

Lightning is discussed here to ensure a basic understanding of static electricity and to dispel any false notions on the matter. Two kinds of lightning strikes are of concern to field installation: the direct strike and the ground strike. Both types can last as little as a few milliseconds and as long as one second. Therefore, the reaction time of the suppression shunt needs to exceed this time requirement. Both strikes result in large amounts of electricity, anywhere from 30,000 to 270,000 amperes, being introduced in that small time frame. These facts being nothing more than physical science, a division of the two strikes must be formulated.[2]

There is generally no protection from a direct strike by lightning. The power of lightning simply bakes most electronics packages regardless of grounding or other safety measures. Electrical flux alone causes irreparable damage to the electronics, aside from the awesome power of the strike itself. Fortunately, as most equipment is located below ground, direct strikes are not common on field apparatus. Installations above ground should always be equipped with grounding rods, Faraday cages, if possible, and suppression. Above-ground electronics are especially susceptible to lightning and every precaution must be taken not only for the safety of the equipment but also for all that surrounds the installation.

In a ground strike, the ground can be considered a conductor of elec-

2. M. Miller, "Lightning Protection for Submersible Pressure Transducers," Water Engineering and Management, February 1999.

tricity. The flux is carried through the ground and affects field equipment at a variety of intensities. The intensity dissipates as it travels through the ground. The travel of electrical potential as a result of lightning is generally referred to as a *transient.* Transients wreak havoc on field equipment because they can come from near or far, depending on the intensity and the conductivity of the ground around it. Transients are not limited to ground propagation. Humidity in the air is also a conductor and therefore a conduit for transients; air itself is a weak insulator. For purposes of this discussion, both air- and ground-conducted transients are referred to as *ground strikes.* Fortunately, most transients can be suppressed at the instrument, or at least their damaging effects can be mitigated. Proper suppression saves field equipment from most transients. The cost of the damaged suppressor is certainly worth it.

High-quality field instruments either come with lightning transient protection or show in their design where the protection should be introduced. Protection from any electrical transient is critical, as lightning is not the only source; it simply causes the most damage. Transient protection is critical over the lifespan of the instrument.

6.6 SAFETY ISSUES WITH CONFINED SPACE

The subject of safety issues is brought up here for completeness. Those who work within the infrastructure every day are acutely aware of the hazards that sewers pose to their health and safety. Each and every utility that performs this work is aware of these hazards and a full discussion on safety issues is probably not necessary. However, steps relevant to equipment installation are brought forward.

At all times, in major lines, the treatment works should be notified of entries so hydraulic actions at the plant that would adversely affect the workers can be rescheduled. The water itself is a threat because it contains disease organisms. The flooding of a jobsite because of plant actions, such as closing the main gate as part of a standard maintenance program, is also a hazard. Therefore, a plant contact person should be notified of any work taking place within a major sewer line. Also, personnel at any major construction site in the sewers along a critical path should be notified so work can be coordinated. Look up and down a critical path to be certain that there are no impediments resulting from the installation jobsite. Construction sites generally cause constrictions and necessitate bypass pump-

Figure 6-5 Confined space equipment is necessary, expensive, and requires surface space to deploy. There should be no expense too great when it comes to the preservation of human life. The extent of such equipment must be realized by the system managers to determine if installation and repairs of the sewer monitors should be handled in-house or via contracted service.

ing. The bypass pumping causes variations in the flow profiles such as would be ordinarily experienced from gravity transports.

A standard body of text for the United States Occupational Safety and Health Administration (OSHA) and for international bodies concerns confined space requirements. Most times, these are far and beyond the actual needs of the workers and they are modified to suit the purpose. Compare the current confined space entry requirements of the utility with the corresponding country's recommendation and determine which should be included in the safety documents. Requirements include recommendations for lower explosion limits (LELs) for gases such as methane and hydrogen sulfide, tolerance limits for oxygen, and other issues of safety for the workers.

There are also issues concerning potential hazardous releases in industrial areas. Substances that might be released into the system should be known and protection issued. Sometimes it is simply a matter of knowing the potential is there. Other times, the treatment works may need to be notified. This is a consideration and should be observed in the site profiling. Many industrial wastes are more a nuisance than they are a hazard. Knowledge of their presence is generally the best defense prior to entry.

The issues of street coordination and lane blockages are of local determination. Laws varying from locale to locale determine how traffic must be detoured in the event of street work. Extra care should be taken not only when workers enter the sewer but also when they install equipment in high-traffic areas. Here accesses with limited street exposure are desirable, of course.

Safety and health are the primary concerns for anyone required to enter the infrastructure.

6.7 TEMPORARY MONITORING TECHNIQUES: SEQUENCING

The availability of resources is always an issue in any management environment. This is the reason for introducing the concept of sequencing.

The following discussion introduces concepts for both temporary and permanent staging of various levels of equipment and, further, to introduce the methods of enabling a vision of development, from the simplest to the most complex forms of implementation. A temporary level monitoring system is appears to be the simplest of the concepts, and the large permanent flow measurement and alarming and control system the most complex. None is of any value unless accompanied by a stepwise procedure for the evaluation and installation of the devices, limited by their particular technology. Sequencing is a stepwise method of introducing equipment and systems into the sewer.

Sequencing is one of many techniques for providing balanced resources for a temporary monitoring program, making maximum use of all limited resources. It also provides structure so the program can be repeated in the same manner and style time after time. This consistency means the results can fairly be compared, and it allows long-term analysis of groups of results. The repeatable method also can provide continuity over personnel changes and movements in administration.

In temporary monitoring programs, the most limited resource is usually equipment to perform the task; the data requests exceed the capacity of available equipment. Locations are generally plentiful, and manpower

Figure 6-6 Sequencing temporary monitors along a critical path. The instrument package will hang from a ladder rung. A small cable will hang just below the vent of the manhole on a stiff wire so the data can be retrieved without removing the manhole cover.

is usually sufficient, even if overtime or extended contracts are utilized to fulfill the request. The equipment, however, is either in short supply or not well matched to the request. Sequencing facilitates maximum deployment with minimum commitment.

The following sections discuss the techniques involved with sequencing for the purpose of generating information about a particular drainage district. The approach is applicable to any data request from any requesting party. The framework can be modified to fit the individual circumstance. It is recommended that tailoring be within the frames described so consistency can be provided over the long term.

6.8 TEMPORARY MONITORING TECHNIQUES: LEVEL SEQUENCING

Level sequencing is the simplest temporary monitoring technique and one that generates a lot of information in a relatively short period. In level sequencing, a variety of inexpensive level monitors are deployed along a critical path in order to determine the hydraulic response of the path to phases of the localized hydrologic cycle.

Level sequencing can be done either by storm event or by time. Storm event is a more common delineation of the study period. Dry-weather steady-state conditions probably will not change much as long as the passage of time is within specific boundaries—say, a period of weeks or even a month. The same conditions probably will noticeably change over months due to changes of season and subsequent changes in groundwater tables and other climatic patterns affecting the localized hydrologic cycle. However, conditions change radically with storm events, depending on the time, intensity, and duration of the event. Therefore, the more common method of study delineation is the storm event because rapid changes can be seen and the equipment moved to another potential site of interest. Reasons why a dry-weather pattern should be studied over time include capacity studies for large sewers and relief structures. However, the greater need is usually to study hydraulic responses to storm events.

The storm event delineation technique requires that a baseline storm and a dry interval be defined for purposes of consistency. The baseline storm is defined to eliminate events that may not be significant to the study. Here the definitions of time, intensity, and duration play a major role. A baseline storm can be defined as one of an intensity of no less than 0.5 inches in any five-minute duration, with a time of no greater than six hours. A maximum event can be defined in order to provide an upper limit. This eliminates freak storms, hurricanes, and other rare occurrences that might skew the intended study. By providing these upper and lower limits, consistent data can be developed over years of studies.

The time delineation is useful for capacity studies and other long-term evaluations. Such studies can track changes in the infrastructure that are not associated with a weather event. As an example, private pump stations from development wet wells, industrial holding systems, groundwater drainage stations, and other contributors cause rapid changes in the system, not expected by the treatment works, that can be studied. Another long-term study involves monitoring for illegal discharges. In small sewer lines, spikes in the system tend to point to illegal discharge or other illicit connection. Depending on the size of the line and the quantities discharged, the disturbance in the normalized pattern is enough to prove the existence of the discharge and may serve as evidence during a court action.

The temporary monitoring solution using level sequencing and time event delineation for this purpose almost changes to the event delineation category as the actual target is an event, not time. Time is only a framework for determining a condition within the structure.

It is possible to have a project delineation with a target of cross-references. It is also possible to have a time event delineation to study storm events over a period and to monitor storm events through a time sequence. This is done to show that what information sought may differ from the goals of the program. Because the program often turns out to be different than expected, expectations are not put on the result. They are only included in the part of the program that is realistically controlled.

Level sequencing consists of a series of inexpensive level monitors arranged strategically along a critical path. Here, the utility performing the study chooses the level devices for the purpose. Unless there is a large amount of in-house talent, the packaged systems supplied by manufacturers generally are the easiest to install, interrogate, and process. Not long ago, it was quite effective to fabricate these devices in house using standard battery-powered dataloggers connected with pressure transducers or inexpensive ultrasonics. Today, manufacturers produce an array of systems that not only are seamless in their integration but relatively clean in the interrogation and data handling. These devices usually cost less than in-house fabricated components, but price is not the only valid criterion. Always investigate to determine what equipment works best in a particular situation.

Once the critical path is chosen, the delineation agreed upon, and the resources set aside, the level devices can be installed. The devices are sequenced so changes within the system along the hydraulic grade line can be observed and documented. A hydraulic grade line (HGL) is the surface of the water traveling under open-flow conditions within the infrastructure. Along the critical path, the HGL is recognizable by the line that the surface of the water makes as a level within the sewers. Rises and falls in the HGL denote changes within the infrastructure that are of note to the project managers. In civil engineering, HGL measurements are generally tied to some elevation. This conversion is easy to perform based on the elevations chosen by the utility. Conversion to any individual system is left to the reader.

A preferred resolution of data points per time series provides the best image of the HGL within the infrastructure; this is the time increment at which the data is recorded. In systems that are fabricated in house, it is possible to develop elaborate schemes of point averaging, as dataloggers have immense capabilities for computing samples and averages. Because of the slow changes within the infrastructure, the question becomes one of necessity. The determination must be made whether or not it is really necessary to take a data point every second and average the points over a minute. In the case of temporary monitoring, the answer is a resounding no. Temporary monitoring is strictly that—a time-limited attempt to determine potential conditions. Battery life decreases with every function computed. Therefore, it is better, in the temporary setting, to take a sample every five minutes or so, bearing in mind the scope of the endeavor. This should provide sufficient battery life or at least provide a clue as to when the batteries will require replacement by the field crews. It is recommended to keep the resolution fairly loose, considering that the weakest component of any piece of temporary field equipment is the battery pack.

The interrogation of the level sequenced units is generally performed using the manufacturer's specifications. The least problematic method is to use the manufacturer's interrogation unit, which automatically segregates and stores the data set. These devices are generally inexpensive and targeted to the task at hand. A laptop computer is a useful tool; it has more versatile functions in the field than single-task devices. Laptops cost more than interrogation units, but their functionality is noticeably greater. The agency can decide which device it prefers. Market awareness of product offerings and available technologies is important.

Level sequencing is a method of generating a lot of HGL data in a relatively short period. The ability to follow points of interest on a critical path provides references for determining responses of the infrastructure to the delineations of desire of the system operator. The goal is to build an index in a database where studies can be compared over the long term as the knowledge base of the agency increases. Such histories can provide system operators with tremendous insights into the infrastructure's responses.

6.9 INSTALLING THE TEMPORARY LEVEL MONITORS

A separate section is dedicated to the installation of the level monitors, because any monitor introduced into the infrastructure is a relational device and the relationships must be defined. An understanding of the structure, pipe, or outfall at which the device is being installed is critical to optimal placement.

The physical dimensions of the site must be known; if they are not, the site must be measured. This is known as *profiling the site.* In carrying out the system inventory described in chapter 3, many of these dimensions were gathered and documented, so little additional work should be required. Documents that provide the physical dimensions of sites should be uniform so the catalog of information can be filed for future reference.

Always, in the case of temporary monitors, the locations that provide the readiest access to the probes should be chosen. The selected technology dictates location and placement. Some compromises may be required, but in the case of structures, a spot should be chosen where the HGL is best reflected and the access is the easiest.

A must for inclusion in any site profile is the grit base, as discussed above. A base profile must be performed to determine if any grit exists. If no grit exists, the installation can go forward with only the consideration of solids deposition and the effects of equipment shape. If grit does appear, the parameters previously noted must be considered before introducing the equipment. Localized depositions can be readily cleared. Extensive depositions need to be evaluated and localized knowledge used in determining whether the base should be included or removed. If extensive grit deposition exists, perhaps no monitoring can be performed until it is cleared. Upstream and downstream depositions are also a consideration in determining the extent of the deposition and are easily obtained from a profiling.

Nonintrusive level monitors are the easiest to install and maintain. Most nonintrusive level devices use ultrasonics to determine the relative HGL within the site of interest. The devices can be mounted right in manhole entryways. Several points must be considered when installing the ultrasonic probe. The first is the distance from the face of the sensor to the bottom of the site; this measurement must not exceed the manufacturer's limits for distance of signal projection. The second is the cone expansion of the sound pulse. The head must be mounted low enough in the manhole so that the cone expansion is not a problem. The head should also be high enough to avoid flooding. The third is the deadband of ultrasonic signal. All ultrasonics have a point near the face of the crystal to which the echo is not properly returned. This is known as the *deadband.* In determining where the head should be mounted, the deadband should be noted and compensated for.

The design of one family of ultrasonic devices effectively eliminates the deadband issue by projecting the sound pulse lengthwise for the calculated length of the deadband, then reflecting it down. These devices look like a tube and may be mounted in the manhole but also within the pipe or structure. These devices tend to be best suited for locations of potential flooding or surcharging.

Another consideration is the distance of the HGL from the ultrasonic head. The distance measured by the travel of the sound pulse from the head to the HGL subtracted from the physical distance from the sensor head to the bottom of the site is the depth of the water. Direct measurement is the best way to verify that the electronics are working properly. Let the device run for about a half hour and take periodic measurements to ascertain that the device is measuring properly.

Verification of the device is not performed by placing a ruler or other measuring device into the water, as one might assume. Such action generates a fishtail and the direct physical measurement becomes inaccurate. It is necessary to measure from the ultrasonic head to the water's surface and subtract this figure from the distance between the sensor head and the bottom of the site. This is the proper method and does not generate any residual in the form of a fishtail or other disturbance.

Installation of an intrusive level monitor, such as a pressure transducer, can be a bit more complex and require some forethought concerning the location and parameters of installation. Anything introduced into the flow becomes an obstruction, which has the potential to collect debris. Also, pressure transducers tend to be sensitive and require protection from passing solids. For these reasons, site-specific profiles are critical.

The usual protection for intrusive level devices is a stilling well. This is a piece of rigid PVC tubing with vent holes drilled so the level in the site can match the level in the stilling well. The name comes from *still flow,* the calming of the passing volume so a tranquil head can be measured. Fastening the well to a flat wall is preferred. Not only does the well still the passing volume's HGL but also it protects the transducer from solids that may cause damage. If the walls are not flat, as in the case of a circular sewer, fabrication of a curved portion to match the wall may be in order. If the flow is shallow enough, a stilling well may not be necessary. Direct mount-

Figure 6-7 A temporary submersible level device installed in a stilling well. Note biological growth on the well. The stilling well protects the sensor from the turbulence of the bubbles and waves.

ing to the bottom of the site is sufficient. This is a discretionary decision based on conditions at the site.

It is possible to offset the sensor to accommodate for grit bases and other obstructions. Also, the level sensor can be mounted on a platform if it is certain that the platform is sufficiently covered with water so that relational operators can properly ascertain the relationship of the sensor to the bottom of the site. The measured distance between the bottom of the site and the sensor is called the *offset.* Offsets can be very large if there is constant HGL in the structure to cover the sensor. The measurements of importance in an intrusive measurement scheme are the level of the water, the location of the sensor within the site, and the offset that the sensor generates by virtue of its mounting. Even a stilling well tends to generate an offset for the transducer. Verification is performed, again, by measuring from a point above the HGL to the bottom of the site to the reference and from the sensor to the water's surface. The subtraction of the two measurements is the actual level or HGL in the site of interest.

It is necessary to secure the cables in any installation. Tie wraps to a stilling well are generally adequate. Securing the cable to the ladder rungs is questionable because of the entry and exit of personnel. It is better to secure cable to the walls of the manhole or to secure them under the ladder rungs so they are out of the way of the access point. However, for temporary installations, fastening the cables to the ladder rungs may be the only reasonable alternative. Temporary cables should be secured well enough so as not to collect any debris in the instance of flooding, which may cause excess weight and damage the cable, associated sensors, or instrument package. Cabling that has the potential to become submerged should be provided with extra protection and all edges minimized to reduce the potential of solids collection.

Level measurements that determine hydraulic characteristics of parts of an infrastructure are particularly used with respect to combined sewers and CSOs. Combined systems have complex hydraulic patterns. Measurement of the three major areas of a combined sewer identifies valuable characteristic information that can be used to understand the actual, not design, function within the sewer system. A level placed in the trunk sewer, a level located at the overflow point, and a level of the discharge to the regulator can generate data that can relate level to loadings and storm events, which can provide excellent diagnosis of the functions of a combined system. It is not important at this stage that the actual sensor be uniform or accurate. It *is* important that the proper depths are recorded and reflected within the profile generated prior to installation. Relationships to any elevation are subjective and unique to the utility that generated the elevation correlations. Knowing when a system is discharging, strictly based on level, in connection with particular storm characteristics, can be valuable when planning not only system operations but also for protection of the environment and dealings with regulators. Temporary level sensors can be placed at any accessible point as long as it can be related to the site profile.

Structures themselves generally have complex hydraulic patterns; even relief chambers do. Levels can provide insight into their hydraulic characteristics and ascertain if structures are performing to design specifications or are being affected by placement. Remember that structures are custom designed to a particular set of parameters and that those parameters change over the many years that the structure is in service because of external changes, system expansions, and modification and shifts in industry and populations. Thus, a level verification can illuminate the performance of the structure. The various hydraulic influences that generate the level do not influence the measurement of the level. For this reason, level is an excellent characteristic on which to base studies and performance testing. By relating the HGL of the structure to grade points throughout the system, performance milestones can be examined and related to the system as a whole. Level equipment can be placed in any portion of the infrastructure as long as the site is related to the profile generated prior to installation.

Combined sewers, structures, and storm sewers are made of similar materials. Level devices are fairly forgiving in their placement and performance. For this reason, as noted above, a temporary installation can be performed based more on convenience than on exact design or HGL considerations so long as relations are made to the location of the sensor. It is always desirable to generate a level reading from a central or maximized location. One of the purposes of temporary monitoring is to determine these locations even if trial and error are necessary to do so. Sometimes the trial produces reasonable results, but it will always build knowledge of the system.

The last matter to be brought forward is the use of mounting and fixture hardware within the infrastructure. Sewers contain hazardous gases, corrosive mixtures, and other compounds that cause many materials to deteriorate. PVC, plastics, or stainless-steel hardware should be utilized in any installation. Most other materials rust or corrode, even if painted or sealed. The hazards of the environment dictate the use of high-strength, high-quality noncorrosive materials in any mounting scheme.

Presented here is a method for generating characteristics of the infrastructure based on level. The ability to record a level and, subsequently, to profile an HGL within the system, can provide significant insights to the hydraulic response of the infrastructure. Using the methods outlined enables the repeatable measurement of characteristics and thus the development of a database that can be used for many years.

6.10 TEMPORARY MONITORING TECHNIQUES: FLOW SEQUENCING

The concept of flow sequencing is similar to the concept of level sequencing. The biggest difference is the addition of the quantity of flow derived from level and velocity. It is generally impossible to calculate flow directly in a manner that a full pipe primary measuring device would, so some relational operator is generated through a cross-sectional area and average velocity scheme. This approach produces acceptable results if the equipment has integrity, the plan is well thought out, and the devices are maintained in the field. In this situation, the HGL is generally not an issue because actual quantities are generated; HGLs are used for level situations only, although novel uses for them pop up in the engineering studies portion of an analysis. Here the emphasis is on flow, or the actual quantity or volume that is passing through the system.

The critical paths are again chosen to link specific points of interest. In the case of flow sequencing, flow requires measurement only at junctions and where there is suspected infiltration and inflow (I/I). There may be a wish to supplement level information at various points for study purposes, but the flow or volume of water being conveyed should not significantly change along a given length of sewer unless there is some inclusion of a structure or junction or an otherwise known introduction of a water source. In the case of a junction, flow should be taken at the tributaries feeding the larger system and mathematically summed. This accounts for the weighted contribution of each tributary and provides a clearer understanding of the hydraulics affecting that part of the critical path.

There are many reasons why the flow should not be directly measured just downstream of a junction in an open-channel situation. The first and the most obvious is the complex hydraulic pattern that the mixing of the tributary systems produces at the point of influx, caused by the open-channel conditions. If the system operated under full-pipe conditions, the confluence of the upstream tributaries could be compensated for or would be negligible. This is especially true if the flow profiles are skewed at the time of confluence. The system then becomes less stable and more difficult to introduce relational devices. Keep in mind that a relational device measures only part of the velocity profile in a fixed boundary volume. A discharge from several perturbed profiles into a greater area causes an irregular pattern to form. This makes it hard to measure flow accurately. Therefore, the devices need be inserted upstream of any mixing effects. If the number of tributaries is too great for one instrument to monitor, the data acquisition and handling program should perform the summations required to obtain a total through algorithms particular to the site.

In flow sequencing, the tributaries to junctions and other points of interest are chosen in order to generate the flow characteristics, or the hydraulic response, expected from natural and generated events. This has significant bearing on the study. Flow is a quantity made up of the parameters of level and velocity. A significant change in one of the parameters can mean that other parameters may not appear affected if one of the parameters is substantially increased. The point here is to be certain that the administrator of the survey is aware of all conditions that can affect the outcome of the program.

An example here is the action of pump stations and the treatment plant along a critical path. Remember that the treatment plant is the largest hydraulic control point along any sewer infrastructure network. If conditions change at the plant, some of the parameters affecting flow could change and disrupt known or otherwise expected conditions. For instance, pumping at the plant or changes in gate position at its influent could significantly affect the velocity of the incoming flow but may not, during a particular set of conditions, affect its HGL. It is therefore recommended that in any flow sequencing layout, known and potential upstream and downstream variables be charted and understood before the test.

The charting and tabulation of conditions known or suspected to influence the critical path is part of a proper program layout. It also serves as a ready-made punch list for notification that studies will take place; notification of any changes from standard operating procedures should be given to the program staff liason. At all times, the treatment plant should notify and be notified so actions there take place with the highest regard for the safety of the workers installing, maintaining, and removing the temporary equipment. In the case of unmanned stations, human intervention is always possible. When the program is put to paper as a plan, the necessary contacts will appear obvious.

The areas of study can be divided along the critical paths because it is unlikely that a temporary monitoring system would be left as a permanent installation. Therefore, the division of study can serve the program, but its administration should be rational. The rationale for division must be repeatable so multiple requests made in the same area prior to the installation of permanent system can share a basis. The plan here is critical and should be heavily documented.

6.11 INSTALLING TEMPORARY FLOW MONITORS

The manufacturers of temporary flow monitors produce systems that are readily installed and calibrated in place. These complete systems can provide accurate results, even with the spacing of data sampling points, which is meant to save power. The installer is responsible for uniform, consistent, and logical placement of equipment. To support this, fundamental considerations should be noted at the time of layout; others can be determined only at the time of installation.

The area velocity device is the type of flow monitor necessary to perform flow monitoring, as described in chapter 3. A level is ascertained and a reference velocity is worked into an average velocity value or into a profile using Manning's equation or other means. In any case, it is important to remember that flow is not a directly measured quantity but one derived from relational operators. These relational operators must be able to measure their respective parameters with as little obstruction as possible.

The first class of installations concerns pipes of 18 inches/0.5 meters or smaller, to which direct entry by an installer is simply not possible. In most instances, installation via a manhole is the only option. A second class consists of pipes up to 48 inches in diameter; these can be entered directly. Sanitary sewers generally fall within one of these two classes, except for major conveyances such as intercepting sewers and other major feeds to a treatment works. The last class consists of pipes above 48 inches; this includes combined and storm sewers.

The area velocity probe or probes should be selected based on a profile and drawings made from the profile. The presence of a grit base should always be noted and profiling should be performed to determine its extent. Once the selection is made, all preparations should be carried out at the top of the manhole. The sensor can then be mounted on a tension band; most manufacturers provide mounting screws. Avoid using tape or other wrappings to keep the sensor in place unless there is no other alternative. Tape does work for temporary locations, but the inconsistent chemical content of the wastewater causes unpredictable wear on the tape's adhesive qualities. It is for this reason, not the quality of the products, that the use of tape is discouraged. As the adhesive dissolves, the tape may become entangled with the sensor, producing erratic readings. Tape is good for generating a smooth surface for the sensor cable. In this case, use a thick tape and be certain to wrap it around the tension band. By wrapping the entire tension band, the installation itself assists in keeping the tape in place. Grout on the edge also helps to keep the tape in place after insertion.

The small sewers, provided there is no grit or the grit is removed, can receive the installation via direct insertion using a tension mounting band. Special insertion tools allow the band to be inserted in these pipes from

Figure 6-8 The level and velocity probes are mounted on a large band, which is compressed and sent into the manhole as an assembly. Surface assemblies allow a reduction of the time workers must spend within the sewer infrastructure.

topside and reduce worker exposure to confined space entry. In small flows, as experienced in small pipes, there are no further requirements to mounting the probe other than securing the electronics package. If heavy or fast flows are possible, a tension bracket should be used to maintain the pressure against the walls of the pipe. The sensor should always be installed downstream of the manhole. This allows use of the water flow as an aid in maintaining the position of the sensor and reduces the potential for debris buildup. Upstream installations can gather debris at the far end of the mounting band, which may not be seen until the device is removed. The end of the band should be sealed with grout or other material to reduce the edge and the potential exposure to debris buildup. Sharp or protruding edges should be removed or smoothed at the time of installation.

Area velocity sensors should not be installed directly at the bottom of the manhole, in order to prevent false levels from being reflected in a flow calculation. Remember that an area velocity device is a relational operative unit. If a backup occurs within the manhole and the hole fills up, the level may greatly exceed the actual size of the pipe being measured. This gen-

Figure 6-9 Improperly installed banding will accumulate floatables and other undesirable debris, eventually causing the electronics to fail and compromising the installation. Note the jagged edges on the band where it was ripped out of the installation from the force of the water against the debris.

erates a flow far outside of actual conditions. This is because the level in a flooded manhole, which may be measured in feet or meters, is being recorded to generate the cross-sectional area, which is measured in inches or centimeters. Most modern equipment has over-range alarms for just this situation. In the event the device chosen for the task does not, the easiest and best solution is simply not to install at the bottom of the manhole. Installation of the tension band in the downstream portion resolves the over-range issue by providing the ceiling for the level sensor.

Larger sewers require more innovative techniques when mounting sensors for temporary use. Because each sewer is unique in structure and design, a different approach may be required for every installation. However, some factors may be consistent for all small sewers; some apply only to structures or to combined sewers. Common sense in the understanding the site and a thorough comprehension of the equipment is requisite in adapting any method of installation.

Larger-diameter sewers and combined and storm systems enable greater access than small sewers do. Also, the requirement to place the sensor downstream of the manhole is less significant because the manhole is no longer the largest structure along the critical path. In the extremely large system, the manhole situation is no longer relevant. Therefore, site identification and characteristics of the observed flow dictate where the sensors should be placed. Profiling is extremely important in understanding the particulars of a point of interest along the critical path. The problem in larger pipes is that the sensors are of orders of magnitude smaller than the sewers in which they are installed and debris and other issues are associated with their placement. This makes the cable runs longer and produces more exposure to hazards than installation in the smaller sewers does.

The exposure of the equipment and cabling to hazards can be addressed effectively with commonsense elimination of protrusions. Using many fasteners to keep the cable close to the wall of the sewer, along with proper grouting and placement of the residual cable, can limit exposure of the equipment to hazards. Cables should be run as high as possible above the bottom of the sewer and away from any structure that may impede the safety of the cabling. The reduction of exposure of the cabling protects the entire instrument from both anticipated and unexpected threats.

The sensor or combination of sensors measuring the area and velocity components of flow must be mounted according to the manufacturer's specifications and the minimum guidelines provided. It is the responsibility of the system installer to enhance the installation by observing conditions that may lead to inaccurate reflections of the sewer.

Generally, the manufacturer supplies or equips the devices with mounting hardware. Tapped holes for mounting are found in the bottom of specific sensors or brackets for attachment as required. Since the actual parameters are so different from site to site, the manufacturer's suggestions provide a basic approach, but individual judgment must be utilized in the actual installation.

The large, rectangular, smooth concrete sewers with no grit deposition are the easiest in which to install monitoring devices. Here, mounting plates can be anchored to the bottom of the pipe and the cable run across and up to the nearest manhole. A piece of PVC tubing with hydraulic grout can protect the cable from passing floatables, especially in low-flow situations. The hydraulic grout should be used to build up a slope in front of the conduit to allow debris to flow over and not become attached to it. Even

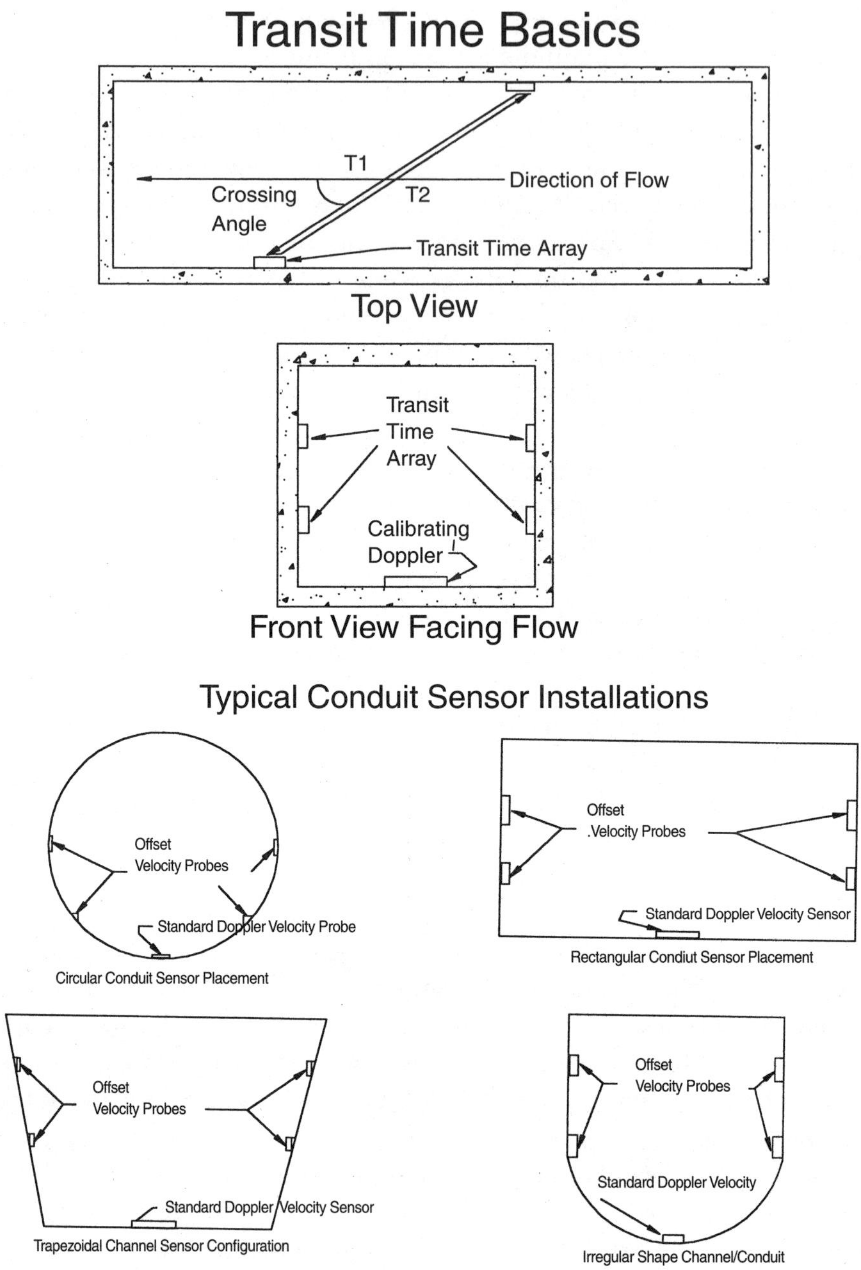

Figure 6-10 Typical installation of sensors in a transit time array. Both major manufacturers' configurations are represented.

though this is a temporary installation, anchoring the cable system is necessary. The sensor itself should be mounted in the center of the pipe to expose it to the maximum flow. Thus the selected technology has a large sample from which to obtain the relational operator and generate a velocity. This sample region is the cross section from which a velocity is calculated. The level device can be installed at any place that accurately reflects the depth, provided that the concrete is uniform. If the level sensor is integral to the velocity sensor, this issue is resolved by its strategic placement.

In the case of grit deposition that cannot be removed, the sensor can be mounted at a near angle. However, the sample taken by the device to generate the relational operator is thereby limited and some of the cross section may be missed. Angling the sensor is not recommended, as most angled mounting schemes produce edges that tend to catch debris. Also, if the side wall is used in a rectangular sewer, only the layers in the immediate path of the signal are measured. Subsequent upper layers or false echoes from the bottom, if the device is placed too low, yield inaccurate measurements. The level readings of devices with integral level sensors may be skewed by angled mounting. This is resolved simply by utilizing external level devices. Therefore, if angled velocity is the only alternative for the conditions, external level sensors should always be employed.

A smooth sewer circular in cross section is another consideration for the larger systems. If there is no grit deposition, placement of the sensor at the bottom of the pipe is most desirable. If the level device is integral, no further considerations for installation are necessary, except for the cabling. PVC does not bend as necessary, so the protection of the cabling is a challenge. Building a layer of hydraulic grout to produce a slope separated from the sewer wall with a piece of fiber gasket is usually enough to minimize solids buildup. It is a bit expensive at times, but the mat used to prevent people from tripping over computer cables in offices does bend in just such a manner. In any case, the cables should be provided some reasonable protection from solids buildup.

If grit deposition cannot be removed, angling the sensor may be possible. Depending on the offset required, the sensor can obtain a good cross section of the velocity profile despite angling. If the level sensor is integral, a reasonable offset can be established to provide accurate information about the flow volume. Grit deposition is a problem here because if the sensor is not adequately mounted above the grit level, it may be and buried in the grit, making readings inaccurate. Be sure that the sensor is significantly above the grit base but still within the flow volume at its minimum. Remember that a minimum level of water—the deadband—is required for all the electronics to operate. The requirements of the deadband, as discussed in previous sections, are particular to the manufacturer of the equipment, which should be consulted accordingly.

Sewers of odd shapes, brick or block, or of other types and materials not readily amenable must be examined in light of the following criteria: the gaps in the cabling and the structure material, the shape of the bottom of the conduit, the hydraulics of the structure, the availability of access, the potential for solids entrainment, and the potential for grit deposition. In older block sewers, grout is required to seal gaps between the cables and the structure material to keep solids from attaching. In egg-shaped sewers, the position of the ellipse and the reason for the design must be considered. Any gaps in the installation will collect solids debris for those portions submerged. Any compromise in the installation results in flawed data and potential loss of the equipment. Here, the goal is to protect the investment in the equipment as well as to get valid readings. Good installations support both of these goals. Flawed or compromised installations undermine them.

The measurement of combined sewers presents distinctive issues. The end of the combined sewer is, in fact, a structure, and any number of parameters may be required. Flow is generally not practically measurable unless it is from the discharge of the regulator. Only this point is practical because of the nature of the structure. If the CSO is in a tidally influenced or watershed area and the discharge is to be measured, the location must be carefully test measured to ascertain characteristics of the site. This requires placement of the probes at points along the dam or discharge to determine what point is adequate or provides clear measurement. Many instrument manufacturers supply equipment with multiple velocity sensors just for this purpose. Note that a combined sewer, because of the nature of its function, should never be entered during a storm event regardless of how mild the event appears. Many consultants note the use of chalk lining or grease lining to ascertain maximum levels. This should never be performed for the safety of the workers just before, during, or immediately after a storm event. Upstream conditions are always questionable without sufficient system knowledge.

Combined sewers do contain interesting points that can be monitored readily. The flow from the discharge is one of those points. Most combined regulators were designed to control flow, which means they are inherently capable of producing a measurement. Placing flow monitors just beyond the influence of the structure at the outlet should provide excellent information on behavior characteristics of the system. The measurement of velocity and level over the discharge point calculated with the flow at the discharge can provide a majority of the information about the hydraulic characteristics of the location. Further, the measurement of the discharge can work in reverse if there is a suspicion of inflow from the watershed. Until the hydraulic characteristics of these systems are better understood from the temporary study, direct measurement of the flow within the trunk sewer should be avoided.

Storm sewers are straightforward in that devices in them measure only during storm events. Device installation requirements are the same as for any other sewer, but storm sewers are a bit more forgiving in terms of solids deposition, though not in the first flush because then the volume of solids in the water tends to be higher. If the same rules are followed as for other types of sewers, the installation should provide the results desired by the operator.

Structures are a unique challenge. Most are custom designed, so there are no hard and fast methods for monitoring them. Some general guidelines determine their characteristics. Flow should always be measured from either an upstream condition or from a discharge. The hydraulic complexities within a structure are generally such that a single sensor is inadequate. A structure is generally a holding or controlling entity, so the utility is concerned with its effects rather than its workings. Keeping this in mind, flow measurement at the inlet or outlet of structures makes sense. The discharge of the structure is the primary consideration, with the feed being secondary, as it still bears residual influences of its upstream components. The influence of the upstream components may, in fact, be an object of the study. However, if the effects of the structure are the object, then the downstream measurement is likely to yield the most information. If the internal hydraulics are desired from the temporary viewpoint, then a level-only measurement should be considered. As with any installation, common sense must prevail. Any direct flow measurement of structures should be deferred until characteristic studies using level can be performed.

Flows from pump and lift stations are handled with primary full-pipe meters. Here no extensive discussion concerning retrofitting open-channel flow metering equipment to a pump station will be undertaken. However, if the station does not contain a primary measuring element, an open-channel flow device can be installed downstream, far enough so that the turbulence and other effects of pumping are negligible. Several manufacturers of open-channel flow equipment carry retrofits for just this purpose. These types of equipment can be AC-powered because pumps usually runs on AC electricity. The quality and design make a lot of difference in the longevity and safety of the instrument.

Many combined sewers and support structures do not perform the task they were originally designed for, nor do they behave as intended. However, they work quite well and to the benefit of the community they serve, for several reasons. These structures were designed at a particular point in time. It is not possible to accurately project all future development, however, and the function of the structures changed over time because the world around them changed. Therefore, do not be surprised if the finest structure is not operating to specific design criteria. Even if the construction were flawless, the world around the structure changed, which changed its conditions as well. There is nothing wrong with this; it is just a factor requiring consideration.

The instrument package should be placed where it can be interrogated quickly and with no confined space access issues. Many times hangers can be placed through a vent in the manhole cover just big enough to admit an access cable, thereby avoiding removal of the cover. The device should be secured to a point that the installer is sure will not be submerged. These are areas of high humidity, regardless of the dryness of the surface atmosphere. Care should be taken in considering mounting height within the manhole for this purpose. Along the same lines, the instrument should be low enough in the manhole so that it is not visible through the manhole vents, at least to pedestrians. Interest generated by such things causes curiosity or alarm in the general public and has interesting effects. It is best to keep the devices out of sight. This requires judgment based on the size and depth of the manhole, the particular characteristics of the installation, and the resources available to the utility performing the installation.

6.12 MIXED SEQUENCES

The critical path method of determining points of interest rarely covers all level points or all flow points. The path method demands that equipment be sequenced to provide maximal benefit to the data requester but minimal deployment of equipment. Mixed sequencing is a reasonable resolution for this situation.

The basic factor to keep in mind is the data request. Sequencing makes it possible to provide the best quality of data possible given the equipment available. A situation could arise in which data is requested with no forethought concerning the difficulty or complexity of the task involved in generating it. Also, many times, the data requester expects particular findings. For this reason, data must always be gathered with considerable planning so that many potential solution scenarios can be matched or otherwise anticipated.

If the basic request is to determine hydraulic characteristics of the critical path, then level equipment can be sequenced in all of the structures and flow in all the adjoining pipes. If the basic request is to satisfy complaints of flooding, then level at critical points is sufficient, plus flow on the main feed to determine if volume alters due to a storm or plant action. If the request is for permanent equipment placement, the best judgment of the system operator is required to determine as much about the hydraulics as possible so permanent positions can be established. Generally, temporary equipment is staged at points where permanent equipment is planned to evaluate overall performance of the installation. Mixed sequencing is discretionary based on the requirements of the data.

6.13 BATTERIES AND SOLAR CELLS

The weakest link in any temporary installation is the power supply. The two most common power sources are solar cells and batteries. Both can provide extraordinary performance, depending on their type and location. Also, with modern low-power chip sets, the drain on these newer design systems has been reduced, extending the length of time between battery changes. However, these power sources also have significant drawbacks that should be considered with any system being installed.

For batteries, two problem areas are shelf time and threshold. The issue of shelf time concerns distribution and storage. A battery manufacturer calls shelf life the time that a battery can remain on a shelf and retain its power integrity. Shelf time is how long a battery stays in storage before it is utilized. Shelf time can be as long as two years because of procurement processes and agency stockpiling. The manufacturer may state that the shelf life is one year; it may be two years before the cell is used. This is entirely possible under the present governmental procurement process. Be sure to purchase batteries with a long stated shelf life. It is also good practice to examine shelf time within the utility and shorten it, if possible.

Threshold is the point at which the electronics no longer work because the remaining potential of the battery simply cannot operate the equipment. A certain potential must exist to push a volume of electricity across an electrical circuit. If the potential drops below a certain level, pressure is insufficient to get the current across the circuit. The problem's source is in a combination of the power requirements of the electronics and the physical limitations of a dry cell battery. Threshold varies with the relationship of each company's manufacturing process and the power rating of the batteries. Batteries may not be strong enough to run the electronics but will work well in a flashlight. Be aware of the threshold issue and check with the equipment manufacturer for input.

Solar cells were, at one time, considered a replacement for batteries or at least a reasonable substitute. However, several shortfalls were discovered. One was the expense. Another was the degeneration of the substrate. The material in earlier cell designs deteriorated over an 18- to 36-month period. A major shortfall was in the placement of the cells themselves. Logistics simply did not permit easy placement of the cells for practical implementation. Climate, sunlight availability, and back current characteristics all play into the use of a cell at a location.

Some of these issues have been resolved. New materials and manufacturing methods have overcome some of the power and deterioration concerns. Solar cells are, however, still expensive and still difficult to place. In many instances, the cost of the cells exceeds the cost of the electronics, so this options should be weighed carefully.

6.14 PROFILING THE SITE

Profiling the site can provide excellent and probably necessary data about key features of the site of interest. It is important to note key features on the site profiles and to maintain them until it is found that the feature is either not necessary or supported with other issues that can provide the same sense of the location. If proper format and structure are noted, then all the points along a critical path can be joined using the profiles and a clear picture of the critical path's hydraulic characteristics will result. Also, uniform profiles provide baselines for future studies and the foundation for the permanent monitoring network.[3]

The site of interest along the critical path should be chosen. From this point, the physical dimensions of the sewer or structure should be measured and recorded. If the point of interest is a sewer, do some up- and downstream research for at least 20 pipe diameters to check for anything that can perturb the flow profile. Obstructions or physical characteristics of note should also be indicated. The requirement for placement of primary elements represents a uniform start point from which a profile can be generated. Grit bases should be included and, if significant, which parts are soft and hard. Manhole and discharge points, where applicable, should be included with their dimensions. Also, include a topside drawing of manhole locations in relation to the curbline or other surface structures. If ladder rungs are to be utilized for hanging equipment or securing cable, note their conditions and number. Ladder rungs in manholes are subject to the corrosive effect of sewer gases and can decompose.

In the final sketch, find one point to which all other measurements can be related. Relate this point to a surface point so the entire portion of the infrastructure of interest can be associated and measured from this one point. After the equipment is installed, note the location of the sensors relative to this point. Note any offsets required to make the equipment functional. The location of the sensors and the equipment in the system can assist in determining locations of optimal performance or less than desirable placement.

The last parameters are the initial conditions. Record the level of the water and the time it was recorded. Then relate the level to the point of reference at the surface. This is for assistance in generating an HGL later. Note the average velocity, as seen with the equipment, and the time it was measured. If velocity profiling must be performed, note this in the sketch with a separate drawing to show the velocity proile. This will document any skew in the pattern. If the instrument can display real-time conditions, be certain that the readings as generated are reasonable for the conditions observed at the time of profile. Dates and times are important for the historical perspective.

3. A. Drinkwater et al., "A Guide to Short Term Flow Surveys of Sewer Systems," WRc Ltd., Frankland Road, Swindon, United Kingdom. Section 3 describes a step-by-step survey procedure that makes an excellent base for discussion.

The deployment of the equipment now is quite clear. Note that anywhere the sensors are placed, they need to be documented for placement and referenced by location if necessary. Many engineering groups insist on elevations rather than levels or depths. Once a proper drawing is made, including the topside landmarks, it is quite easy to convert from depths and levels to elevations. Many computer modeling groups are particular about this, as they use the HGL in generating simulations of the system during various conditions.

The key here is to be consistent in the parameters that are taken. Be certain that the same parameters, where possible, are taken for all the points of interest along the critical path. The ability to relate all of the study parameters to one point on the surface allows the translation to elevations as needed by engineers if they so desire. Generally, however, the goal is to determine the level or flow within the system. Recording uniform parameters makes the translation of data that much more valuable to the system operators.

6.15 COLLECTING THE DATA

The retrieval of the data from any temporary network deployed with any scheme has particular issues that should be noted and properly attended. Data can be obtained in many forms and a uniform method of tabulation should be employed to generate consistency as programs evolve or are weighed over time. This section discusses some of the finer points to consider with respect to data collection and processing. Actual interpretation of the data is the subject of other discussions.

There is no standard time increment over which to collect data from the system. This is a function of the electronics package data storage, the requirements of the user, and the display to be generated. Experimentation must determine the best data collection interval, using frequent collections at first and decreasing the frequency until an ideal is achieved.

The ability of small electronics packages to interpret and store data has increased exponentially in recent years. These massive changes, driven by the personal computer industry, have also made the use of the equipment simple enough to be operated by individuals with little training. Complex mathematical algorithms can be constructed with minimal programming abilities. Such advances also allow the collection of data and its transport and processing to be equally simple. Little space is required here to describe data collection techniques.

The first method involves using a manufacturer's data interrogation unit. This is a targeted device that serves no other purpose than to extracting data from the field installation. Generally, the device finds the beginning and the end of the data automatically and extracts it accordingly. The human operator need only plug the unit in and wait until the download is complete.

The second method combines a laptop computer and a manufacturer's proprietary interrogation and processing software. This approach is more common but requires more human intervention than the other. The amount of intervention is strictly a function of the software. Most software finds the beginning of the data set and extracts it accordingly. Some requires the operator to find it and to note the boundaries within which to extract. In both cases, the data is extracted using the manufacturer's proprietary protocol. Some specialized service providers have developed their own data extraction routines. For this reason, and especially when there is any mixed sequencing of equipment, a procedure should be in place for handling the data once it exits the field instrumentation and enters the data handling area. This allows for efficient processing of data assuming the variety of protocols used by individual manufacturers for data handling.

Data gathered from field equipment should be moved to a central repository. Before removal, be certain that the interrogation program can convert to either a spreadsheet program or to standard delineated text. Agree on a uniform naming convention for files generated. Usually a site code, year, month, and day can be included in the file header. Determine if standard delineated text will be utilized or if a spreadsheet of choice is preferred. Once these choices are made, the data can be converted, moved, and labeled. It should then be examined for holes or other issues that suggest a site deficiency requiring correction. No other analysis should be performed at this point other than to ascertain the integrity of the equipment. The data should be filed for further analysis and action.

The temporary network is now profiled, installed, recording data, and being interrogated. Because the goal is to obtain usable data, data availability is critical. The polling of data from the site and its handling once back at the central facility make the data processing task straightforward, assuming a protocol for processing the information once the files are retrieved. Temporary locations are easy to relocate in the event that the data obtained is not satisfactory to the end user. The integrity of the data is critical and its handling is of the utmost importance. This is why the protocol agreed to must be adhered to and why any changes in the chain of custody must be made uniformly among all the files in the database.

The collected data should be reviewed and verified through spot checks of the equipment until the personnel involved are comfortable with the installation. An established protocol should verify the equipment installation at planned intervals so that the data can be matched for accuracy. Such verification provides quality control to all the data generated.

6.16 PERMANENT MONITORING SYSTEMS: OVERVIEW

The in-place or fixed monitoring network is the complete temporary installation plus data communications and other utility services. It provides all the functions of the temporary network, but with the convenience of not requiring weekly or monthly site retrieval of the data. Also, on better devices, it provides the ability to watch the hydraulic actions within a system during a storm event or maintenance procedure. All permanent installations should begin with a temporary survey and profiling. Then the best equipment can be chosen based on the individual characteristics of each location.

Currently, few options exist for permanent monitoring systems. They are either adaptations of temporary systems used as permanent installations, or custom blends of equipment integrated for the purpose. At the time of this publication, only two companies manufacture permanent sewer infrastructure monitoring systems as a total solution. Permanent monitoring systems are broken down, for this discussion, into two categories: with and without utility service. These are called, respectively, full permanent systems and partial permanent systems. The discussion begins with partial systems, as most aspects of these systems have already been presented.

6.17 EVALUATION FOR PERMANENT INSTALLATIONS

The success of any temporary monitoring system and the interest the data generates determines if progress should continue toward making the system permanent. The best method for this determination is to evaluate the experience from temporary monitoring and to develop a model or grid that outlines the resources necessary to take the next step. Grids are generally simpler to generate than models and easy to read. Model development is certainly a better option, but models are difficult to construct and calibrate. The discussion here is limited to grids, which combine the results of a use study with firsthand experience with equipment deployment, installation, and maintenance.

A use study should be performed based on the information generated from the temporary monitoring system. Temporary monitors generate their own interest and care requirements based on the interest of the owners and operators of the sewer system. The interest is in the data that the instruments record from the infrastructure and its relation to the effects from mishaps. The care requirements, when met, keep the data sound and viable. Because each point along the critical path where equipment is deployed is unique to the system, city, and nation in which it resides. Generalizations can be made, but the site has unique characteristics that set it apart even from those along the same path. This fact creates a case for individual examination for the issues relevant to deploying long-term equipment there. The ability to get equipment into the site and provide reasonable service needs to be weighed against the need for and use of the data. This appears to be a backward method of evaluation, but it makes the most economic and reasonable sense. If the issues involving installation and maintenance are so great that the site will require continual maintenance far in excess of what the utility considers reasonable, is the data of sufficient interest and value to the utility? Only the utility operator can answer that. Each point must be evaluated for the balance between the information generated and the maintenance involved in keeping the site operational.

A grid can be constructed for each site ranking the interest along with its installation and maintenance issues. Remember that this grid is being developed now with firsthand data generated from the temporary deployment, including input from maintenance personnel, the engineering group, and those parties with an interest in infrastructure issues quickly generates a grid ranking sites of increasing interest with those if increasing installation issues along with increasing maintenance issues. This grid becomes the basis for choosing points and critical paths for permanent installations. It also assists in calculating costs associated with installing and maintaining equipment. Some sites are imperative to install regardless of

Figure 6-11 A permanent stand-alone enclosure. The vent at the top releases sewer gas to the atmosphere. Utility service is provided from the nearby utility pole.

cost or difficulty. This is understood; such sites should be given priority. The grid ranks them for classification purposes and presentation to the authorities as necessary.

The grid can be expanded to include the needs and wants of the parties involved. Once consensus is reached concerning resources and integrity in the data distribution, needs and wants can be pared into short phrases and added to the grid. It does not matter if the utility is private or governmental. It does matter that the grid be framed with respect to the uniqueness of the entity for which the utility operates. For this reason, no hard and fast rules can be presented.

Once the grid is generated, the parameters of the points of interest can be inserted. Several waves of analysis occur at this point; it is a straightforward task to eliminate during the first pass all the parameters that appear unlikely to be measurable. Reexamine that which is left. If the resources available do not match the amount of work needed to install the system, make another pass and eliminate more parameters. Repeat this analysis as indicated.

The result is a layering of interest in each point. Several layers, each requiring a commitment of resources, are now available. These, in themselves, generate options once funding is clear. Through this analysis, sev-

Figure 6-12 Profiling of a site is necessary to ascertain the total physical conditions and the flaws that would impede installation, operations, and maintenance. Such issues as this decaying wall can compromise the integrity of any installation, even those with the best equipment. Note the grit buildup at the base of the decaying wall.

eral options can be presented even if sites are compromised. Once the analysis is complete, a decision can be made to move the program forward or consider another method. This approach yields two results. One is the several levels or layers that provide options for permanent installations. Another is the conclusion that the options generated are not acceptable and another method should be considered.

The second analysis also begins with aligning all points of interest along all critical paths with the parameters of interest. From this, all points along one path are chosen with all parameters to be installed in order to generate experience and expertise in this area. The idea here is that the installation will expand over a period of fiscal years or financial periods. By installing one section with all the parameters, the utility can ease into a full system. This is an excellent method for generating both experience and understanding of installation, operations, and maintenance of this technology.

Yet another method is to pick points along several critical paths, provide all the desired parameters, and develop the experience base from this range of situations. This approach is logistically difficult. If the sites are scattered, maintenance planning is at the mercy of the very parameters that the instruments seek to measure in order to rectify. Traffic, weather, and other matters of public interest are always of concern. In the development stage, sites should not be so scattered that more time is spent driving between them than actually working on them. There is an advantage to developing expertise on several critical paths at once, depending on the terrain and localized hydrologic cycle. If this method is selected, sites must be picked strategically. If operators can develop expertise in several critical paths at once, maintenance costs will be reduced and questionable data can be detected quickly. This saves time and money.

Whichever analysis method is picked, the target is obviously the same: to minimize the street time and maximize the information obtained. The ultimate result is to shorten response time for sewer maintenance activities and decrease inconvenience to the citizens who rely on the sewage system or the commuter who is detoured by substreet activities. This is always considered along with the protection of health, safety, property, and the watershed.

This development is absolutely necessary before steps can be taken to install permanent monitors. A permanent system of some form cannot be effectively installed without this planning, as shortcomings and oversights are difficult to correct once the project is under way.

6.18 PARTIAL PERMANENT MONITORING SYSTEMS

A partial permanent monitoring system is a temporary system that has been converted to a permanent installation. Little has to be done to convert a functional temporary deployment to a permanent system if the temporary system was installed correctly. If the temporary system produces the required data, then it need only be converted to a permanent installation and the data collected at set intervals. Such partial permanent systems are flexable enough for modification and change, while they provide a permanent reference for hydraulic actions within the pipes.

The partial permanent monitoring system is more organized and elaborate than the staged temporary locations. A temporary location is designated out of interest or necessity, one point at a time or along a critical path. The time frame is based on a need and the placement of equipment among points is accepted. This is the reason for the discussion on sequencing. Any system that no longer has definitive deployment times—that is, it is targeted for use over years and not months—is, for purposes of this discussion, considered a permanent network. Permanent systems include rain gauges and other peripheral devices as needed or currently used by the utility.

Some installation requirements are necessary to make the temporary system permanent. Most involve the securing of the instruments, the hanging of enclosures for the devices where necessary, and instituting regular maintenance schedules to keep the equipment in optimal condition. The subject of maintenance is covered at greater length below. Some maintenance issues are outlined here because of their relevance to the topic.

The goal of a partial permanent monitoring system is to generate data on the continued hydraulic response of a location. Knowing the equipment is in place provides an information source for the system operators, who can disperse the information as necessary. Data requests are handled soon after an event of interest or when the need to poll the site is noted. Various parties who want the data are aware of the deployment of the instruments, the relevance of the data, and specific information based on a particular storm or maintenance activity. However, the information is delayed until the data can be removed from the location of interest.

The first step in converting a temporary deployment for permanent use is to reevaluate the installation for the data it produces. If the data are acceptable to the parties that requested it and the site requires little maintenance, given the conditions, then the site should be reinforced for a longer term; it may require modification so that maintenance and replacement can be performed. Maintenance, for a temporary system, means removal of the equipment and deactivation of the location.

Stilling wells for pressure transducers should be remounted so the transducer can be readily accessed for cleaning and replacement. The well itself can be retrofitted to be removed with the transducer that is mounted within it, or it can be bracketed to accommodate just the sensor removal. There is no formula for such retrofitting because of the diversity of products available. The manufacturer of each device should be consulted about mounting requirements that will elicit maximum performance from the device. Once these requirements are understood, they can be easily integrated into the mounting. A pressure transducer, depending on the specifications, requires cleaning, recalibration, and inspection every three to six months, depending on its location.

Ultrasonic level heads pose unique challenges under long exposure in the sewer environment. The accumulation of water from humidity and from flooding negatively affects poor-quality devices. With respect to devices of any quality, hanging debris from floatables and other solids significantly affects the signal and probably prevents it from being transmitted altogether. For this reason, periodic inspection and cleaning of the head is required. Also, because of the potential effect of air temperature variances on signal propagation, the head should be inspected and calibrated, or at least verified, to ensure signal integrity. The long-term mounting requirements of ultrasonic heads also may change. Some devices already have fixed mountings and no additions are required to secure them for the long term. Others require a permanent mounting, which should be installed as part of permanent site deployment.

Cabling should be tailored to the distance between sensor and electronics package at the location. In many cases, the maximum cable lengths are requested at instrument purchase so the device can be deployed in any number of circumstances. Cables thus can be two to three times the length required for the point of interest. Excess cabling has the potential to collect static noise in the transmission of the reference signals. Therefore, cable lengths should always be as short as possible for the location. It is recommended that this type of equipment be retrofitted with cabling of the precise length needed when it is being converted to a permanent installation. Any cable that is exposed to water should be contained in a conduit system to protect it from floatables and other solids. The conduit does not have to be sealed; it simply acts as a shield.

The issue of velocity sensors requires special attention. The velocity sensor is generally the most sensitive part of the equipment and the device most subject to fouling. Many times, depending on the manufacturer, the level sensor is integral to the velocity sensor. The location of the velocity sensor compounds the issue further. Depositions from grit, solids collection, and grease will foul the probe, as it has full exposure to the flow and cannot be protected because of the nature of its work. Many attempts have been made to design low-profile, limited-exposure, and so-called self-cleaning devices. The fact is that the sensor is still directly in the flow and directly exposed to the maximum hazards of the environment. Low-profile and limited-exposure devices provide the best solution given current technology. The self-cleaning sensors are integral to the installation, amount of debris expected and other factors otherwise unexpected in the deployment that makes the cleaning effects questionable at best. This outlines the situation facing any long-term exposure of potentially temporary equipment adapted to a permanent environment. The issue is maintenance and the interval sensor required to provide valid data.

The issues surrounding velocity sensors can be remediated with the assistance of a proactive maintenance program. The system operators should schedule cleaning of the sensor at the same time that the level sensors are scheduled for maintenance. A maintenance procedure as simple as brushing the sensor with a broom to clear it of debris, including debris that cannot be seen by the naked eye at the time of maintenance, extends the life of the installation and helps to ensure accurate data over the long term. Thin layers of grease or debris, which are not visible but can be detected by touch, can form over the sensor and interfere with signal propagation. Brushing is enough to clear the grease and permit free propagation of the reference signal. Better flow devices preamplify their signal at the sensor, but they too require this kind of cleaning.

At no time is it recommended that an unshielded hand touch a sensor that has been exposed to the sewer environment. It needs to be remembered that wastewater is a medium for biological growth. The microorganisms and other living entities in it attach to any stationary location, including a sensor, and reproduce there. The living biology in wastewater is evident in the linings of many sewers and fixtures in wastewater plants. The point is that the growth is everywhere in a waste stream and extreme care should be employed because of it. Maintenance activities are more critical in fixed equipment than in devices left for temporary study purposes—they are a large, if not the largest, factor in the successful continuation of the system over a long time—and even sporadic cleaning and inspections keep down debris accumulation and biological growth to some level and help to maintain continuity of the site.

The next step is the integration of additional devices, such as rain gauges, biomonitors, and samplers. The most likely addition is rain gauges, as most weather station information other than rain now comes from the Internet. Biomonitors and samplers are discussed below; note that they are easily integrated with partial networks. Adding rain data to flow and level can provide a complete picture of the hydraulic response and the long-term hydraulic characteristics to the localized hydrologic cycle of a given drainage district. Rain patterns provide the basis for understanding the response of the sewer system to various conditions as they are presented to the infrastructure.

The establishment of a small sample rain gauge network was suggested in previous chapters for the purpose of obtaining samplings of the size, frequency, and magnitude of storm events characteristic of the drainage district. Now the equipment is integrated with the temporary monitors being converted for a long-term application. The ability to combine flow, level, and rain data in order to develop systemwide pictures is a tremendous boost toward characterizing systemwide hydraulics. The rain portion provides the feed or input for the response, which in itself is a result. Thus, the integration of the rain gauges adds the final touches to a basic permanent monitoring system. The partial permanent monitoring system now can address both cause and effect with respect to hydraulic activity in the sewers.

It may be necessary to reconsider the original placement of the rain gauges in light of the experience accumulated from generating the data from the temporary sites and making correlations based on the localized hydrologic cycle and the terrain in the drainage district. The experience base generated at a local level is critical in placing the devices and in repositioning them as needs dictate; many people now have opinions about where the devices should be placed. At all times, keep in mind the manufacturer's requirements for spacing and clearance. Also, remember that rainfall is not an exact quantity, as noted in previous chapters. Relationships and totals can only be considered approximate.

Biomonitors and samplers within the network may be included in the partial network. Many samplers have attached water quality probes; the data they yield can provide useful insights to the content of the wastewater along with serving as part of an alarm system for contaminants. These devices can be easily integrated, as they are manually polled after their deployment. A sampler always must be attended, as the samples have to be removed. This makes the inclusion of these devices straightforward.

6.19 FULL PERMANENT MONITORING SYSTEM

The full system differs from the partial system in several distinct ways. The full permanent system has data communications abilities, utility service, and a central computer for automatic polling. The utility service is a permanent fixture, as is the mounting and assembly of the equipment. A full system has support and a complete maintenance program. The full system does not require any manual extraction of information (unless a data communications link breaks down); it can be accessed via a computer communications protocol. Each of these factors is discussed in this section.

The full permanent monitoring system is not converted from a temporary installation. However, its roots are certainly in such a deployment. A full system is thought out as a whole, taking into consideration all the critical paths within the jurisdiction of a utility's authority. Installation may, however, be accomplished in stages, deploying all the critical paths first in one drainage district and then in another. In a full system, utility services are fixtures of the point of interest and are installed as a permanent placement. The requirements for permanent utility service, whether electricity, phone, or water, include code compliance and securing situations more thoroughly than in a partial system. Each step in the planning of a full system is critical to its success and to the information it is intended to generate.

As discussed above, a use study should be performed to determine the potential uses and requirements of such a system. Temporary monitoring yields site information that translates into grid factors for analysis. The only addition to the grid aimed at developing a full system is the availability of utility services. The system engineer must be intimately familiar with local requirements and codes for the addition of utilities and with their associated costs. Utility costs associated with even the simplest location can be staggering.

The complex structure of a full permanent network requires the integration of electricity, the installation of data communications capacity, and excavation, all of which involve requirements that must be addressed before the service applications are filed. Questions concerning deregulation and privatization and the resources required in establishing these relationships must be answered as well. The major issues are brought forward here for evaluation.

6.20 MOUNTING REQUIREMENTS

Instruments may be permanently mounted either above or below ground. Permanent installations below ground tend to be cumbersome, are subject to corrosive sewer environments, and require tremendous space considerations. However, they are much less susceptible to vandalism and other

Figure 6-13 Permanent monitoring installation. Note the presence of the fixed conduit and solar cell at the top of the pole. Permanent systems are vigilant in the monitoring of the infrastructure and the watershed.

street hazards. Above-ground installations are easier to access and maintain, and many repairs can be implemented without the necessity for confined space entry precautions. This can certainly limit the number of personnel required for a service call. The choice can be made based on individual site circumstances.

Below-ground installations, especially where live electricity is planned, require special grounding and floor considerations. This is due to groundwater conductivity and the transient effects of lightning. Also, dirt and debris have a tendency to collect in the enclosure or chamber. Venting of the equipment may be an issue, depending on the heat dissipation and absorption of the selected devices. If the devices tend to give off a lot of heat, venting is probably required, especially if the access doors are in direct sunlight for a good portion of the day. Low-level vents may cause obstructions unless they can be run to a pole or other fixture out of the way. Conduit runs always must be sealed to prevent condensation and water runoff from entering the box and equipment. If these considerations can be properly addressed, below-ground installations may be of interest. Otherwise, below-ground permanent installation with live utility services should be revisited and reconsidered. Rights of way are generally not an issue with below-ground installations because the utility tends to hold them already.

Above-ground installations come in two types: a freestanding structure near the point of interest and a mounting on a pole, building, or other structure near the site. Bridge abutments serve as great points of attachments, as an example. The freestanding box needs to be big enough not to cause a tripping hazard if mounted along a footpath; on the other hand, width is an issue for boxes placed along a curbline or in a similar narrow space. Also, automatic venting of the box must not cause noise or odor problems for the surrounding area. An advantage is the limited amount of conduit required, as the box is generally located near the point of interest. If space is available near the site and the installation is practical, then the freestanding structure is a viable option to consider.

The pole- or structure-mounted box is a smaller version of the freestanding box; the conduit is run to a higher location. Such sites are less frequently vandalized than are freestanding installations because they are more out of the way. The instrument boxes can be wider and smaller and provide more ambient airspace in front of the instruments. Pole mountings generally are high enough to require ladder access; this simply adds a piece of equipment to the crew truck, which may already carry one. There is an additional requirement for conduit runs to the box. One major issue concerns rights of way, which are usually required for above-ground installations. Even interagency installations require some form of permission letter to complete the task. Rights of way are a big concern, especially in the trenching for conduit burial and placement of equipment.

The instrument of choice may not require ambient airspace or venting. This changes the requirements of the enclosure and the overall space requirements, along with the costs associated with the installation. Sealed electronics do not require any ambient airspace, except for venting of the pressure transducer where necessary and if required. However, conduit is still required to run from the instrument box to the site, if not for general craftsmanship then for local codes. This is where the planning grids become useful, if not indispensable. The evaluation of the equipment per these requirements can readily be added to the grid and the best choices become obvious.

6.21 UTILITY SERVICES

The placement of utility services as a fixture is the hallmark of the full permanent monitoring system. In any industrial operation, the utility service is the single greatest operating expense over payroll and overhead. In the light of recent deregulation and privatization in utilities and services, it pays to gather information and shop competitively for the best service and price. This discussion provides an overview of determining utility costs and requirements for the full permanent monitoring system.

Power to the location provides an electrical source far superior to batteries and solar cells. The service also allows repair personnel to use the power rather than portable generators to run their tools and equipment. To determine the cost of power to the location, first identify the nearest point from which power can be extracted for the location. Determine the cost of running the service to the site of interest, either through standard estimates or utility quotations. Here, it is important to know the policy and procedure of the local utility provider. Generally, the utility provider is required to run service within a specific distance of a customer's service connection at no charge to the customer. Calculate the maximum power that would be drawn per month by the equipment; this figure is available from the individual manufacturers. Modern electronics generally makes these draws minimal. The electric utility usually charges a fee per month or billing period. Expect the minimum charge to exceed the maximum power draw. Generate cost figures for three- and five-year periods, or the service life of the equipment. This figure can now be compared with the cost of solar cells and batteries.

Utility grids are a useful method of calculating total costs and rating technologies. Standard spreadsheets can be employed and provide excellent results. In the case of phone service, cellular technology may be more expensive than phone lines, but if the number of systems required is great enough, competition among service providers may allow for discounts not offered for single units. Deregulations in the power industry mean that electric service will also have this competitive feature shortly. If the sewer system operator can shop for its utility provider, it may obtain sizable discounts simply by showing the numbers and letting the utility marketing people tailor their service to the monitoring requirements.

The addition of live electricity to a location changes the mounting and insulating requirements of the installation. The temporary locations using low-power DC batteries need transient protection. Power installations, in comparision, are subject to electrical codes, which protect the safety of all citizens. The code generally requires additional insulation, lightning protection, and shielding for the equipment as well as insulation from the sewer environment. A sewer is classified as hazardous in any country and conduit requirements are specified in each jurisdiction.

Figure 6-14 Weatherheads for utility service. The heads keep water from weather and runoff from entering the instrument package. This case includes a separate weatherhead for a tone telephone line and standard electric service.

Phone costs can be computed the same way against cellular communications and radiofrequency (RF) transmitters. A full discussion of data communications equipment follows, but one need not master the technology to know its costs. Build the grid for the costs of construction and three and five years of expenses for the phone lines, cellular service, and RF transmitter, with license and site survey. Because rates vary per service provider, government regulation, and jusridiction, actual recommendations cannot be made here. It is quite possible that a mixed bag of equipment will be the best choice for the site. Estimates may have to be computed for each location, but a base computation can be generated for the grid so technologies of greater or lower preference can be priced accordingly. The determination of costs for data communications technology is considerably more difficult than for power providers. This is because not all technologies work in all environments and designs. Many times, combinations of data communications technologies are necessary to make the system functional and to avoid inordinate expenses.

The cost of utility service must be considered separate from the actual technology, although the technology is certainly a part of the system. This cost can vary greatly depending on location of the site and the technology selected. Both are variable and quite site-specific. It is best to avoid making an irrevocable commitment to one technology, and always expect the technology to change. Adaptation of this part of the system is necessary for its long-term continuance. Most high-quality instrument manufacturers do make interfaces for the technology of choice. Many of these features are already installed on the processor boards of the devices. Check with the equipment manufacturer to see if a particular utility service works best with its devices.

6.22 DATA COMMUNICATIONS: OVERVIEW

The discussion now focuses on data communications technology rather than on the utility service provider. Data communications has two aspects: technology and utility. At one time, these were entirely separate. Today, with deregulation and privatization, the two areas have moved closer together, and many techniques used in the selection of a power vendor are applicable in the selection of a data communications vendor. The more popular technologies are analyzed in this section, as are methods of evaluating them.

The feature that makes a system permanent is its ability to be polled from a remote location via in-place data communications. Some of these price and technological boundaries are dropping with the advent of low-powered cellular service. However, for now the classification presented provides enough background to accommodate the changes as they manifest.

Any time a site on a critical path can be accessed reliably without a field visit, a direct cost savings and multiple indirect benefits accrue to the utility operator. In any management scheme, the ability to accumulate information quickly and make informed decisions promptly maximizes the services offered by the utility and makes it more competitive. The ability for a site to communicate directly with a central or remote operator, as long as the data have integrity, means that it is possible to detect problematic situations before they cause property damage, health threats, or other concerns to the community. Any time a potentially hazardous situation is detected before its manifestation, a cost savings is incurred. Also, the public relations value generated by demonstrating this efficiency does not have a price in money. The number of sites with data communications abilities should increase over the years as these benefits are better understood.

Up to recent years, field equipment utilizing data communications equipment was prohibitively expensive. Cellular technology had not yet matured and RF was heavily licensed by national and international carriers. This made the implementation of data communications impractical. The personal computer revolution has driven many of these technologies quickly to maturity, and now they are both low in cost and reliable. It makes sense today to incorporate data communications into even temporary devices, if only to learn and understand data communications basics.

This is an area of great excitement in the sewer management field. Any time information can be obtained without a field visit, benefits accrue and the image of the sewer operator in the community improves. The next three sections outline the more popular data communications technologies and how to determine if they will integrate with a system.

6.23 DATA COMMUNICATIONS: TONE PHONE LINES

The term *dial-up,* which is synonymous with *tone,* comes from the earlier days of telephone-switching circuits where the switch would rise in a stack to make a particular connection. The term was then applied to rotary phone service. In any event, dial-up phone service was historically the most cost-effective and reasonable choice for field equipment in relation to other options. The term *tone line* is much more appropriate; however, it is almost unknown to the general public.

Ever-increasing pressure from the cellular industry is closing the cost and technological gaps between tone lines and cell phones, and so tone lines may not now be the choice in many areas. It certainly is the easiest to experiment with, as most personal computers are equipped with modems that use tone lines and many people do not have cellular phones. Dial-up tone service is still the easiest to debug with the least amount of diagnostic equipment or assistance from the service provider.

A basic dial-up system consists of field equipment serviced by a standard household tone line, a modem on the instrument end, and a modem on the computer. The computer is equipped with a program that can dial up and interrogate the field instrument through either an automatic polling sequence or operator intervention. The simplest systems have one modem in a computer that polls all the sites of interest, one at a time. Data is taken and stored until an operator wants to examine it. More elaborate systems have card cages where multiple modems are addressed and sites are polled simultaneously. Remember that for each modem on the rack, there must be a physical dial-up line to service it. Communications switches at the central location can minimize this requirement. This can be expensive, so the option must be weighed carefully first.

Standard dial-up tone phone service uses a pair of copper wires to carry the signal between the instrument and the computer. The carrier potential is generally between 50 and 90 volts DC (as defined in the United States), which should be measurable using a standard multimeter. This carrier transports the tones that are the voices at each end of the transmission. The voltage range should be consistent within an area—that is, all the potentials within a range around the site should be the same as the potential at the instrument under consideration.

Loadings are the next issue for the carrier voltage. Two potential problems that load the system, making communications difficult or simply not possible, are ground loops and line quality. Ground loops are a serious issue and can be detected only after equipment provides a load to the phone service. This can be diagnosed by measuring the voltage as soon as the equipment is attached. If the drop is too great, a ground loop is possible and the phone service provider should be contacted. First, however, another check should be made. The field equipment itself can generate a loading problem if defective. This is always the first check because the phone service provider usually notes the field equipment as the issue. Test the field equipment on another circuit to be certain that the equipment is not the source of the problem. Be certain, before contacting the provider, that the equipment is not at fault and was tested.

The quality of the line is the next concern. A line disconnected from the equipment should carry no measurable AC voltage; check using a standard multimeter. A standard phone plugged into the line should be clear and without crackle and other noises. Finally, the jacks used in the connection should be crimped properly to the cable. If all this checks, the data communications equipment is the final link in making the connection clean.

A note on fiber-optic services: Many phone service providers are converting to fiber optics and no longer use copper wires. This poses a serious issue if the data communications mode of choice is a dial-up tone modem. In many areas, the conversion to a fiber-optic trunk does not produce a carrier voltage strong enough to make the electrical connection between modems. If checks turn up no in-house problems, the carrier should be contacted about the implications of fiber optics. The telephone service provider can make corrections so tone equipment can be utilized without impairment.

This brings up an interesting point. The standard dial-up tone phone line was designed to carry voice communications. A French communications engineer, Emile Baudout, calculated the bandwidth capable of being transmitted across a standard tone line. His calculation, referred to today as the *baud rate,* was 3,000 cycles per second. Much more is expected from a tone line today than Baudout calculated was possible. This effort is a great tribute to human ingenuity in adapting the materials utilized in these endeavors. The point here is that most carriers still guarantee a transmission rate of only 3,000 cycles per second *bits per second.* Equipment simply will not function if data is pushed across the copper wiring too fast. Many coding algorithms make speeds of 115 kilobytes/second possible and reliable over phone lines. Question data transmission rates for field equipment because it is subject to many transients that a home computer is not subject to. Uses of the technology that exceed its designed capacity should be field tested not only for reliability but for time and exposure.

The issue now is connecting the physical line to the instrument. The low power that a tone line uses does not require any code compliance except to local ordinances. The issue is getting the service from the nearest phone junction to the instrument package. This depends on the type of mounting that was chosen for the site. Most services are above ground and the physical line has to be deployed from them. This requires weatherheads to seal the cable run and conduit to the instrument package. In this case, there are local ordinances, telco requirements, and the needs of the utility requesting the service which must be considered. This differs from power requirements, which are usually already in line with local ordinances. These requirements must be compared to the site requirements to ascertain the difficulty of the installation.

The use of tone service for data communications is now a tried and true approach. There are many good sources for equipment and services and, with deregulation and privatization, extra attention is paid by the phone service agent trying to get the business. The ability to utilize standard personal computer parts and the manufacturer's ability to retrofit or fabricate devices to readily accept this technology make tone service worth considering.

6.24 DATA COMMUNICATIONS: CELLULAR COMMUNICATIONS

The recent explosion in cellular communications has made this once cost-prohibitive technology suprisingly affordable and versatile. This utility is entirely driven by service providers, unlike tone phone and electricity, which were originally either governmental agencies or public utilities. Independent agents compete for the service and can be extremely flexible in the service pattern, per-call provisions, and volume considerations. Cellular communications may well become the communications medium of choice because of its immense flexibility.

A discussion of batteries and power supplies is first in order, as the battery, as noted earlier, is always the weakest link in the electronics package. The low power draw associated with cellular communications systems opens a host of potentials and blends not possible with tone lines. It allows temporary monitoring systems to be converted to almost full permanent systems with the addition of a cellular phone. It allows strategically placed alarm devices to monitor for contaminants and to provide alerts in the event of discharges. This is all dependent on the battery pack. Given the commitment to maintain the batteries of a system, then cellular communications without live electricity is an exciting option. If the batteries cannot be easily maintained, then external power is required. The low-power units do not require much more than a trickle charge against the batteries, but live electrical service is subject to the usual codes for installation. Modern low-power circuit design here can make a supplemental system feasible.

Battery-powered cellular sites should be considered as supplements to the full monitoring system where utility electricity simply cannot be reasonably provided. Places where it is simply too expensive to provide utility service can be so supplemented. Such a mixed system will complement the full monitoring system and provide adaptability that a completely fixed system cannot. Such temporary locations can be added and moved for greater versatility. They give a well-rounded design to the network and help to provide better service to the data requester.

The cellular phone and the electronics package must interface. Most cellular manufacturers already plan for this and either offer their own manufactured solutions or purchase aftermarket options and adapt them. Some equipment manufacturers have adapted their devices to accept cellular communications. This makes interfacing with the cellular network simple at the instrument level. The blending of communications equipment, because of adaptation in the market and competition among service providers, can be a fairly painless task and one rich in rewards.

The only issue, other than the power, is the location of the antenna. Equipment located below ground is not suitable for direct cellular communications. The antenna must be strategically placed so it can find a cell with which to communicate. The cell and the cell phone communicate their ID codes and maintain a connection as long as a call is taking place, such as for data retrieval. However, terrain and structures can block the ability of the cell phone to make the communications link. Hence, the antenna must be placed within the communications range of the cell. This is determined by a site survey that can be performed either by the service provider or with equipment they can provide. A good test is to take a standard cell phone from the provider of choice and watch the indicator lights or display to see if it finds a cell.

The antenna for the cellular communications device must be in signal range of a cell. Better commercial antennas can point in the direction of a cell. This can be difficult for street underground installations and where antennas cannot be readily mounted. Sometimes a quarter wave antenna can be mounted in or near the vent of a manhole cover, just below the surface, and still be able to communicate with a cell. In the case of designs incorporating boxes mounted on structures, placement is not a problem and the antenna can be strategically mounted to the best area in a field for reception. In the case of an equipment box that is itself a structure, the antenna can still be mounted per the manufacturer's recomendations. Several solutions are available through the cellular provider, including site surveys for signal strength, for temporary systems being converted to permanent application and whose equipment is not necessarily fixed.

In the case of box installation to equipment packages, provisions for electricity resolve the battery issues and make the location free of one utility service that requires hard installation. Those locations to which power is provided can accommodate industrial cellular communications devices, some cell phones without any keypad or other less tamperable device, working at slightly higher powers and efficiencies than standard consumer units. The increased power requirements are negligible, but heat is a major issue in device operation. The equipment enclosure can accept the antenna mounting readily and obviate to the battery charging concerns. This appears to be the best possible arrangement.

The premise is that supplying electricity to a permanent system can allow cellular communications to be utilized and eliminate the additional connection to tone phone service. The power requirements do not cause excessive draws on current or skew an estimated bill. The only concern is that cellular services tend to be compromised during lightning storms. As long as this issue is recognized and addressed, then cellular is an excellent option.

6.25 DATA COMMUNICATIONS: RADIO FREQUENCY (RF)

The use of RF communications has waned with the increasing popularity of tone and cellular services and the shrinking availability of frequencies. However, RF is a viable option and many manufacturers provide ingenious methods of circumventing the frequency shortage, some of which are discussed here.

Radio frequency communications require more power than the two technologies discussed above. The single greatest advantage is that this equipment is always working and transmitting, so real-time continuous communications are possible. Modems and cellular technology require initiating a call and can simply hold the line open for usage, but charges apply. RF devices are always on, but no charge accrues because the activity is simply transmission of radio waves. In cases where real-time data is critical, RF transmitters are a long-term, cost-effective solution. The disadvantages of high cost at purchase and the extra power requirements can be quickly offset by RF's constant transmission ability.[4]

Most RF equipment transmits in the 902 to 928 MHz range, designated as the industrial, scientific, and medical band, or ISM bandwidth. Although ISM bandwidth is designated for unlicensed equipment, Federal Communications Commission (FCC) licenses are generally required to operate

4. "Site Surveys: Who Needs One?" Grayhill White Paper, Bulletin #748. Grayhill does an excellent job of delineating classfications of RF devices.

such devices over large areas. A designated frequency is either assigned or shared for transmission purposes. Most manufacturers of these devices can cross-license frequencies for customer use. The device transmits first an identification code and then the data. Any receiver can recognize the code and accept or reject the transmission accordingly. The code also determines the size of the data packet. The approach is similar to computer communications except the medium is radio waves. The devices are applicable over a physical plant or entire cities. Satellite communication is also possible with these devices if the system is set properly.

The advantage, as noted previously, is that the devices are always transmitting with no expense other than the license fee and the power to operate the units. The disadvantages are capital expenditure and noise. RF devices are highly susceptible to interference from the same technologies as noted above and from many devices used in any modern society. Electrical transients and transmission noise from cell phones, laptop computers, and other forms of communication constantly interfere with the signal. If the frequencies are the same, the signals will collide and potentially disintegrate in a physics phenomenon called *destructive interference.* At best, the signal will become so noisy that it no longer makes sense to the receiver and is rejected. Pagers operate in the 930 to 931 MHz range, which is dangerously close to the ISM band. Anytime the signal loses integrity in the higher band, it can cause the data to be carried over the lower-frequency bands.

Another issue is the flat spot in the RF range. A flat spot is an area in the transmission range where radio signals simply do not work. They are a result of high structures, valleys, and other variations in terrain that block the propagation of the radio frequency transmission. Other than boosters or other methods of redirecting the signal, the technology that can be adjusted to get the signal into a flat spot is very limited. The size and location of flat spots in a given area is best determined through a RF site survey performed by qualified communications technicians. If the flat spots are small or negligible to the success of the system, RF is certainly a consideration for full permanent monitoring systems. RF may be questionable for partial permanent systems based on the expense at purchase in comparison to other alternative communications technologies. If the flat spots are large, RF should simply not be considered.

Many utilities successfully employ RF transmission equipment to monitor and control pump stations and other sites in the infrastructure and at the treatment works as well. The technology is reliable and gets better with each wave of technology innovation. Weighing its advantages, pitfalls, and customer requirements in the analysis grid is the best way to determine whether or not the technology will serve the needs of the utility operator. RF's biggest advantage in this application is for when real-time communications are needed.

The better RF devices act as repeaters for each other. In extended systems with many nodes, when a coded signal is heard by a member of the network, the signal is brought in and rebroadcasted to improve signal clarity and integrity. This constant boosting of signals provides for maximum performance under the duress of noise and interference from other communications devices. Also, if a node drops out of a loop, the problem is quickly detectable and service can be promptly dispatched to the location. Many times, the startup requires that the antennas be repositioned for maximum reception given the terrain. All parts of the system are thus tuned to each other as well as to the transmitter and receiver. This is easily done on structure-mounted or box-mounted antennas.

Few, if any, RF devices operate on batteries of any size. Even low-power devices require a power source of greater integrity than a battery. For this reason, the only practical operating environment for an RF unit is utility-provided power. Using RF for deployment of any temporary site or conversion to a partial permanent network is probably impractical with the current technology offerings. However, if a full permanent system is being considered, then RF technology is certainly a possiblity. A full site survey for RF flat spots is recommended as a first line of analysis prior to any investigation of equipment or compatability.

The ability to generate and monitor live data is a goal of the best monitoring networks. RF communications technology can provide this easily if the terrain permits and the capital expenditures allow. RF technology is a solid performer with a proven track record and, if the network design permits, is readily incorporated into the system at installation.

6.26 SYSTEM INTEGRATION

System integration is the process of making all the decisions for the devices and services of choice—technology, central computer, maintenance commitments, field hardware, and other aspects—and bringing the system together as a functional, usable infrastructure monitoring system. System integration is also the point in system development at which the system engineer selects the additional parts of the network to be included. Possibilities include rain gauges, biomonitoring, samplers, and pump stations. The overall decision is which parts should be included and which remain separate.

The critical paths are again chosen for the points of interest along them. Every point of interest along a critical path is delineated and labeled according to type. Parameters of flow, level, rain gauge, pump station, sampling, biomonitoring, and automatic control are identified. The decisions are based on the needs and requirements of the utility operator and those who require the data. However, guidelines are provided for all the points and parameters that should be included, for those that may be included, and for those that do not need to be included.

Rain gauges should always be included; there cannot be too many rain gauges in a given vicinity. Even with the best methods for deploying them, system operators find vast differences among the gauges per storm event for time, duration, and intensity. Using too few gauges may not yield enough information to determine the effects of a storm. Therefore, include as many gauges as possible.

Pump stations can be as small as a sump pump providing occasional lift services or as complex as a multipump facility lifting hundreds of million of gallons per day. The key is to identify those whose hydraulic effects are big enough to be of interest to the operators. Having a flow measurement at the next point along a critical path may be enough to provide the necessary information to the utility operator. If signal outputs from the controller of the pump are accessible, they should be identified to provide flow from a pump curve or primary device, if so equipped. Flow measurement devices may have to be installed so the hydraulic characteristics are better understood. Each pump station must be examined to see if its inclusion will benefit the system at large. Generally, some on-or-off condition of the pump station is desirable if flow is not available or logistically possible. Storm stations have the same criteria and also require evaluation for integration. Inclusion of all pump stations that have a significant effect, as determined by the system operator, on the critical path is recommended.

Samplers do not necessarily require inclusion, as they are portable devices and require attention every day or so in order to change the bottles. However, if a sampler has a trouble bottle or can determine toxins through chemical analysis, it should be included so the network is aware of an alarm it may be broadcasting. Samplers, which are 24-hour devices, demand operator attention within that time frame to remove the samples taken and reset.

Biomonitors can be critical in determining the quality of the wastewater being transported by the system. Continued monitoring for quantities such as temperature, pH, ORP, and DO can provide useful information. Devices should have ranges so alarms can be transmitted in the event of a value outside the range. This can alert the treatment facility to undesirables that are en route or about to be discharged into a watershed. The facility, so warned, can respond appropriately.

The subject of automatic control is one of growing interest as the technology of wastewater processing improves. Any automatic control is considered a hydraulic control point and should be included so its actions can be monitored and the hydraulic characteristics it generates better understood. Gate positions and other control functions should be able to be seen by the system operators.

Computer systems are required to poll, store, and process the data; potential purchases must be evaluated. Several years ago, this would have been the toughest part of the operation, as the personal computer was still maturing and mainframe systems were expensive. Today, the technology is available to everyone and no special skill set is required to make these integrations. Most manufacturers' software is designed to configure the system with limited user input. Third-party communications software is generally not difficult for regular computer users to configure and operate. Therefore, what needs to be determined is how much equipment is to be utilized and whether or not vendor software will accommodate each point of interest and each critical path. If the commitment is to remain with one vendor's hardware products, then the accompanying software solution is probably adequate. If the grid shows that a diverse selection is necessary, then third-party software should be considered. The decision showing the diversity will be obvious with the number of variations per location.

Use this information to form a final grid of those functions of the monitoring system beneficial to the individual utility. From this, the decision can be made to roll over devices, add devices, or combine devices.

6.27 SUMMARY

The most important part of any monitoring system is understanding the desired result from the installation and the data generated. This chapter covers a matrix of considerations that should be addressed before any action is taken in the procurement and turnover of any infrastructure system.

Note what is important in this section. The first consideration is the selection of temporary equipment, which can be of any style or kind as long as it provides hands-on experience with the site of interest. The second is the inventory of current equipment, such as the rain gauge network. Third is a general inventory of the system of interest and the quantities sought from the sites along the various critical paths. Fourth is the decision-making process that determines the kind of system to install. Fifth is the choice of utilities and services that make the system functional. Sixth is the data communications choice. Seventh is a review of what should be included and what excluded. Eighth is the installation considerations for each type of site. Ninth is the final development and installation parameters based on the experience generated from all these activities.

It is only now that the next step—the choice of equipment for each location—can be taken. The preparation, which may at times seem tedious, yields a matrix into which only certain devices can fit. Without these steps, in one form or another, maximum yield from the system is not possible because some consideration will be missed, thereby causing a rethinking of the entire program and its applications. This decision-making process keeps the program on track and brings forth all the needs, wants, and desires of the agency.

Planning is the most difficult step in equipment selection. Part of the planning process is the generation of experience in site installations. Preferences for technologies and manufacturers can be developed based on field experience and not sales presentations. No consultant or service provider can adequately tell the utility operator what it needs; they can only suggest and compile the information. Only the utility agency has the total package of information for making this decision. Remember that the infrastructure is the responsibility of the utility, which daily works to serve and protect the public. Obviously, no group better is equipped to make these determinations than the system operators.

CHAPTER

7

Equipment and Service Selection

7.1 INTRODUCTION

Once a full understanding of all the factors required to make the integration and to plan from the proper experience base is achieved, next comes the task of selecting the instrumentation that will be introduced into the infrastructure system. If preparation for this step was adequate, the selection process will be straightforward. If an area was missed in the planning, the omission will show up as soon as equipment is selected.

The steps of the planning process have been numerous. The basic concepts of modern sewer management systems have been reviewed. The technologies and their applications have been identified. A method of classification has been established and identification of points of interest and the critical path technique has been developed. The systems approach has been discussed and thoroughly outlined. A rain gauge system has been installed for generating information on the localized hydrologic cycle. A series of temporary monitors has been systematically installed; this has been building an experience base and parameters for permanent installations. The entire system has been mapped on a grid that designates the parameters desired, the relative difficulty of installation, potential utilities nearby, and the ranking of importance to the critical path. Through an iterative process, sites have been added and removed so only those sites that are important remain. Choices have been made for data communications. The decisions have been made as to which portions of the temporary system will

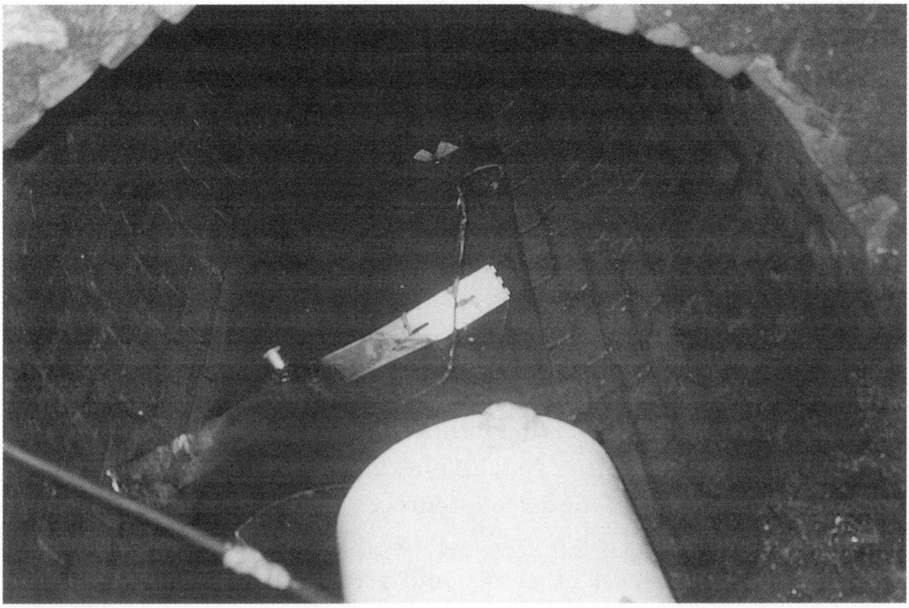

Figure 7-1 It is important to completely understand the application and the conditions of the infrastructure to be monitored. Improperly installed equipment will have a short lifespan in such harsh environments. This particular failure was a result of improper understanding of wet-weather flows and their force.

remain temporary and which will be converted to permanent. All, at this point, leads to the selection of equipment and the methods of installation.

The evaluation and selection of technologies to be introduced into the infrastructure begins with firsthand knowledge of the infrastructure, its characteristics, its potential hydraulic responses, and a fair idea of the localized hydrologic cycle. Only with this battery of information at the fingertips of the system engineer can a successful system be constructed.

The subject of this chapter is the evaluation and selection of technologies. The varied systems around the world and the uniqueness of each does not permit definitive equations for this procedure. However, certain parameters are key and can be readily adjusted by the utility, including reliability, life cycle, applicability to the specific site, cost to purchase, and cost to own. Consideration of all of these and more is required to generate the best purchase and fit.

It is important to realize in the beginning that the purchase price of a field instrument is the least of the cost. The cost to purchase can be a small fraction of the cost to own. The cost to purchase is the capital expenditure for buying and installing the field devices. The cost to own is the long-term expenditure required to make the devices useful over their life cycle. The cost to own is often a surprise to utilities, as it is seldom factored into the cost of the system as a whole. This long-term support can be listed as direct expenditures or made part of existing programs, depending on the choice of the utility. If the cost to own is not a factor in the financial plan for the monitoring system, the system will probably have a short life cycle, fall into disrepair, and cease to be useful to the utility operator.

Another major concern is the justification of the resources to purchase and support the monitoring system, but this is the subject of chapter 10. The efficiency aspects of the selection process are the topic here. Remember, to the uninformed, this purchase shows no promise of direct economic benefit. Its ultimate cost efficiency and return on investment must be proved to the utility managers. This makes the correct selection of technology the first time around critical, and it is one valid reason for the intense research and long study periods described above. The methodology provides a structure for justification when presenting the plan to the management teams. Direct structured methodology is absolutely critical in presenting solid financial plans. Of course, direct economic benefits do accrue to a sewer infrastructure management system. The ability to convey this information to decision makers is essential.

This chapter describes the method for selecting equipment for long-term use based on grids, as outlined in chapter 6. Choices of equipment for short-term use can be based on individual site requirements or on a piece of technology that fits some loose guidelines, even if only for scientific interest. Long-term capital investments require more analysis.

Most of the analysis presented, at least in the first several iterations, will not be specific to the measurement or other aspects of the desired result. Development and planning involves sifting through a range of applicable devices and manufacturers best suited to the task. Many products can do the job; the goal is to generate more than the minimum. For example, temporary monitoring devices simply do the task. This is acceptable if that is all that is desired. But the limited information these offer will inhibit future expansions and reduce the potential automation can offer to a utiity operator.

The operators of the sewer infrastructure are the only ones able to determine the best technology for their particular application. This is the reason that an adjustable framework is provided for the evaluation, rather than equations for selection. The frame has to be tailored to the particular utility's policies, practices, system size, and available resources. The frame provides all the tools for making the equipment evaluation as long as the craftsmen involved provide the necessary energy in building the monitoring system.

7.2 AN OVERVIEW OF TECHNOLOGY EVALUATION

The concepts applied to the selection of instrumentation can be used in deciding any technology preference. Data communication, as an example, can be used for any application that requires remote communications—for example, traffic systems, medical monitoring, and library services. Construction and integration are relevant to many other areas as well, and they form databases that can be utilized over and over and can be adapted to fit many settings. However, let us focus on technology that can be applied only to the sewer infrastructure.

Sewer infrastructure technology can be divided into two distinct areas: technology designed specifically for sewers, and technology that works in the sewers as well as for other applications. Depending on the grids used in the analysis, the choices used in developing a sewer management system may be from either category or from both. Historically, every combination has been utilized. Reasons for selection include local politics, resource allocations, court orders, and the utility's efforts to modernize operations. In any case, unless a service provider performs the task as a whole, seldom is one area of technology applicable throughout the system. This complicates the evaluation.

The process of evaluation is constrained by the product offerings by the manufacturers. Sewer infrastructure monitoring is a very thin slice of technology; only a few very big players supply this equipment. Among these manufacturers, the competition is stiff, which actually works to the utility's benefit. Competition drives technological improvements and enhancements, much as the drive for faster computers drove the expansion of the personal computer market. However, the applicability of these technologies to sewers is a function of sewer conditions, unique designs, the imagination of the parties involved, and resources. When imagination is employed, wonderful results may be achieved. But common sense and the experience of other users must be used to limit experimentation.

The limited equipment market puts stress on the evaluation process. The tendency is to reach too far in the effort to find devices that satisfy all of the dreams of a utility. Dreaming is wonderful, and dreams are what makes many concepts a reality. The practical dreamer, however, relies mainly on technology with a track record. To illustrate, picture a system engineer evaluating the top five product offerings of the top five manufacturers. Feeling they will not meet design specifications, the system engineer then reaches into other areas and blends technology that appears to work but is not either tested or certified for the hazards of a sewer. A reasonable degree of success may be achieved; however, the engineer now has a one-of-a-kind entity requiring many more resources for its support than the standard product offering. The chances of long-term failure are greater when experience is lacking, and this method is prone to it. Therefore, employing standard product offerings work even if they offer less than total dream fulfillment at the time of installation makes sense.

The risk of failure in the long run from lack of planning is another justification for the extensive planning required to bring these systems on board. The desires of a utility or other interested parties may exceed the capacity of current technology or product offerings. By using the iterative process and matrix analysis to match objectives with technology, expectations for the system can be brought in line with available products.

The best introduction to the various technologies is through sales representatives. Sales teams hold the information about the technology desired. It is therefore valuable to maintain contact with them. There is information in sales material; unfortunately, it is hidden behind sales hype. Nonetheless, the sales team and the literature that they produce are invaluable. Get on manufacturers' mailing lists. The product literature updates they send provides food for comparison as the system's experience base expands. Understand that sales material is biased toward the target product. However, the material contains information that can enable further understanding of the project.

There is another source of information for independent product testing. The United States, England, and Europe all have independent laboratories, funded by equipment manufacturers, that test and rate devices for purchase. However, these services generally are expensive and difficult to obtain. Information on independent testing is much more readily available in Europe than the United States. It may make sense to either join these agencies or purchase their results.

It can be worthwhile to initiate product sales meetings, but do so with caution to avoid an onrush of salespeople into the facility. Explain to the sales teams that a preliminary analysis is under way or that a first cut of technology is taking place and invite them to present an introduction to their product line. Do not exclude manufacturers of what seem to be familiar products, as technology is changing quickly and new products should be reviewed as soon as they are available. Be sure, however, not to waste the time of the parties involved.

The first steps in evaluating technology will open the door to a flood of material concerning current product offerings. Apply the utility's experience base to screen the sales literature. Facts are buried in the sales hype and must be dug out. The sales literature of reputable manufacturers is easier to decipher than others. One of the first notes to add to the compilation of sales material is the relative difficulty of getting facts from the material.

Data not immediately required for the analysis should be cataloged and the product binders updated as new products are introduced. By tracking the families of products offered by a manufacturer, a utility can determine the stability, commitment level, and forethought exhibited by the manufacturer. Products that appear to be evolving continuously and have backward compatibility can be considered well thought out. Products that appear unchanged for many years or that change constantly with no provision made for older versions already in use should be carefully scrutinized. Tracking products and their family lines becomes an indicator for the manufacturer themselves and the staying power they possess in the marketplace.

This compiling generates an initial body of information confined to sewer product offerings. It establishes a system for tracking product introductions and updates as well as families of equipment offerings. The result is a solid foundation on which to base specifications for equipment best matching the utility and its unique sewer infrastructure.

7.3 EVALUATION OF TECHNOLOGIES: SECOND-TIER CONSIDERATIONS

The term *second-tier considerations,* for this discussion, indicates a closer look at current product offerings applicable to the infrastructure and the support groups from which these products and services are generated. After this analysis is completed, development of the plan for procurement and implementation can begin. Much more can be gained from companies that manufacture instruments for the sewer infrastructure than simply the equipment. Manufacturers have a mature experience base with respect to the deployment and utilization of their instruments. This knowledge base can be tapped by asking intelligent and informed questions. Reputable companies accept the challenge of answering them because the manufacturers realize they learn as well. As long as the exchange is mutual, and the questioning reasonable, the manufacturer generally is gracious and accommodating in configuring particular sites to their product offerings. By partnering with the manufacturer, the utility obtains the best fit and customized solutions.

Because of its uniqueness, each sewer system poses its own set of problems. This is the reason for the matrix, as previously described. The task of the system designer is to use the concepts discussed, along with local knowledge, to generate the most pertinent questions.

The concept of partnering the manufacturer is introduced, seemingly early in the analysis. At this point, a number of reputable manufacturers may still be in the running for the project. The working partner is critical to completion of the endeavor. The utility personnel must be able to relate and work with the personnel of the manufacturer. Any issues in which this may not be possible should be identified early on. If the issues are correctable, which they generally are, then they need to be identified and resolved. Many times, however, the administration of the utility may not be compatible with the way that a specific manufacturer does business. These issues are generally beyond the ability of the system engineer or other utility personnel to correct. These issues are procurement, finance, and the support agencies that administer the business end of the system operations. Partnering brings out these issues so the unresolvable ones can be identified and the relationship dissolved or corrected. In the above cases, the manufacturer usually has to change for the sake of the business endeavor. Squarely facing problems, regardless of their difficulty, makes resolution possible if the desire to do business is strong enough.

Companies can be grouped roughly in terms of the products they offer. Full-service providers manufacture and provide the entire package. Some manufacturers team with service providers to produce full working systems. Aftermarket service providers fabricate and supply systems of choice based on the matrix or other needs assessment. Some suppliers simply find

Figure 7-2 Retrofitting existing flumes and other structures can be difficult, especially those exposed to high flows. Verification of the structure and features that would inhibit instrumentation should always be evaluated as part of the screening.

equipment and the utility is responsible for making the system work. This last type of supplier is divided into two categories: those that supply, service, or support equipment for sewer infrastructures only, and those that supply process instrumentation applicable to a range of processes.

7.4 TECHNOLOGY CONSIDERATION: FULL-SERVICE PROVIDERS

A full-service provider is a company that manufactures, markets, installs, and services sewer infrastructure equipment as a turnkey operation. It is a one-stop solution. The company has developed technologies based on their experiences, provides installation services, and has a staff for long-term maintenance for the operation. There are very few of these companies; those that exist are quite specialized and draw on considerable expertise to make the system successful.

A full-service provider has developed a range of technologies and offers a solid body of experience with the package. The products meet some or all of the specifications of the utility operator. Their own developed solutions for measurement and alarming have levels of success based on the uniqueness, specification, and cooperation as received or experienced from their utility customer database. Because sewer infrastructures are the only or core business of full-service providers, the product offerings are narrowly targeted and generally not useful for other applications. This concentration on sewer infrastructures is to the benefit of customers.

The biggest advantage of a full-service provider is the fact that highly specialized core businesses yield an array of products whose design can take into account the uniqueness of each sewer infrastructure. The systems install readily and produce results in short order. If the matrices outlined in previous chapters are considered, the utility simply turns them over to the full-service provider, issues a contract, and awaits the results. The full-service provider approach should certainly be considered for large-scale installations.

One disadvantage in this approach is that the equipment is generally of a closed architecture and does not adapt readily to later expansion. (This may not be an issue if there is no potential for expansion to biomonitoring, sampling, or control as the knowledge base matures.) By the same token, existing equipment may not be able to be integrated into the system. (If no equipment exists, or if existing equipment is simply to be removed and reassigned for temporary service, this may not be an issue.) This is relevant if the utility operator is considering the conversion of temporary locations to permanent monitoring. Such a plan may not be possible using the full-service provider. Also, the range of equipment that may potentially be utilized is diminished because the utility is limited to the offerings of the provider.

The advantages of a full turnkey system delivery and operation must be weighed against the integration of existing devices and the limitation of product offerings. If there is no preference or the utility has little experience, then the full-service provider is probably the best choice. If the utility feels strongly that existing equipment must be a part of the network, the vendor must be consulted and the possibilities for integration of these devices analyzed by the providers' representatives for best fit using existing or developed protocols. Because these providers handle everything, spare parts and supplies are taken care of by means of a service contract. A dedicated contact person and a work order are needed to arrange repairs.

The evolution of electronics and advances in product development are rapidly changing the offerings of the manufacturer. Products that previously were unintegrable can now be modified and reintroduced along with new products and service. The manufacturer is possibly the only source of information about these possibilities for integration. Full-service providers are always developing more open architecture. Only they can tell the utility where they are in the process of development.

In sum, choosing the full-service provider approach means that individual system components need not be evaluated; one decision takes care of all. It is critical to evaluate the provider as a whole and decide on the level of involvement that the utility intends to extend. The less involvement planned by the utility, the greater the appeal of a full-service provider.

7.5 TECHNOLOGY CONSIDERATION: MANUFACTURERS AND TEAMING PARTNERS

The early days of the personal computer revolution spawned competition among companies that saw the possibilities inherent in the rapid movement in computing technologies and incorporated modern software techniques in modular electronics packages. This allowed many utilities to construct and maintain systems in house. Manufacturers seeking to tap this share of the market developed an approach whereby the utility took control where possible and a team of partners provided support services otherwise.

This is an interesting combination of business entities that sometimes are on one team on one job, and are on opposite teams on another. The placement of companies, does, however, circumvent many of the local political and economic obstacles encountered in municipal or other governmental and utility procurement processes. Here, the knowledge base of several companies is presented to a utility and a broad spectrum of integration possibilities is available. Much of the experience brought to the table here involves related businesses, which provides a different perspective than a solid core business approach, such as with the full-service provider. As with the full-service provider, factors concerning the applicability of this blend to the task of preparing technology for introduction to the sewer environment must be weighed.

The core business of most manufacturers is the constructon of equipment for the sewer infrastructure. Tried and true technologies are blended with new approaches, yielding an array of product offerings for many applications. This has the advantage of less integration when applied to difficult or troublesome locations. Manufacturers, however, unlike full-service providers, generally do not offer direct field support for installations and startup. Most reputable manufacturers maintain service phone lines so the equipment can be debugged and installation problems resolved over the phone. The point is that the utility is responsibile for getting the equipment into the ground, which is not always possible. Therefore, the manufacturer has agreements with one or several providers of services to perform installation, initiate startup, and address operations issues that are completely handled by the full-service provider.

These third-party providers are usually known, reputable companies rich in experience working within sewer infrastructures. They are versed in the manufacturer's product offerings and have working relationships with them. The utility can select which service, or part of the service, is appropriate to support and underwrite the resources available from the utility's management policies. With the broad experience that underpins the manufacturer's offerings, the part of the service that the utility chooses to purchase can go forward, with the utility providing the rest of the support. This approach leads to a sense of ownership by the workers and buy-in from the utility's technical staff and first-line supervisors. It can be a difficult blend, but the result is a wider range of products than is offered by the full-service provider.

The major issue is the potential for substitution or incorporation of products that are not from the main manufacturer. At times this substitution may be by choice, depending on the type of equipment or the installation required. An example is the use of specific data communications technology. If the preferred data communications technology is not supplied by the main manufacturer, the utility can purchase that specific technology separately. If the utility decides on a particular level measurement technology, this can be integrated by the third-party supplier with the primary supplier's equipment. Note the flexibility this allows the utility operator. This can be valuable if there is an established base of experience from which the utility can draw its information. Generally the products supplied or otherwise integrated are designed specifically for the sewer environment. This ensures the integrity of the equipment being installed.

The advantage of the teaming approach is the ability to choose from a broad range of products and services and to benefit from a large experience base. It allows the manufacturers to do what they do best—manufacture—and the service providers to exploit their specialty. It allows for flexibility in the complying with local governmental and administrative requirements. The approach also provides a sense of ownership by the employees of the utility, who will be expected to work with the equipment. It allows for non-manufacturer-specific equipment to be directly supplied, if desired. In general, however, the manufacturer and teaming partner approach tends to streamline the replacement parts process, minimizing the stockpile of diverse spare parts required to keep the system functioning.

The disadvantages involve the integration of several groups of potentially incompatible interests. Whenever companies are brought together and set to work on the same endeavor, problems may arise. Many of these concerns are addressed through the agreements made between the manufacturers and the service providers. It is simply issues with people that tend to slow the process down if not addressed at the outset. Another downside is in the case of partial support from the utility. Whenever partial services are offered, it is easy to shift blame when problems arise. Careful contract writing is required to eliminate, or at least minimize, this tendency. Finally, a non-uniform system may result if care is not taken in the procurement and administration process.

In sum, manufacturers and teaming partners can provide a broad base of product offerings in a potentially greater variety than the full-service provider can supply. The experience base also tends to be broader because of the multiple resources. However, the same factor inhibits the process because of the added personnel issues in the administration of the project. If there is an experience base with a range of manufacturers or preferences for specific technologies, then it is reasonable to evaluate this option as a potential choice.

7.6 TECHNOLOGY CONSIDERATION: AFTERMARKET SERVICE PROVIDERS

An aftermarket service provider is an entity wherein the service provider—rather than the manufacturer—is the main component of the system development. In this evaluation, there is no teaming or loyalty to any particular manufacturer, technology, or supplier. The system can be selected based on knowledge of the market and the products offered and the ability of the service provider to integrate chosen products.

The scenario presents a large base of product offerings still specific to the sewer environment but much more broad than the manufacturers' targets. The ability to pick and choose instruments allows the best blends to be introduced with closer adaptation to specific applications than one manufacturer or one full-service provider could generate.

The lack of loyalty to one manufacturer can have two effects. The first is fierce competition among manufacturers to supply equipment for the job, many times with additions and upgrades at reduced rates. The second is that integration issues tend to limit cooperation among the various manufacturers. This means that the utility is completely reliant on the integration expertise of the service provider. Such services work well if the service provider is experienced with all the equipment being provided and if it works with reputable manufacturers, which want to show up well wherever their equipment is used. It does present problems in obtaining full cooperation and problem resolution during the trying time of initial system deployment.

A serious downside here is the issue of maintenance. In all cases, maintenance is handled according to manufacturers' directives and either service contracts are written for the task or a streamline parts stock supports the system. It may be necessary to maintain large stocks of replacement parts and supplies because the equipment is so diverse. The diverse parts stock means smaller quantities of each part are purchased, leading to the loss of quantity discounts. The parts handling issue arises from the desirability of devices whose replacement parts are small. Even the best devices, however, require replacement parts. Small stock issues are sometimes more difficult to manage than large stock because of lag times in procurement and potential shortfalls in program funding. Even if contracts are issued for service, there may be problems managing them if they are not all given to the aftermarket service provider.

The advantages are the diversity of equipment that can be incorporated in the system and the ability to exploit the best features of current technology. The downside is the lack of general support and the reliance on one major source of expertise for a diverse body of equipment. Maintaining diverse lines of equipment and replacement stock can be a complex and time-consuming endeavor. Spare parts and supplies issues should be weighed thoroughly, especially with respect to the maintenance issues.

7.7 TECHNOLOGY CONSIDERATION: EQUIPMENT MANUFACTURERS: TARGETED

The technology evaluation process is one in which the utility intends to integrate targeted equipment and make the blends necessary to generate a full monitoring system based solely on manufacturers' equipment. The targeted equipment is made by manufacturers who build devices strictly for the sewer infrastructure, and therefore hand direct application to the environment.

Programs in which the utility takes full responsibility for integrating equipment and implementing the system generally are less than successful unless there is extraordinary commitment to the program by the utility agency. Utility personnel are not unqualified for such integration, but problems arise when they must handle it in addition to their usual workload. In such cases, the minute an emergency or preferred project comes up, the monitoring program is postponed. This causes lag time in the installation and, generally, a system with little continuity. Again, a high level of commitment, probably far in excess of the other options, is needed.

Many times, some great talent reside in the utility agencies and large bodies of expertise, combined with local experience, appear to have the best blend of talent for such integration endeavors. While this is quite true, any program in a utility environment that is not the focus of a day's work is difficult to complete just by the nature of infrastructure management. This is brought forward in defense of the people who generate large skill sets working day after day in the field. They know the personality of each site that requires monitoring. The nature of the sewer infrastructure business means that the workload these people are expected to maintain tends to exceed reasonable expectations. Adding this type of complex systems construction to their workload, along with the inherent replacement parts issue, constitutes a recipe for failure. It is therefore recommended that extensive consideration be given overcoming the workload obstacle in such endeavors.

7.8 TECHNOLOGY CONSIDERATION: EQUIPMENT MANUFACTURERS: GENERAL PRODUCTS

The selection of non-sewer-specific products for sewer-specific applications needs to be approached cautiously. This approach was popular in the days when few product offerings were designed exclusively for the sewer environments. In this strategy, all manufacturers of equipment that can provide the required measurements are evaluated even if their products are not specific to sewer environments. This includes manufacturers of process instrumentation and datalogging equipment. The approach requires broad knowledge of process instrumentation, data communications, computer integration, and system startup.

Infrastructure monitoring systems prior to the growth of the instrumentation market and the personal computer revolution were constructed in just this manner. If a full-service provider could not fulfill the task with specially targeted devices, plant process instrumentation was adapted. Many of these general products were less than suited to the sewer environment and required extraordinary maintenance. However, at the time, this was the only reasonable method for construction of large permanent systems. Some were built by in-house teams and others were released to general contractors. Because each was unique and based on the equipment experience of the system designer, there was no uniformity among systems from which to draw accumulated knowledge. However, general experience was shared with the industry, and both the utilities and manufacturers learned from such projects.

It may be necessary today, in certain circumstances, to step out of the targeted product market to obtain equipment for a particular site or critical path. Manufacturers can adapt their devices to most situations or construct equipment for specific applications, expanding their own knowledge base in the process. The construction of entire systems with equipment not designed for sewers is a poor option and the one with the greatest likelihood of failure, though this is not to discount those small, specific projects into which a one-of-a-kind aftermarket product fits and works quite nicely.

7.9 COMPILING RESULTS OF THE CONSIDERATION PROCESS

The various methods described above all require a complete understanding of the measurements to be taken at each point of interest along a critical path. In almost every case, the utility operator must select the technology with which to measure the parameters. A perception that the choice of technology belongs to the service provider is largely incorrect. In most cases, the utility's system engineer is responsible. Even if the responsibility is assigned to the service provider, the system engineer should be on guard and possess a thorough knowledge of the available technologies as a result of the evaluation process.

The evaluation methods presented in the previous sections allow the system engineer to determine the best fit for the utility. The goal is to match technology to the utility's principles and practices, but this is an iterative process and may need to be repeated several times until there is a clear understanding for the best fit per the utility's resources. After the first set of iterative evaluations is complete, a more specific set of steps can now be taken to determine the best technology to be utilized by the agency.

7.10 PRACTICAL EVALUATION OF TECHNOLOGY: STUDY AND VENDOR FEEDBACK

The most difficult task is the consideration process—making the evaluation that identifies the vendor, technology, and service level most appropriate to the job. Once this is done, the point at which actual equipment will be selected is near. However, in many ways it is more important to be comfortable with the firm (or firms) that will be providing and supporting the equipment than to have the perfect type of measuring device going into the ground. Remember that the expert on each type of equipment is the manufacturer. Teamed service providers have experiences to share, but it is the manufacturer who is ultimately responsible for the success of the hardware.

Specific steps can be taken to narrow the evaluation to a single case for final consideration. Following these steps puts all the current technological offerings on an equal footing and provides specific parameters for the final analysis.

It is difficult to evaluate all products and services equally because the issues are always changing, as is the technology itself. A base point, a place from which the evaluation can proceed with no further influences and changes, must be identified. Only then can the choices be made with integrity and clarity.

Freezing the technological advances means that, past a certain date in the analysis, no further technology and no improved or new products are considered. Suppliers must be notified of this date so they can submit final proposals for evaluation. If the utility waited for technology to produce the ultimate personal computer, it would never procure any. The technology is always changing and improving; the reason for the artifical freeze is to stop the bombardment of change long enough to comprehend current equipment offerings. The freeze should not be implemented, however, if a major technological breakthrough is anticipated within a short time frame. When providers are notified of the freeze date, they should make the utility aware of any pending major announcements. The utility can judge the significance of this information and proceed with or halt the freeze accordingly.

The freeze must be enacted internally as well. Further development of the system plan must stop in order to halt the constant new demands for its performance. Notify personnel of the date on which the freeze will be implemented.

The freeze must exclude new suppliers of instrumentation and services from submitting proposals. If the efforts of the utility were adequate, every potential vendor had an opportunity to present its material. The issue here is that late introduction is a sales tactic not of the highest standard. Even if a late solution appears viable, the vendor must be rejected because there is no limit to the number of potential vendors behind it.

There is always the risk that the latest technology will be excluded because of the freeze. The question is how much was really missed. The likely answer is that even major technological changes were not significant enough to warrant the confusion including them would have caused.

Once the plan and proposals are frozen in time, technologies can be reviewed on an equal footing. Specifications of like equipment can be compared and the best, most accurate, and most cost-effective instrumentation selected. This comparison is performed by building matrices based on level measuring devices, flow measuring devices, data storage requirements, and support equipment. For example, level devices have minimum and maximum ranges, operating temperatures, MTBF (mean time between failures), recommended maintenance intervals, and measurement and error ranges that all must be considered. Flow also includes velocity and noise parameters. The support equipment is that which is required to get the system operational. Support equipment also needs to be identified based on the derived system requirements.

A lot of preliminary work is involved in planning the selection process if the procedures outlined in this book are followed. The goal of the preliminary work is to save a lot of evaluation time. The evaluation teams receive grids and make recommendations if the team could perform the task based on the location of utilities, choice of data communications, and other resource commitments of the utility. This is almost the same as answering a questionnaire. From the returned documents, segregate the technologies to see which solutions are recommended for which type of site. This is important in determining if one solution is being forced to fit many points of interest or if a spread of product offerings more tailored to specific environments is materializing.

The grids now contain the solutions provided by the vendors. These solutions, per the points of interest, include specific hardware recommendations. These say a lot about the spread of equipment offerings and the application of the technologies. The manufacturers' specifications must be compared to the manufacturers' recommendations for installation, operation and maintenance. This verification process ascertains whether a generic application of one type of device or a critical placement of like instrumentation is provided. Generate another grid of these solutions, which can easily be segregated using standard spreadsheets, and review the hardware as specified. Compare the specifications of the vendors against the potential solutions.

The grid now shows the recommendations and the specifications of each supplier. It is easy to cross the information and compare the various solutions recommended by each group. Most of the time, the vendors' recommendations are well thought out. The issue is whether parameters such as range, deadband, grit bases, and utility sources match the hardware solution the provider recommends. The experience generated with temporary monitoring now enters the evaluation. Utility personnel can identify recommendations that will not work as the manufacturer states because of the particular conditions at the site. Many questions will arise concerning the application of specific technology to unique or troublesome locations and should be noted. Such questions generate not only interest but also respect from reputable manufacturers and service providers.

The questions prompt a variety of responses. The more reputable manufacturers address them all. Most of the responses are clarifications based on the experience of manufacturers and service providers who can illustrate their success in similar conditions. If no similar conditions exist from which the vendors can draw, they can provide evidence that the technology will work efficiently in the circumstances described by the utility. Even though all situations are unique, other systems do have similarities. Reputable manufacturers can state that their devices work effectively under

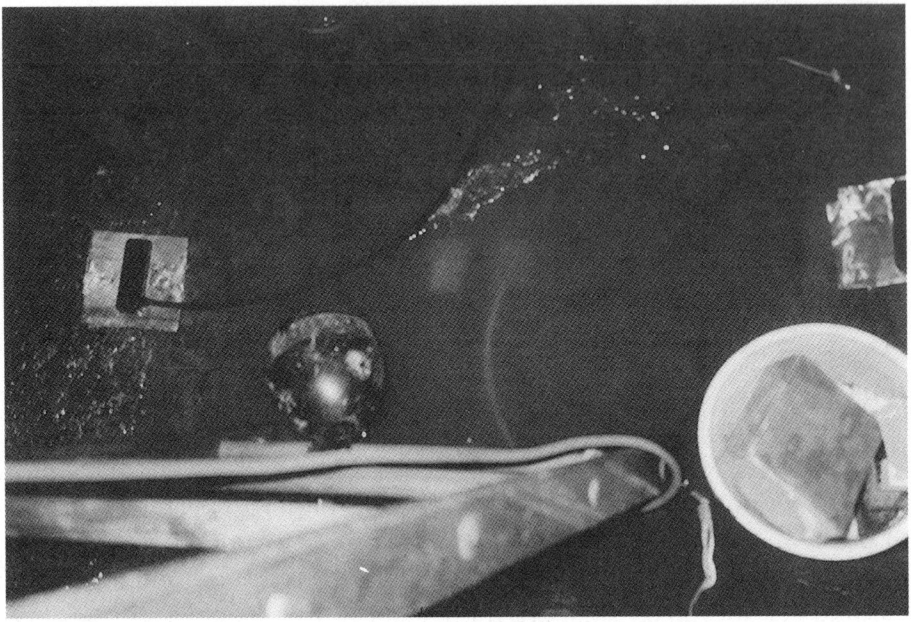

Figure 7-3 This is a multiple Doppler installation on an extremely wide sewer. This was achieved with the assistence of the manufacturer, whose understanding of the equipment overcame a difficult problem. This particular manufacturer's feedback provided an excellent solution to a difficult application. (Courtesy of American Sigma. Used with permission.)

characteristics comparable to those described by the utility; they can back up the claim through a warranty at the time of purchase. Always refer to the vendors' specification sheets when checking the solution proposed for a utility site condition.

The responses from the vendors now need to be coupled with the field experience of the utility's maintenance forces. The review of the equipment should be presented and input requested from the people who work daily within the infrastructure. Most of the evaluation will be agreed by consensus, and there will be some significant information from these individuals concerning specific details of the equipment. This invaluable hands-on experience adds another layer to the analysis of the behavior and response of the equipment. Certainly the field experience is limited to whatever the utility owns at the time of analysis. However, the voice of experience can shed light on specification questions and issues related to particular vendors' product offerings that cannot be obtained in any other way.

Seldom does this analysis form the final iteration in the evaluation. Generally, another step is required before the final decisions can be made. All these steps are required to evoke the expertise and background knowledge needed to ask the difficult questions and draw the tough conclusions.

7.11 PRACTICAL EVALUATION OF TECHNOLOGY: DEMONSTRATION PROGRAMS

There comes a point in the evaluation of equipment when all the specification analysis and comparisons provide nothing further of merit. The only solution is to try the devices being considered in places where they can be fairly evaluated. The places should be of equal settings within the sewer system so the performance can be observed and measured properly against the competitor devices for the performance specificaitons claimed.

Demonstrations are difficult but provide valuable firsthand results. They are labor intensive and require daily maintenance to be completely successful. The suppliers should be made aware that their devices are to be compared to and rated relative to their competition in the utility's environment. No claim will be made for anything other than success or failure at a particular test; this should be made clear. The testing parameters should also be supplied to the vendors so they know ahead of time the basis on which they will be rated. After the test, be prepared to provide full documentation to the manufacturers or those who provided the devices to justify the position taken by the testing group and the choice made on the basis of performance.

Clearly, one site cannot yield enough information with which to accurately determine the performance of an entire line of equipment. One site also cannot provide nearly enough data for all the installation, startup, and maintenance issues that may accompany the installation of a full monitoring system. The test is designed only to indicate whether or not the equipment matches the specifications listed in the product literature. This is all that can be expected. It is interesting, however, to see how many devices fail when installed. This is one of the benefits of the demonstration.

The first consideration is the time frame of the demonstration—generally at least 30 days but no longer than 90 days. Schedule the test during summer and winter in order to stress the instruments under extreme conditions. Because the devices are meant to determine hydraulic response during both wet and dry weather, one to three storm events of a certain intensity and duration should be recorded during this test; the time of the program should be either expanded or contracted based on the number of storm events recorded.

Location is the next consideration. Option one is to place all the devices in the same site and compare the results. This approach has advantages and disadvantages that should be carefully analyzed. No previous profiling of the site or any temporary flow data is required, but a site profile is necessary so the vendors can provide a directed solution. Similar answers and skewed data are then obvious. The program parameters also allow for one site visit by the crew, who looks at the data and reports on the devices. The disadvantage to the approach is that the sewer could get crowded and interference from other equipment may result. Measuring positions per the devices' specifications may also be compromised. These are reasonable complaints. The issue must be considered properly prior to deployment to avoid generating mistrust.

Option two is to find a number of sites of the same complexity and size, preferably along the same critical path, and give each vendor an entire location to work with. This approach is a bit labor intensive; it does, however, generate the best results. The sites must be equal on most counts so that all vendors have the same opportunities to demonstrate their products. The technique keeps the vendors segregated, which is valuable for this type of trial to maintain uniformity. The sites should be sufficiently challenging such that the manufacturers' best efforts must be employed. Reputable manufacturers enjoy such challenges and like to present their offerings under these conditions as long as there is the potential for business at the end of the demonstration.

Manufacturers may wish to use their assigned location as a sales site, bringing in other communities to see their installation. This is to be expected, as the companies have expended their own resources to demonstrate the product. If requests do not strain the utility, this should be encouraged. The sales opportunity also provides an additional incentive for manufacturers to do their best, as if the installations do not work well or if problems come up, manufacturers cannot use sites for this purpose.

It is unreasonable for a utility to ask for free demonstration equipment. Also, the final contract should be substantial enough to attract the better vendors to the demonstration. The size of the potential job must be balanced with the amount of equipment and services requested. Reputable manufacturers provide at least one demonstration unit to be installed at the utility's expense. This is a good balance that should be appropriate to many systems that require instrumentation. If both the task and the volume of business at stake are large, more services can be expected from manufacturers. Some vendors may be eliminated by lack of participation.

A final consideration is the storage of data. The utility needs to identify a time slice that reflects the conditions within the infrastructure. Thirty minutes is too long for this purpose. Entire pump cycles can be missed and storm events can pass by in half an hour. A five-minute time slice is recommended. With a time slice comes the method whereby the data is read and compiled. In a battery-powered system, it is unreasonable to assume that the device is always measuring; this wears out the batteries in no time at all. If utility service is used, then the time slice can be defined not only as a reporting period—say, five minutes—but also in terms of how often the data is read—say, every ten seconds, averaged over the five-minute period. Remember, the power source limits the ability of the device to make the measurements in a reasonable manner. The technology has limits that must be respected.

A demonstration program, when organized correctly, is an excellent method of comparing technologies under actual conditions. It is also a method of determining applicability of technologies to the infrastructure, based on the sites selected. Lack of participation in a demonstration can eliminate a manufacturer from the bidding for the work. Getting all the information from the program is the key to making the final equipment decisions.

7.12 PRACTICAL EVALUATION OF TECHNOLOGY: DEMONSTRATION PROGRAMS TECHNIQUE

The demonstration program, to have the maximum effect, requires that the same parameters be measured at each of the sites equally, with no bias for or against any brand or type of equipment. Even if the utility has broad experience in one or several brands of equipment, it is always advisable, and in the best interest of the utility, to review all the product offerings. Do not assume that in-house experience with a brand of equipment means that brand is the solution for all the sites of interest along every critical path. The nature of changing technology dictates that all choices within a range should be periodically reevaluated so they can be repositioned and reclassified to reflect changes and revisions.

Evaluation of the demonstration program itself is critical for the utility, which must be absolutely sure the guidelines that they generate meet the goals for this endeavor. Each of the major parameters to consider is discussed in this section. Guidelines may be added or deleted based on the requirements of each utility. This baseline of these guideline parameters serves as a frame for discussion within the utility and modification to the individual performance standard is encouraged once proficiency in the products is gained.

The demonstration program should reveal how easy it is to install and remove the devices. The reliability of the data should be tabulated, along with the number of times gaps appear in the data and their types. Downtime and uptime of each device should be noted, as should the ability to debug the instrument in the field. Maintenance requirements should be rated by service calls made versus the manufacturer's claims. The accuracy of the device should be determined in place, even if no historical data is available as a guide. Data accuracy should be compared to the manufacturers' claims for the instrument. Discripancies between product literature and actual field experience with respect to calibration requirements should also be noted.

The factory support and the teaming partners' followup on problems should be recorded and comments on performance and response made. Data communications should be rated for their ability to communicate and adaptability to the environment. The quality and construction of each device should be rated against the others. This includes tolerance of the sewer infrastructure environment and a list of the features that contribute to tolerance. If the system requires software for interrogation and operation, a rating of the ease of use, functionality, and compatability with the current or projected computer system should be made. Finally, the cost to purchase and the cost to own should be analyzed.

The ease of installation and removal is a primary issue in equipment evaluation. The device should be able to be readily installed and removed. The cables should be easy to hang in an organized manner. Note whether the sensor is integrated with the instrument package or if it has to be fabricated as a package. This can cause problems at permanent sites if all the cabling has to be removed in the event of an electronics failure. Note the lifespan of the sensors and the mounting requirements. It may be difficult to comparatively time installation processes because the structures in which devices are installed differ, but some conclusions can be drawn. Note whether or not the sensors appear to be susceptible to fouling, and follow up on the fouling particulars up during the inspection process. Draw no conclusion concerning the result; simply note any issues that the sensor generates because of whatever occurs in the field test. This can include protrusions, mounting requirements, and anything that does not appear straightforward. It may be a design that works well in heavy fouling. Observation here is simply a way of recording the learning experience.

The rating of installation and removal has to be done bearing in mind the configuration that the utility wishes to use. This tentative configuration must be presented to the vendors so they can understand the requirements. Here, vendor feedback is critical because the mounting and enclosures preferred by the utility may not in fact be the best. This is one way in which vendors can be helpful; they can suggest alternatives that the utility should consider. However, for the demonstration, not all such alternatives can be implemented, as this would compromise consistency among the participants. They do not have to be final as long as the lessons learned from the installation are incorporated in the final specification document.

Data has to be interrogated from the location. The first consideration is whether or not the data is complete and without gaps. Continuous data is a function of accurate datalogging and storage. Complete data is the function of integrity of the data. The data should be received, stored, and presented for interrogation in a straightforward manner with any of the desired parameters attached in its data set. The desired parameters include time, data, and site location, as agreed between the manufacturer and the utility. It should also be determined whether or not the device will substitute values in the event of a missed reading. This should be known up front and the data substitution feature turned off for the balance of the demonstration. Substitution is an excellent feature to have, especially in the event of erroneous or overrange data. However, the demonstration project requires raw data as it comes from the device, without data massaging or substitution, to ensure fair comparison.

The way that the instrument samples and stores data should also be outlined. If the instrument at the site must rely on battery power, it is unreasonable to ask for a short data point in the sample average. If utility electric is available, the short step and sample size are reasonable and should be made known to the vendors. Here, it is a matter of reasonable expectations.

Data retrieved from the instrument should be as accurate as possible.

The manufacturers provide specification sheets with their parameters for accuracy if the installation is made according to a certain set of guidelines. This is quite important; the most accurate measuring device cannot be accurate if not installed according to the manufacturer's instructions. For this reason, care must be taken not only in the installation but the effort it takes to make the complete installation with all resources necessary for its success. This part of the analysis should concern the ability to move an accurate data set generated by a device and move it into storage with minimal data manipulation. The amount of data manipulation may affect the result that is sought as a result of the reading from the site.

The downtime and uptime of the device is a critical issue in the evaluation of the equipment. An instrument that is constantly experiencing downtime cannot generate reasonable data to extract, even if the data handling is exceptional. The instrument must work consistently and be able to withstand the sewer environment. Causes of downtime include lack of or intermittent power, measurement overhang, exposure to conditions for which the device is not designed, and ground conditions causing equipment failures. If a device experiences significant downtime, it is imperative that the cause be determined and relayed to the manufacturer.

Any device introduced into an environment as hazardous as a sewer requires debugging. This is necessary in order to set parameters that optimize the operation of the instrument and to adapt it to the sewer. However, the debugging needs to be easily performed so as to minimize time spent within the infrastructure. The amount of debugging that is reasonable is a judgment call based on the experience of the field crews. The device should be easy to debug, in the opinion of the maintenance crews involved. List the features that affect ease debugging. For example, can the device be debugged in place or must it be brought up to the street? Devices that are easy to debug but must be removed each time debugging is performed are less than desirable. The instrument should be able to be debugged in place with as little diagnostics as possible to keep the site functioning. Here, a decision must be made as to the reasonability of effort and resources to be expended in the debugging process.

This also calls for a judgment call based on the relative experience of the maintenance crew performing the inspection. Instruments that are readily removed without confined space entry should be considered easy to debug. Any time a confined space entry can be avoided, it certainly should be. This portion of the evaluation should be written to explain why the devices are more or less desirable with respect to debugging and how these opinions were generated.

Maintenance is the major factor in evaluating an instrument in place. Maintenance issues are difficult to discern in the sewer environments as they are unique to treatment plant and cannot be compared accordingly against each other and manufacturers' specifications. Many argue that the standards for the maintenance of wastewater plant equipment applies equally to equipment intended for sewer environments. This is not so. The comparison is a valid initial step where the environments are comparable. However, treatment plant equipment is confined to the plant boundaries, where the traffic and access can be controlled. Sewer environments cannot control traffic, access, or other natural and man-made occurrences. These issues distinguish the maintenance evaluation of sewer instruments.

The reason for maintenance analysis is to determine the amount of effort required to keep the devices measuring and to repair the devices when that is necessary. (The contracts, spare parts, and expertise required to expedite these repairs should also be evaluated.) These two issues have associated costs that are part and parcel of the cost to own.

The level of in-place maintenance needed should be verified along with the data generated. Periodic site inspections determine whether or not the installation is susceptible to debris buildup, grease accumulation, and other problems that may require confined space entry to correct. The profile of any sensor introduced into the flow has characteristics that make it more or less susceptible to these hazards. Only through direct observation can these issues be properly analyzed. It is important to make few generalizations concerning sensor design, as manufacturers generally are careful to address maintenance concerns. The evaluation task is only to observe and record how the sensors respond in the environments in which they are introduced. Pictures are credible evidence in making claims to the manufacturer of either success or failure; they should be included in the final report.

The repair and upkeep of the devices' electronics is a critical issue over the long term, especially when numbers of them are considered for purchase. Some of the devices will not require spare parts, except for sensors and related mounting hardware, or a contract with the factory. Other devices will require parts stock and qualified technicians to expedite repairs. A true feel for this situation can be obtained only through field experience, where difficulties and failure rates can be readily assessed. The utility must determine if it is worthwhile to maintain in-house support or simply to obtain electronic swaps for the instrument packages. Field experience provides the information necessary to make these determinations. The cost-to-own models provide the financial data to complete the decision process.

The issues concerning factory support are critical to the evaluation. The initial concern is whether the factory representatives understand the task as described. If a lack of understanding is uniform across all vendors, then the assignment was not written properly and should be rewritten and redistributed. It is imperative for the utility to know what is desired from a monitoring system and to leave the door open for more measurements to be added. The utility must understand the goals of the system clearly enough to convey them to the competing vendors.

Feedback from the various suppliers can now be assessed. Many of the questions generated indicate the level of interest and understanding possessed not only by the company but also by the individuals assigned to the task. The people assigned, and the interest they show, are critical to the overall success of the project. Note how complaints are handled by each vendor. The common sense of the response and methods for correction should be analyzed and documented. Factory support for any piece of specialized instrumentation is critical, as nothing is foolproof. How a vendor supports its products is a vital characteristic of the company and an indicator of the value of its sales claims. This portion of the demonstration adds that piece of feedback from the various suppliers to the total picture.

The maintenance support provided by a company during the demonstration is also an indication of what can be expected after the installation. There will always be problems beyond the expertise of the utility's field crews. It needs to be determined if a support team, independent of the sales force, will support the endeavor in the long term. Sales forces generally provide exceptional support to make the sale. The issue is what happens after the sale. This part of the analysis must be made separately. It is easy to determine the presence of a support function within an organization by the existence of toll-free numbers, contact people, and other indicators in the product literature. Such information must be segregated from the sales effort during the term of the demonstration, as there is no other reasonable source as accurate from which it can be procured.

The company should understand the various infrastructure systems in place for best fit of their products. Remember that instrument manufacturers do exactly that—manufacture equipment. The utility's main business is the management of the infrastructure. In order for manufacturers to make excellent products, they must understand the nature of the utility's business. They must understand the difference between a for-profit privatized utility and a not-for-profit municipal agency, along with the concept of a regional authority, which blends the two. This knowledge should be reflected in each and every conversation, letter, and field visit. Manufacturers do not have to know how the utility does its job. Equipment manufacturers and vendors need to know only what will assist their efforts as a function of their business plan. These concepts have to be forefront in their presentations. Vendors must be able to understand the nature of the work in relation to the business plan of the utility. This knowledge must also be reflected in all they do.

The general quality and construction of the instruments should be examined in terms of aesthetics, toleration of sewer environments, and external functions, where possible. The device should clearly be of high quality. The features that make it amenable to sewer environments should be examined. If the device has local indications, their functions and ease of understanding should be examined. If the device is not equipped with local indications, a reflection in the price to show the cost savings should be noted based on the elimination of visual displays. If the device has contact closures, their insulation from the sewer environments should be examined and noted accordingly. Utility connections should be noted for compliance to local electrical codes. The quality of a device is apparent to the trained eye. The guidelines presented assist the untrained eye to learn.

The instruments of choice generally have two types of software accessible to the user: an interrogation and setup program and a data acquisition and utility program for data handling and display functions. The data acquisition program also can interrogate the instruments if so equipped and desired. Software makes for an interesting evaluation as the computer knowledge base grows. Individual preferences generally are for what is familiar, and so the large amount of software on the market makes for interesting combinations of experience within a diverse population group. Be sure to evaluate software not on its degree of familiarity but rather on its ease of use and the features offered. It is difficult to remove personal bias from this evaluation, but it must be done.

The examination of the software should begin with its ease of loading. The software should not be hard to load on the computer of choice, assuming the computer meets specifications required for the software's use.

Few keystrokes should be needed to accomplish the task and minimal effort required to return to either the previous menu or the main menu. Generating graphs and charts should be relatively simple and the movement of data through the system straightforward. The look and feel of the software should not be complex. An export function should easily move the data to a spreadsheet for further analysis. Software should not require a special type of machine; however, this may be an internal desire at some point. The software should be able to coexist with other programs without difficulty.

The nature of technology expansion demands consideration of accessories and upgrades issues. As the demonstration program moves forward, questions and concerns about events that may or may not happen always arise—hence, accessories and upgrades. The manufacturer should be able to project a genealogy of products, services, and accessories that will support and provide paths for future system expansion where applicable.

An area of interest is the allocation and division of memory. The ability to partition memory or otherwise maximize its use through either static or dynamic allocation based on site conditions and setup functions is an asset to the device. Such features as these should be noted on the evaluation because they enhance instrument performance.

This section has addressed the peripherals rather than the core of the verification. However, all these functions are necessary to determine whether or not the equipment has value to the utility. Assessing them provides a basis for fair comparison. If a device cannot rate satisfactorily at this level, the next discussion is not necessary. If a device does well to this point, then the next parts of the analysis—verification of data, and the cost to own—assist in making the final cuts.

7.13 PRACTICAL EVALUATION OF TECHNOLOGY: SITE VERIFICATION

Site verification requires extensive fieldwork. This is only for close and intense evaluation because of the expense involved in a confined space entry. Investigating the devices in the field and observing their behavior under sewer conditions are the best ways to learn about their life expectancy and performance. This section describes the steps required to verify the accuracy of devices, and the issues that arise in the analysis.

The calculation of open-channel flow is as much art as science. This is due to the unsteady and ever-changing nature of the flow itself and the environment in which it is transported. The sewer is the place where undesirable materials accumulate, and the material in itself makes the calculation of true flow difficult. In performing the analysis described in this section, it is necessary for these issues to remain at the forefront of attention.

The program for analysis should be of a definitive time frame. Regular site visits, including confined space entries, should be scheduled at the convenience of the maintenance crews. They should be held at random times of the day to yield a spread of verification times to compare. Regular and frequent visits are required in order to make the functional determinations of equipment operations. The data retrieved also assist in developing the cost models, as frequency of maintenance is also a factor in cost.

Length-measuring devices, such as a tape rule and string, and a portable velocity meter of known accuracy, will be required. If the instruments have not been certified for verification purposes, send them to an independent and recognized calibration laboratory and have them certified, providing the certification document to demonstration program participants. This establishes a ground for fairness and proof of that intent.

Schedule as many field visits as the field crews can manage over the term of the demonstration. Also, schedule visits such that the locations can be observed as soon after a storm event as it is safe to enter the sewer system. Establish a method of tabulation, including time and date, and record the data desired for each location. These documents must be kept so that participants can review data pertinent to their own devices if they like.

Site verification, in this discussion, includes velocity and level, plus a flow calculation. This approach reduces the time spent within the sewer system at each stop. Level is measured using the technique described in the section on temporary meter setup. A subtracted difference is used in order not to introduce a fishtail in the level measurement. A single point of reference is utilized, whether an ultrasonic head or a ladder rung, for all the measurements, and it should be known to all parties.

The velocity is a different issue, as the velocity probe must be, in fact, inserted into the flow. The point of reference must be the same for all test locations, which is just beneath the surface of the water. By keeping to this standard reference point, the measurement can be applied equally to all test locations. Velocity is the parameter that is to be verified based on the velocity measurement read by the vendor's equipment.

There are many approaches to measuring velocity in an open channel. The method described here provides a reasonable representation of the velocity without extensive profiling of the various layers. This method is recommended in order to minimize the time spent in the sewer. The reading can be made and then the crew can exit the location quickly. This allows many sites to be evaluated in the shortest amount of time.

The following technique applies to devices that measure the entire velocity profile, such as Doppler area velocity instruments. The idea is to position the velocity sensor just below the surface of the water, as close to the centerline of the sensor as possible. That velocity reading should then be reduced to 0.9 of the reading as observed in that layer of velocity. The value obtained is a reasonable working equivalent of the average velocity as generated by the entire velocity profile. If a point velocity measurement is used, the probe must be placed at the same depth and relative location as the installed sensor in order to read the same layer. Then the velocity probe is inserted just below the surface so the computed velocity can be determined. Thus, for Doppler devices, a one-point measurement is required, and for the point velocity probe, a two-point measurement is required. To measure transit time, the probe should be inserted in between the acoustic couplers during the measurement phase so the swath of velocities is recorded with no reduction factor, then again just below the surface for the average measurement.

The procedure follows in this format. Be certain that the device can be programmed to record the various velocities used in the computation. The measured velocity, as seen by the sensor, is referred to as the *raw velocity* and the computed average velocity as the *final velocity*. Doppler devices directly compute final velocity as an average of readings in a range of frequency shifts. The result is a single value. This device, then, has only one reading to display. Most others display both the raw velocity and the final velocity used in the flow calculation. Both values can be applied for the verification. The instruments should be programmed, where applicable, to record raw and final velocities for interrogation and display. Make the time-stamped readings either one point per Doppler or two points for the others, as noted. Compare these readings to the time-stamped printout after the device is interrogated.

This method provides a time-stamped record of the velocity measured by the instrument and a time-stamped value for the velocity measured by the verification crew. The comparison of these velocities is an issue for discussion. The readings have little chance of being exactly the same because of the constant surges and troughs in the flow through sewer systems.

Therefore, leeway is required in the verification process to account for the time difference between the instrument sample and the hand reading. Also, handheld velocity probes tend to have some dampening, which limits the amount of movement seen at the probe head. It is therefore difficult to hold a manufacturer to, say, a +/– 2% range. Use common sense. If a reading of 1 m/s is recorded on the instrument and the handheld probe registered 0.9 m/s, this may well be accurate enough for the verification even though it shows a 10% error.

The reason for this allowance of error is because the many layers are constantly slipping around and over each other in an open-channel environment. They tend to be non-uniform in their boundaries. The relative level of the water volume is constantly changing over the range of measurement and because of the constant mixing of these layers. In contrast, laboratory conditions are controlled, and the reference point will be uniform over a large variety of tests. Because the sewer has no controlled conditions, the relative error appears larger than it actually is. Also, because of the range of numbers being considered, a small change can result in large apparent error.

Taking into account these considerations, extract the final velocity and perform a flow calculation based on the cross-sectional area. Use the continuity equation. This can be done by computing the area of the cross section, then verifying the flow between the calculation and the instrument reading to ascertain if it is relatively consistent. Such tables can be generated using standard spreadsheets. Tabulate the instrument readings from the calculated values, and generate a difference table. There should be relative agreement of the values read by the instrument and the calculated result. Note the term *relative.* Observe for other issues as well, including velocity direction and radical drops and rises in velocity that have little hydraulic basis.

The analysis should then examine the numbers generated. These should appear to agree with known site conditions. Obviously, the closer the numbers are to known quantities, the better the device. Exact values are always the goal, but this is certainly not expected. The examination of the comparison points should provide assurance that the instrument is properly reflecting the conditions of the sewer. The results should be charted to show their relative closeness to or distance from the data obtained. A percentage of error can then be generated.

The table can be part of the final report used in the evaluation. The approach taken to compiling the values is up to the utility performing the analysis. However, the ability to take a point or two per day and make verification claims enhances the process of making equipment choices, especially of equipment that has substantial associated costs.

The process also includes observing site conditions as a result of the introduction of equipment. As the velocity and level verifications are taking place, the crew should also take notes on the conditions of the instrument, such as debris accumulation, grease buildup, and any other condition that would inhibit accurate measurements. From the examination over time and the observation of debris buildup, an estimate of the number of maintenance calls required per month or year can be made. If the sensor is susceptible to deposition and solids accumulation, then the number of potential maintenance calls increases. If the design resists such buildup, maintenance requirements decrease.

The validity of this method for verification appears a bit tentative. However, the nature of open-channel flow is so variable and dependent on conditions that these must be accounted for in the final analysis. Once the variability is understood, then the analysis can commence with closer tolerances based on localized site conditions.

7.14 PRACTICAL EVALUATION OF TECHNOLOGY: COST TO PURCHASE/COST TO OWN

The discussion now turns to cost. This is the major interest of the financial appropriations team. The ability to show these costs in a standard format increases the potential for system approval and for proper cost allocations over the life of the field equipment. Finance is often overlooked by the field engineers and technical people, who are mainly concerned with the quality of the program. Cost-to-purchase/cost-to-own models quickly segregate the long-term expenditures for monitoring devices and are invaluable for other types of life-cycle evaluations as well. The variation of this model presented here is referred to as the *simplified cost-to-purchase/cost-to-own-model.*[1]

The simplified cost-to-purchase/cost-to-own model is provided because it addresses many issues that must be incorporateed in financial analysis. The particulars of local government and issues regarding international governing agencies, along with risk management issues, are unique

1. The Cost-to-Purchase/Cost-to-Own Model was developed by the author for American Sigma. It is the property of American Sigma and used by permission.

to each model situation and can be added to the model to provide a full and complete financial analysis.

It costs more to operate and maintain field equipment than it does to make the purchase. The majority of the costs are indirect, that is, costs already spent and allocated within the utility. However, indirect costs can grow into direct costs if extra resources must be committed to the endeavor. Minimizing total cost is the goal of employing the cost-to-purchase/cost-to-own model.

The cost to purchase comprises the actual costs associated with the procurement, installation, and startup of the monitoring devices—the contract cost for the final purchase. The cost to own comprises the costs associated with the short- and long-term maintenance of the devices as a whole system. The parameters associated with this portion derive in part from a generalized understanding of system operations and in part from site-specific requirements. This means that costs such as labor and materials are required and expected, but the procurement, overhead, and labor contract requirements are local costs that must be independently accounted for in the analysis.

Understanding the concept of models is a prerequisite to constructing them. A model is a representation of the potential full actual costs or operating concept cost of construction which illustrates or estimates total costs. It simulates the conditions of interest; conclusions drawn from the analysis represent projected outcomes of actual conditions. Models represent that which is known about a situation and provide working parameters for development of full programs. The advantage of using models is that hypotheses can be tested without constructing the live system.

The advantages of models are also their drawbacks. A model begins with what is known about a situation and enables analysis of additional factors introduced into the simulation. The drawback is that it cannot address unanticipated conditions. This is the reason that many medical and biological models fail; the vast number of complex chemical reactions yet unknown cannot be accounted for in the model. The point is that the results of analyses bases on models must be carefully interpreted and correctly understood.

The task here is to generate a simple model as an illustration and a guide for developing situation-specific simulations. By stepping through a series of concepts and examples, a model can be constructed that takes into account the individual utility's requirements and regulations.

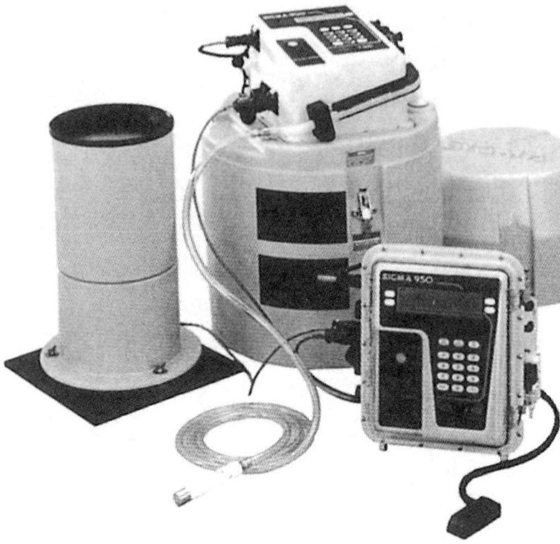

Figure 7-4 An integrated solution. In this situation, all the hardware is matched and can share data in the same format. Such systems are desirable from a maintenance and repair viewpoint.

The financial model described here concerns the cost to purchase, own, and operate a system of flowmeters in a small sewer district. The identification of the parameters is important because they structure the analysis and inform the interpretation of the result. The less a result has to be interpreted, the greater its value to the utility. This is due to the fact that the resultant has less a margin to be misinterpreted and fitted into the table of desired results.

The model begins with one standard street crew and the necessary resources for safe operations. In all these examples, standard U.S. safety regulations and funds are utilized. They can be translated into any system once outlined. Under these regulations, four workers are recommended for confined space entry: top person, radio operator, flagman, and entry person. However, most municipalities still use three (top person, radio operator, and entry person), so three are utilized here. Any other personnel required by local regulations can be added accordingly.

The typical sewer maintenance laborer is assigned a wage of $12/hour for this exercise. The foreman of this crew is assigned $14/hour and the sewer crawler $10/hour. A set figure of 30% overtime and a multiplier for overhead, workers compensation insurance, and benefits of 1.5 times the salary expense is assigned as well.

The following resources support the crew: The vehicle and equipment calculations describe a standard crew truck of 10 tons, with complete confined space entry equipment. The vehicle itself leases for $1,500/week with an additional $500/week in operations and maintenance cost, including standard tires, batteries and accessories, and fuel not covered by the lease. A flat rate of $80,000 for equipment and supplies per year is used to describe all equipment, upkeep, gas meter repair, and loss replacements. This calculation yields the weekly figure:

Vehicle/Equipment

Operations	$1,500/week lease + $500/week operations	$2,000/week
Tools/Equipment	$80,000/year / 52 weeks	$1,540/week
Subtotal		$3,540/week

Resource Allocation

Foreman	1 × $14/hour × 40 hours × 1.3 (overtime) × 1.5 (benefits)	$1,092/week
Workers	2 × $10/hour × 40 hours × 1.3 (overtime) × 1.5 (benefits)	$1,560/week
Vehicle	<calculation above>	$3,540/week
Subtotal		$6,192/week

Many of these costs are not actual but assigned for the calculations. However, because they are held constant for the entire analysis, the assumptions affect only the fit of the result. The $6,192/week figure can be applied across all of the equipment to be purchased for uniform analysis. It only requires adjustment for the local adjustments of wages and peripherals. The calculation yields a uniform cost of putting one crew on the street for a week. The multiple of a three-worker crew on the street using the model value of $6,192 is valid for discussion. Every time a new crew is assigned to the street, under this model, the time frame is per year and the model period is per week.

A maintenance crew responding to the pre-existing conditions described requires setup and breakdown time. This model assumes 15 minutes to set up or break down a job site in any area. It is understood that heavy traffic areas require up to 40 minutes and that off-road low-flow or shallow areas require less. The details include cones, gas detection, and equipment setup. The average work crew can do four minor inspections and cleanings per day, including setup, breakdown, and transit between job sites. These figures allow another frame for the analysis. If the costs associated with public liability insurance are ignored, as most municipal agencies are self-insured, a cost per site visit can be generated:

Site Visit

Inspection/Repair	\$6,192/week / 20 jobs/week	\$310/job

This figure represents the uniform cost for performing one minor inspection as noted in the parameter set. This cost can now be applied to the analysis of the cost to purchase/cost to own. It also represents the maintenance cost to perform minor work at a site; this cost is likewise carried over to the analysis.

The equipment is now considered in terms of cost to purchase. A selection of equipment is tabulated for consideration at various prices.

Cost to Purchase

Meter System A	\$4,500 × 6 units	\$27,000
Meter System B	\$4,300 × 6 units	\$25,800
Meter System C	\$2,000 × 6 units	\$12,000
Meter System D	\$4,200 × 6 units	\$25,200
Meter System E	\$8,000 × 6 units	\$48,000

This shows a range of what the specification might be in regard to similar equipment. System E is four times the price of system C. Three of the systems are similar in price. System C is the extreme low. Immediately the specifications should be reviewed to determine if any errors were made in the analysis. If it can be confirmed that all of the specifications are equal, then the process can continue.

Because, for the purpose of this analysis, all the devices are considered equal here and the points of interest similar, the installation costs can be held equal as well. Cost to own includes the maintenance and repair of the devices. Their life cycle is held constant for three years, in this example, to focus on the maintenance and repair issues.

Periodic Maintenance

	Interval	*Quarterly Site Visits*	*Cost of Maintenance*
Meter System A	quarterly	1 × $310	$310 × 4 = $1,240
Meter System B	weekly	12 × $310	$3,720 × 4 = $14,880
Meter System C	monthly	4 × $310	$1,240 × 4 = $4,960
Meter System D	weekly	12 × $310	$3,715 × 4 = $14,880
Meter System E	monthly	4 × $310	$1,240 × 4 = $4,960

The figures are estimates for periodic maintenance, which excludes any equipment failures or unplanned maintenance per year. The interval is based on manufacturers' recommendations and actual experience in keeping the equipment functioning. Note how quickly the costs increase for a device that requires large amounts of maintenance, regardless of the purchase price. The periodic maintenance per meter is evidently a large determinator of cost. The more site visits required, the more resources must be commited and the greater the cost to own the device. It is clear that the purchase cost of the meters will have little impact on this calculation.

Next, the cost to purchase is added to the partial cost to own.

	Maintence First Year + Cost to Purchase
Meter System A	$1,240 + $27,000 = $28,240
Meter System B	$14,880 + $25,800 = $40,680
Meter System C	$4,960 + $12,000 = $16,960
Meter System D	$14,880 + $25,200 = $40,680
Meter System E	$4,960 + $48,000 = $52,960

Notice how Meter System C is in a class by itself and is no longer comparable to the rest. Also, A and B are no longer similar in cost. This is only the first year. Here is the calculation over three years.

	Maintenance over 3 Years + Cost to Purchase
Meter System A	($1,240 × 3) + $27,000 = $30,740
Meter System B	($14,880 × 3) + $25,800 = $70,440
Meter System C	($4,960 × 3) + $12,000 = $26,880
Meter System D	($14,880 × 3) + $25,200 = $69,840
Meter System E	($4,960 × 3) + $48,000 = $62,880

Observe how System B is the most costly due to the periodic maintenance costs. Systems A and C are now comparable; C is no longer in a class by itself. Systems B, D, and E also are comparable. The lesson is that the cost to maintain the equipment far outstrips the cost to purchase it. Systems that appear the most costly at first may not be if their maintenance requirements are low.

Now the electronic maintenance costs are added. This will also change the system costs quickly. A three-year period is considered for maintenance and repairs. Systems A, C, and D have a one-year warranty. System B has a 90-day warranty. System E is sold only with an on-site maintenance contract. Again, uniformity must be ensured. The three systems with one-year warranties have no assigned first-year maintenance costs and 6% costs thereafter. The system with the 90-day warranty is assigned 6% costs. The contract is assigned 20%, based on empirical information. The System D units carry a $200/year calibration cost in addition to the repair costs.

3-Year Maintenance and Repair Costs: Electronics and Calibration

System A	$27,000 × .06 × 2	$3,240
System B	$25,800 × .06 × 2.75	$4,257
System C	$12,000 × .06 × 2	$1,440
System D	$25,200 × .06 × 2 + ($200 × 3)	$3,624
System E	$48,000 × .2 × 2	$19,200

Note that System E was a service contract, so the number of entries required to do the maintenance was added. Therefore, an estimate is required to bring all the other systems into line with the contract price. The calculated cost of a confined space entry is $310. A certain number of these entries must be added to the electronic maintenance cost of each system. System D requires three because the device must be removed for calibration, as noted. Systems C and D have sensors that are part of the meter; therefore, any repair requires that the meter be swapped. For this case, two confined entry costs are assigned. If the electronics can be disconnected from the sensors, one confined space entry per service call is allocated. Because all electronics fail periodically, a certain number of calls is required per year. Two calls are assigned per year per device. Systems A and B have only one confined space entry because their sensors can be disconnected from the electronics. Systems C and D have two costs per call because of the added work to pull both sensor and electronics from the site. System E has no costs because of the service contract. System D has two extra calls per year because of the calibration required by the manufacturer. This is performed for three years.

3-Year Site Visit Costs

System A	\$3,240 + (\$310 × 6)	\$5,100
System B	\$3,750 + (\$310 × 6)	\$5,610
System C	\$1,440 + (\$310 × 12)	\$5,160
System D	\$3,624 + (\$310 × 12) + (\$310 × 6)	\$9,204
System E	\$19,200	\$19,200

Observe that no longer does the least expensive system have the least cost. The model shows that Systems A, B, and C are about equal when the costs of maintenance are considered. The costs associated with System E are difficult to interpret because maintenance contracts whereby all service is covered have advantages in manpower savings and other intangible benefits that are difficult to reflect in these models. However, if the budget of the utility is geared to contract operations, the full-service or contract area may well be cost-effective by keeping other resources free for emergency maintenance. The policies of the utility must be taken into consideration, as the cost for the System E contract may be in line with a contracting policy.

The break point, as designated here, is the threshold where the maintenance costs become so large that extra responsibilities cannot be added to existing maintenance personnel and more must be hired and equipped. If the model crew can enter four sites per day each day, excluding holidays, and an allowance of 10% is assigned for entries failed because of weather, hazardous gas, or equipment problems, the following occurs:

$$4 \text{ sites/day} \times 5 \text{ days} \times 52 \text{ weeks} - 10 \text{ holidays} - (10\% \text{ failed entries}) = 900 \text{ site visits/year}$$

Potential Site Visits per Year

	Interval	*Site Visits*	*4 Quarters Yearly Site Visits*
System A	quarterly	1 × 6 meters	6 × 4 = 24
System B	weekly	12 × 6 meters	72 × 4 = 288
System C	monthly	4 × 6 meters	24 × 4 = 96
System D	weekly	12 × 6 meters	72 × 4 = 288
System E	None <Contract>		

	Site Visits/Members	*Meter Purchase*	*#Meters per Crew*
System A	900/24 = 37.5	6	37.5 × 6 = 225
System B	900/288 = 3.125	6	3.125 × 6 = 19
System C	900/96 = 9.38	6	9.38 × 6 = 56.25
System D	900/288 = 3.125	6	3.125 × 6 = 19
System E	Contract Option–No Entries		

The analysis shows that, based on the maintenance intervals, a new three-person crew and equipment is required for every 225 meters purchased for System A, 19 meters for System B, 56.25 for System C, 19 for System D, and none for System E, because of its full-service contract. Field maintenance time has the biggest impact on cost to own. The greater the failure rate of the device, the increase of the field service and the increased cost. This is why the demonstration program is so important, as hands-on maintenance is experienced and those empirical values can be accurately added to the model.

The break point identifies when additional crew is required for budgeting purposes. Any time a crew is added, the yearly cost is ($310 per visit 900 visits = $279,000) for the three-worker crew, truck, equipment, and resources. This is why the break point is significant. If, with the purchase of lots of 19 devices, the associated costs rise by $279,000, this makes the devices relatively expensive. If the costs rise $279,000 for every lot of 225 devices, this is relatively efficient. Note the term *relative.* This analysis also puts a different focus on the contract option, which now appears quite economical.

The numbers presented here must be adjusted for the particular utility. Once this is accomplished and any contract estimates are included, the model takes on a greater relevance to the utility operator. Confined space entry costs, the cost of vehicles and the manner in which they are procured, resource allocations, and other factors affecting the systems and field crews affect the uniform and variable costs. The amount of investigative work performed to generate accurate numerical sets determines the accuracy of the cost model.

The cost-to-purchase/cost-to-own model is a valuable tool in generating ratios and projections per planning for the cost to purchase and operate infrastructure monitoring systems. The ability to determine which costs to make uniform and which costs to vary is the key to accurate interpretation of the results. If contract work is to be considered, all the functions that a maintenance crew would perform should be accounted for in the contract price in the same manner that the in-house calculations are determined. Such calculations provide values that can assist in the decision process.

7.15 SUMMARY

The selection of any equipment bound for use in an environment as hazardous and as unpredictable as the sewer infrastructure must be well informed and carefully analyzed. There are many methods for making these evaluations; few have definitive structure. The goal of this chapter is to provide a structure for a systematic analysis.

It is important to know exactly the system into which the equipment is to be introduced. It is equally important to realize that one kind of device will not work in every environment. Although many manufacturers claim this, it simply is not true. Too many variables exist at each and every site and in each and every sewer system to allow such a claim.

The variety of available devices and peripherals, and the uniqueness of the systems themselves, tend to require demonstration of the devices and their applicability to these environments. Manufacturers are generally willing to participate in the demonstration as long as they are treated fairly and logically and as long as the utility displays a genuine desire to understand and learn. Manufacturers treat such projects as learning endeavors themselves and shape their product lines accordingly.

Hardware evaluations can be performed methodically. By partnering the supplier during the demonstration, the utility learns what its desires should be based on the best available technology or the most appropriate technology for the particular system. Depending on the size of the job, much of the up-front engineering may be provided by vendors in their own quest to improve. This cooperation assists all concerned; the result is the best blend of products and services for the utility and its customers.

The cost-to-purchase/cost-to-own simplified model is a way to consider the financial aspect of the selection. The financial picture is of great interest to budget professionals who neither want nor need to know the technical aspects of a project; their concern is the bottom line—that is, cost versus efficiency. The model presented in this chapter enables technical and financial experts to communicate.

There are many facets to procuring a sewer infrastructure monitoring system. All are critical to the process and none can be left out. The technology application is such a small and specialized area that few resources are available to provide guidance. Another issue is the uniqueness of each sewer system, which leads to the modification of every concept to accommodate these characteristics. This makes the bottom line a complete understanding of the utility system, so all the uncommon aspects of each point of interest can be properly accounted for in the final product. It is therefore important to plan, test, budget and model all facets of a monitoring system to find the best system fit and reduce the exposure to failure from a less than adequate design.

CHAPTER

8

Use and Applications of the Data

8.1 INTRODUCTION

The result of all the effort is the use and application of the data generated by the newly installed system. For this discussion, it does not matter if the system is temporary or permanent. It *does* matter that data are retrieved, delivered to a central storage location, and processed for use by the utility on behalf of its customers.

The information obtained justifies the system that produces it. If the data meet the needs and expectations of the user, then the system is perpetuated. If the data have no meaning to the user, the system will be abandoned and the expenditures considered an inappropriate use of funds. The best data, with the highest integrity, prepared in impeccable style, are valueless if not structured to meet the requirements of the user. Data presentation requirements may be simple so that fancy or flowery presentations simply get in the way. This often happens if the data are targeted for other need-driven analytical programs within the utility. Managers, however, often prefer presentations with flair and style.

Regardless of style, data should embody good, accurate information. If the data is compiled and tabulated well but is inaccurate, the presentation is irrelevant. The line between accurate and inaccurate data is so thin that it is difficult, at times, to discern. This makes the recording interval, time average, and other factors in which data are collected as important as the presentation.

8.2 TIME SLICES AND RECORDED INTERVALS

It is necessary here to determine how the data were recorded and why. The installation itself is not an issue. This discussion covers only the movement of data from the sensor to memory for interrogation—the electronic handling of data. (Particular time slices and recording intervals are possible only for some technologies; this is explained below.)

Under current technological limitations, there is no such thing as a continuous recorded reading. All readings are discrete, which means they are independent of one another. The boundary between the start and end of a reading is defined because a computer can hold only discrete points in its memory. With chart recorders, the pen never leaves the chart, which is always moving. The reading is never interrupted, even if the instrument itself goes off line. This is a continuous reading. The advantage of a continuous reading is that every potential event is noted as it took place in the field. The most serious drawback is that the recording requires interpretation but the data cannot readily be entered into an analytical software package. Therefore, the data is only as good as it appears and cannot be manipulated without significant operator intervention. Discrete readings over specific time frames or intervals are necessary for computer analysis to be possible.

The time slice here is defined as the interval over which the reading is taken. The recording interval is the period over which the information is written to memory. These are potentially different readings. The instrument can measure a time slice of one second, with a five-minute average of those time slices written to memory. Both are required for proper memory management of the device. It is possible that the time slice and the recorded intervals are the same, which indicates that one point is moved with no further work required. However, the slice and the interval have particular functions applicable to the condition being monitored, and are uniform with current computer standards. This system, as is universal across all fields of instrumentation, provides discrete points for analysis or for entry into other monitoring or control schemes.

The time slice is the raw instrument reading before analysis is performed. Within the reading, one or many measurements can be taken and stored in an intermediary section of instrument memory. Generally, this is considered short-term memory; the better devices have management systems that can allocate memory on demand. At the time of programming,

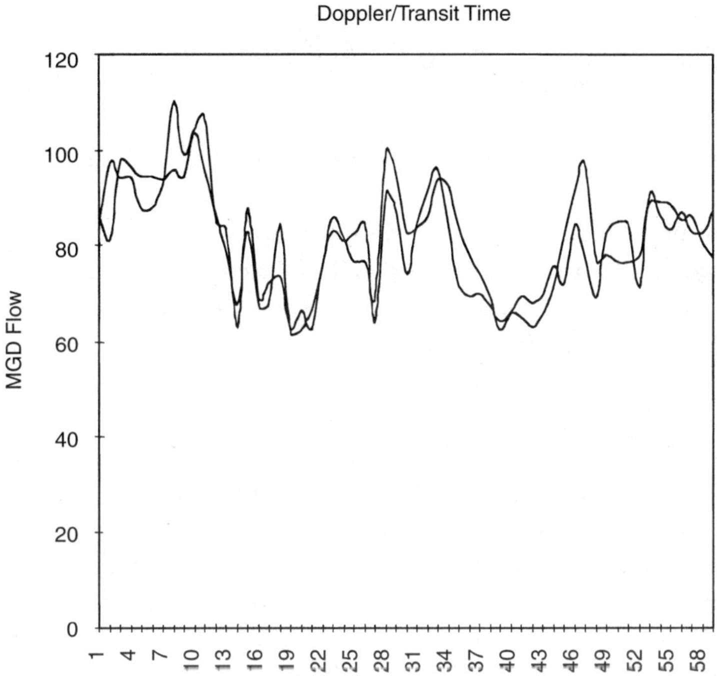

Figure 8-1 Comparison of transit time and Doppler under uniform conditions. The perceived difference is a result of the sample time and time slice.

the instrument's setup function allows the time slice to be entered. There may be many references to the time slice; therefore, the manual or technical support team must be consulted for the specific terms the manufacturers chose to use. The time slice is a function of allocated memory, power source, and accuracy. These are discussed as separate topics.

The recorded interval is the consolidation of time slices so the data is mathematically computed into the data point of preference. In many instances, this is the sampling interval, or the time that all of the slices are recorded in. *Sample* refers to the data as measured by the instrument. Samples, once converted to a value that is functional given the circumstances, have some statistical calculation performed on them to best represent the total time slices of record. This is an excellent method for recording rapid changes in hydraulic conditions while saving space in instrument memory.

In the example above, the time slice is a one-second data point and the recorded interval is five minutes. Because the interval is the same, the slices all have the same weight; therefore, a straight mathematical average can be performed to determine the point over any given five-minute time frame that accurately represents the interval. Some process schemes take slice measurements at predetermined steps or intervals. This is not the ususal approach in sewer management because of the limitations of field electronics.

Does more data storage provide more accuracy or simply generate a higher potential for error? It is always desirable to gather as much data as possible given the technological limitation of the equipment. As noted above, a pen recorder transcribes continuously and every activity is viewed. It is difficult for electronics to do this because they store only discrete storage points. Therefore, a compromise must be struck with respect to the amount of data to be stored. Averaging, peak recording, and other methods provide vital information with little threat to the storage area and no loss of important data. Knowing the outcome the utility wishes to see and the purpose of the data helps in reaching this determination.

Many field instruments are designed to take one time slice in the recorded interval. This is a power-saving feature that does not limit the functions of the device. With batteries the weakest link of the system, manufacturers work to economize on the electrical use of the device to lengthen the life of the cell. Therefore, in many cases, there may not be a clear delineation between the time slice and the recorded interval.

The data being generated need to be considered for proper determination of an appropriate time slice and the recorded interval. A recommendation is that all the recorded intervals remain the same at all sites. This is so the data sets are of uniform size, which may assist in the handling and presentation of the information. The time slice can place tremendous drain on the device, depending on the type of location and equipment. Therefore, the time slices can and should be varied based on the accuracy of the information required and the task being monitored. In the case of level measurements, the time slice can match the recorded interval and provide reasonable results. In the case of custody transfer and other critical locations, the smallest possible time slice should be employed to incorporate as much data as possible. If the site has utility electricity, the size of time slice and other storage concerns are not necessarily an issue.

The only circumstance in which the time slice is a major consideration is when power allocation is critical for the longevity of the location. The proper allocation of the memory is a function of the number of points stored and not necessarily a consideration for power. The ability to allocate primary and intermediate storage areas then can be seen as a great advantage to maximizing the resources within the electronics package.

8.3 MEMORY AND MEMORY ALLOCATION

Memory is generally allocated based on the operating parameters and software utilities that a manufacturer has incorporated into its equipment. Space for memory is limited by the physical area of the chipset used by the device. The memory is usually divided into three areas, although this varies by manufacturer. These three areas are called primary, intermediary, and final memory. The generic outline given here explains the convention and theories used by vendors. Generally, the same memory set used to process data is also used to contain the program that runs the instrument. For this reason, the method by which a device allocates memory is critically important to the function of the device.

The first region in allocated memory is a primary storage area, where all the raw readings are placed as they come from the sensor. This area stores only the readings for particular time slices. The readings, as recorded, are relational operators for the value desired. In the case of Doppler devices, the data are represented as a series of frequency shifts. In the case of transit time instruments, the data are represented as a series of time shifts. In the case of electromagnetic velocity, they are potential voltage. Algorithms applied to this data set generate intermediary values that bring the data representations more in line with the requirements of the final data product and the needs of the end user. In other words, the relational operators are converted into usable readings. These values are moved into the intermediary memory area. The size of this allocated memory area must be consistent with the programmed settings. The primary storage area may need to be much larger than the intermediate storage area. This is especially true in Doppler readings, as many shifts per reading are moved into the primary area for conversion.

The intermediate storage area is where values are organized. In many cases, it is the place where the time stamp is applied after the statistical manipulation of the values. Generally, the only function that occurs in the intermediary storage area is the average or tally, which is written, in turn, to final or long-term memory. Depending on the architecture of the device, the values may reside in the intermediate memory for a specified time frame until blocks of them are ready for movement into the final storage area.

Final storage is the area directly accessed in interrogation. Information taken to the data depository comes from this area. There are two methods of final storage tabulation of the data. The first is wraparound, where there is a constant movement of data, the most recent replacing the oldest. The data flows, or wraps, so a historical trend is constant within the device. After interrogation, there is no need to erase or otherwise clear the memory. The data pointer is generally a time stamp or other identifier attached to the information. Wraparound is probably the most common form of final memory assignment used in field instruments because the small backup of data means that in the event that a problem arises in data transfer or central processing, the device need only be reinterrogated.

The second form of final memory allocation is fixed or fixed placement, whereby when the memory allocation is filled, the device no longer records data. After the instrument is interrogated, the memory must be cleared and reset. Generally, a data pointer tracks landmarks in the memory stack. The operating software requires knowledge of the pointer in order to know where the data stack begins and ends. The filled memory either simply stops taking data or issues a "memory full" message. Fixed storage is useful for time-sequenced events where a sequence of events of interest is defined. The fixed-storage record retains the readings when there is a time delay in retrieving the information.

Memory allocation should always be performed by the instrument software and not by the end user, because the operating program and mathematical routines must also be placed in memory and only the manufacturer knows how big these programs are. The better devices allocate memory automatically, so little operator intervention is required. The operator generally does not have direct access to either primary or intermediate storage.

8.4 STORAGE OF THE DATA

Field interrogation is always determined by the manufacturer, who provides the proper tools to interrogate and reset any parameters of the device. The data, once returned to the central depository, must be labeled by site and date so retrieval can be performed readily and efficiently. A naming convention must be used to denote each site uniquely. The naming of a site by its cross street tends to lack uniqueness as there may be many points along that street or on another critical path, which may lead to duplicate names.

The naming convention can consist of the street that runs along the critical path, plus a corresponding number. This is acceptable as long as there are no other paths on the same street. Some conventions simply increment the points by number with a cross reference. Others use the first initial of the receiving waters with incremental numbers. Others utilize the nearest local street with incremental numbers. The convention selected does not matter as long as it will expand to accommodate the growing system.

A feature on some software allows the header, or site information, to be placed directly on the data file at the time of interrogation. This is a valuable help in maintaining the organization of the data and simplifying the interrogation process. An automatic file-naming provision also simplifies the process. The site name and identification are entered to the instrument at the time of setup; this header is extracted with each interrogation. The ability to maintain large file structures becomes imperative as the extent and coverage of the system increases.

8.5 DISPLAYING THE DATA

A pictorial representation of the information is critical in the initial analysis of the data retrieved. The graphical interpretation provides the basis for pattern interpretation and the development of characteristic information about the location. However, a style and type of generic graph should be chosen for the base analysis that best suits the data displaying requirement of the task at hand and the utility's overall preference.

The manufacturer's proprietary software is one source of graphic formatting. If the manufacturer's software is equipped with a graph generator, the data is presented in a manner potentially fitting the data in the best possible venue for the circumstance. Vendor software is geared for ease of use and requires little technical ability, thus placing it within reach of many users. It is generally the safest and most realistic approach if the utility is committed to one system. This provides a more complete system match, allows for a more holistic approach to system management, and is often the most economical solution. The software should be able to export data to commercially available word processing and spreadsheet applications.

An extensive and diverse system may not lend itself to a manufacturer's proprietary software solution unless the internal protocols are known and can be included in the data processing algorithms. In this case, a third-party package is required. Many excellent solutions exist, including packages that run on any computer platform. The requirement is that the protocol for the interrogation can be obtained from the instrument manufacturer and successfully incorporated into the product of choice. The graphing functions and interrogation systems of these third-party packages require more expertise to operate than do manufacturer's software solutions and may even require specially trained staff to manage the effort. The dedication of such personnel may or may not be possible under the current managerial scheme of the utility, so this option should be considered with this in mind.

Data can be displayed consistently; the uniform approach encourages parties of interest to obtain, view, and utilize the data retrieved from the field.

8.6 TRENDING AND TREND IDENTIFICATION

The initial method for identifying hydraulic characteristics of the point of interest is pattern identification. This is a tried-and-true method in the process control world for determining steady-state conditions, then observing the changes from steady state. Pattern identification can distinguish the characteristics of a system and its responses to weather and man-made events.

The ability to identify a characteristic trend of a location based on its hydraulic response is the beginning of complete understanding of the infrastructure. The change in the steady state, as recorded in the pattern, reveals the distinguishing aspects that make each site a unique entity. These trends allow the system operators to plan, modify, or reconstruct the location in the event of deficiencies that may cause problematic conditions within the critical path.[1]

The goal is to collect a series of trends, study them, then apply specific rain data to observe the response. Storms of similar time, duration, and intensity should produce similar trends within the point of interest. Observe also, if possible, the upstream and downstream conditions to determine the effects on and of the critical point; correlations are possible.

Here is a trend analysis procedure:

1. T. Day, "Sewer Monitoring Information for Proactive Infrastructure Management," New England Water Environment Association, Marlborough, Massachusetts, September 1996.

1. Identify a trend in the data from a 30-day period.
2. Note all the rain events over the same period.
3. Correlate these two sets of data.
4. Note the following information to identify lag time:
   - The time each rain event was first recorded
   - The time each rain event first registered in the sewers

Lag time is important in understanding the characteristic loadings to the sewer infrastructure, especially with respect to the first flush after a prolonged dry spell. The lag time, called the *loading lag*, can be calculated along the critical path to the treatment plant. The loading lag can provide the treatment works with adequate time to prepare for severe weather. Loading lags become loading leads in the event of snow meltdowns and hydrant abuse, as no events are recorded by the rain gauges.

The loading lag can be of assistance in describing issues involving the localized hydrologic cycle. It also provides substantial documentation to regulators of events not driven by precipitation. Most regulatory guidelines are driven by localized precipitation and this is defined in the localized hydrologic cycle. This becomes an issue when the treatment works are attempting to justify excessive flows via snow melt (there is a time lag between the snow and snow melt). The ability to document the changes in key points of interest along a critical path, along with evidence of temperature and snowpack, can assisy both the regulators and the treatment works in developing policy, regulations, and process modifications. Dry-weather responses demand different protocols than wet-weather responses for both the sewer infrastructure and treatment works. The pattern characterizing the hydraulic response to dry-weather events can be matched to that of similar wet-weather events.

This example is just one way in which trend identification can yield valuable predictive information. The match can have far-reaching effects on every aspect of the wastewater system. It is therefore necessary to be confident of the measurements made by instruments in the field. Trend identification also provides documentation of conditions that may not appear linked to the definitions used in regulations. In the above case, a snow meltdown could introduce flows to a treatment plant that exceed the dry-weather limit. In many cases, a rain threshold must be reached before an event is considered wet weather. Therefore, flows exceeding the daily average dry-weather flow require explanation. Pattern identification demonstrates that the flows experienced due to meltdown are comparable in intensity and duration to those generated by rainfall.

8.7 PATTERN IDENTIFICATION FOR HYDRAULIC RESPONSE

The first analysis of the data concerns the nature of hydraulic responses on the critical path, as noted above. Accurate precipitation information, readily correlated by date and time, is essential. The ability to line up these events by date and time is critical to identifying loading lags. By determining the loading lag at each point along a critical path, the loading lag to the treatment plant can be calculated. Determining the hydraulic response is also an excellent introduction to more complex applications of the data.

Each area of the infrastructure generates a characteristic response in both dry and wet weather. This response generates a pattern that is recorded by a monitoring instrument. This pattern is a representation of the hydraulic response of the point of interest. Study is required to completely understand the particulars of each location and the conditions that generate the response. For example, a location may possess characteristics that can change the profile of the critical path if certain conditions are introduced. A wet well near a pump station introduces flows if the well fills too high. A dam provides relief if the system experiences too much volume. Excessive volumes have effects that are not readily predictable by surface analysis.

The in-place equipment provides information about the hydraulic response to these conditions. The system operators expect that conditions introduced into the system have particular effects and responses. However, the changing infrastructure system introduces conditions that may not have been planned for in the system design. Changes to the local point have hydraulic effects up and down the critical path that must be considered at other points of the system. The goal of trending is to track hydraulic responses in the changing infrastructure as the system grows and develops.

The method involves developing a regular review of the patterns that the system generates, hydraulic point by hydraulic point. Whether the review takes place daily or weekly does not matter as long as the interval is consistent, for the general pattern for each location of interest will become apparent. Dry-weather steady-state conditions are the easiest to observe. The steady-state response is continuous and repeatable, and generally can

Figure 8-2 Flow measurement at a channel junction. Always measure the flow from the contributors and sum for the results. In this case, the hydraulic patterns below the junction at the point photographed make flow verification difficult.

be charted for a week on one chart. The changes over seasons based on the localized hydrologic cycle should always be considered and understood. This can be performed either by observing the weekly charts over a year or by sampling charts for weeks of known steady-state conditions. Note the level over the time frame and normal diurnal variations as they occur. Treatment works activities, which show up as disturbances in the general pattern, are easily identified and verified through plant records. Note also variations to the diurnal cycle that do not appear logical. The regular review and analysis of these trends is critical to understanding the dry-weather hydraulic response.

The wet-weather response is the next condition targeted for analysis. Consider that the dry-weather steady-state condition is now understood. The first major variation is wet weather; the wet-weather response should be examined in light of the severity of the storm recorded. Simply grade the recorded storms from less severe to more severe for this analysis. It is advisable to change the time boundaries to match the storm recorded so

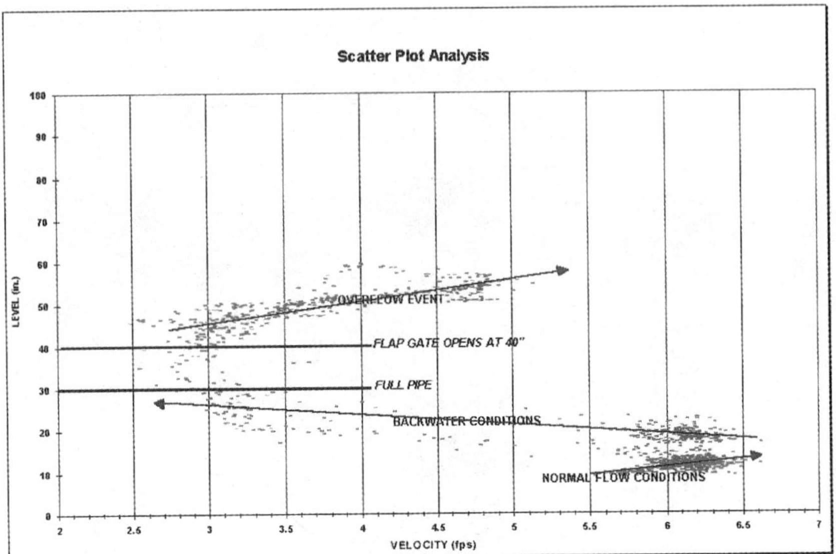

930 FM VERSION 1.02 PRIMARY DEVICE: AREA VELOCITY LOCATION: O/F # 25
SHAPE: ROUND PIPE SITE ID: 00000001
DIAMETER (ft): 2.5 SERIAL #: 000

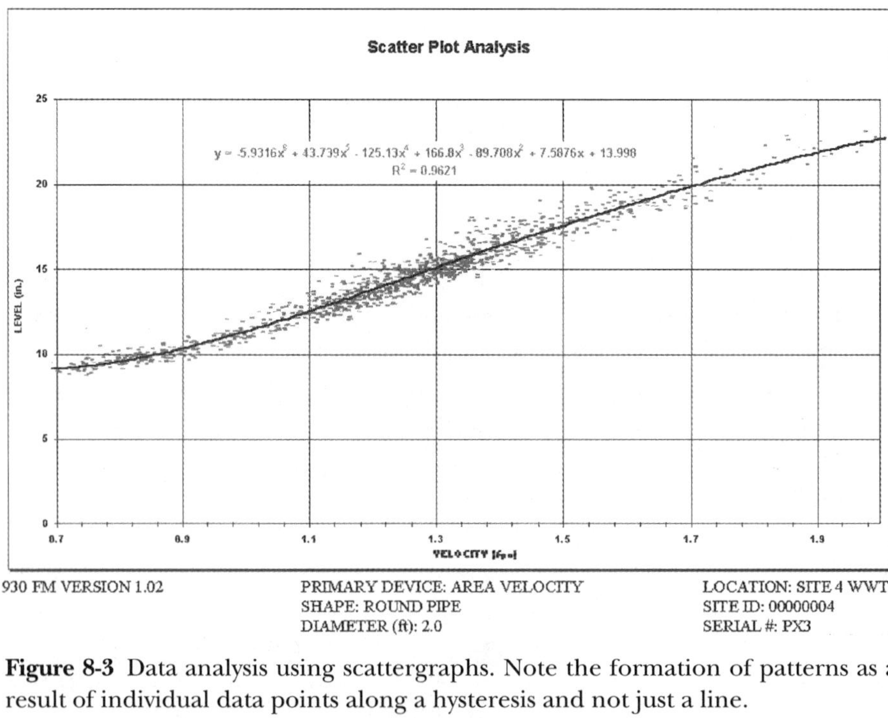

930 FM VERSION 1.02 PRIMARY DEVICE: AREA VELOCITY LOCATION: SITE 4 WWTP
SHAPE: ROUND PIPE SITE ID: 00000004
DIAMETER (ft): 2.0 SERIAL #: PX3

Figure 8-3 Data analysis using scattergraphs. Note the formation of patterns as a result of individual data points along a hysteresis and not just a line.

"dead space" intervals are eliminated in the data. The loading lag must be considered here; this is the gap between the time the precipitation starts and when it is first measured within the infrastructure. In sum, the data are analyzed using the time frame of the storm and with respect to responses to the storm conditions and intensities.

Identifying responses to wet-weather conditions is a formidable task. There are many ways in which an agency can classify a wet-weather event; forethought is required. It should not be necessary to classify every type of event, but simply to place them in some general category arrangement. The utility's engineering department may already have developed classifications that can be modified or used directly, depending on the application.

Tracking the changes in the structure of the sewer infrastructure is essential because the physical changes made to the system can alter the hydraulic response of the entire critical path in both dry and wet weather. When such changes are made, the resulting new patterns must be identified.

Identifying of dry- and wet-weather patterns is an excellent way to begin understanding the behavior of the sewer infrastructure as a result of changes, additions, and current conditions. Many patterns of interest are so idiosyncratic that generalizations are impossible. This is why it is critical for the response of each point of interest to be carefully understood and then grouped where applicable. The initial analysis is best performed with the dry-weather steady-state identification. The loading lag can then be calculated based on the beginning of the wet-weather event. Also, events that cause loadings to the sewers but are not tied to weather can be identified and tracked for regulatory and operational purposes.

8.8 PATTERN IDENTIFICATION FOR SEWER MAINTENANCE

The ability to determine and predict maintenance requirements is one of the bigger selling points for any sewer monitoring system. It also generates the fastest return on the investment in the system and its support resources. The ability to track changes in the steady-state response and to answer these changes with proactive maintenance activity leads to direct and positive investment returns and the most effective utilization of sewer maintenance forces. These activities can help minimize street searches and provide quick remedial action before relatively minor problems become bigger issues for the ratepayers.[2]

A proactive sewer maintenance operation requires the commitment of targeted maintenance crews and the tracking of their activities. Some graphs will be misinterpreted and crews may, in the beginning, be sent to the wrong sites. This is to be expected. Therefore, it is advised that one crew be targeted at first to undertake these endeavors and introduced to the concepts; later additional crews will be brought in one at a time.

Workers often resist monitoring systems out of fear that the systems are meant to eliminate jobs. The automation of the sewers does *not* mean a decrease in manpower. It simply means that the work is more efficiently assigned and is completed sooner, minimizing backlog. In any sewer maintenance system, there is a large backlog of work for any number of reasons, including finances and logistics—but not, usually, the inefficiency of the workers. The goal is to assist workers in targeting their activities while protecting the watersheds and the ratepayers of a community. This message has to be carried to the workforce as the system goes on line. There may, in fact, be no further growth in the maintenance crews, but this can be due to any number of factors, including automation. There certainly will not be a decrease in the workforce as a result of automation. The confidence of the workers and their ability to work with these systems is the key to making the maintenance system function properly and to the best interest of all concerned. Therefore, worker buy-in is critical to the success of the system.

A work order generated from the field instruments is known as a *computer pick*. A service plan is needed to implement and track the computer picks for work order generation from the interpretation of the patterns as printed on maintenance charts. It should be a simple matter to designate a field on the work order system, where applicable, for this type of tracking program. Once the maintenance crew is chosen and the work order system modified, then work can be assigned based on the computer-generated trends. The crew should be assigned work orders in the same quantity and manner as if they were generated from inspections or field complaints. At the end of the workday, the computer picks are tallied for according to the percent of correct picks, percent incorrect, and percent inconclusive.

A successful computer pick is a good interpretation of a change or

2. T. Day, "Applying SCADA-Type Technology to Sewer Maintenance" (paper presented at the Urban Wet Weather Conference, Quebec City, Quebec, Canada, June 1996).

other skew in the daily dry-weather trend. The deviation from steady state generated sufficient interest to send the special crew to investigate. Such deviations in the pattern would be obvious changes and would require little interpretation. Such changes include sudden rises in dry-weather levels, unexplained increases or decreases in a flow measurement, and fluctuating HGL measurements. All tend to be obvious and require immediate attention.

An unsuccessful computer pick should not be treated as a failure. This is a learning process and one in which judgment calls are based on observed behaviors of the hydraulics never before observed in this manner. Therefore, the unsuccessful pick should be investigated by the field crew to determine what caused it. Upstream considerations, changes in dry-weather flow, and structural modifications are all legitimate reasons for issues in the system. The tally should be cataloged so the pattern can be identified as *no issue* and filed for future training purposes.

An inconclusive computer pick is one in which evidence from the trend shows that there may be issues in the field and the investigation shows there may have been a problem, but evidence is insufficient to validate the issue. This happens when floating debris clogs a location but eventually breaks free. Large washdowns of solids are common, especially in the first flush after a prolonged dry spell. These blocks may free themselves naturally in the constant flow rate. Thus, the chart accurately recorded a problem, but on investigation the block does not appear. Evidence of trash and other debris may show that it once existed. These indicators are also important, as the continued washdown or repetition of such blocks suggest upstream conditions that may eventually require correction. A successful computer pick is one where the condition observed on the graph matches the condition at the point of interest. These also need to be tallied for analysis and comparison.

It should now appear evident that all three types of issues—successful, unsuccessful, and inconclusive picks—are important in discerning pattern recognition. Obviously, everyone prefers the correct pick. However, many times assumptions about circumstances are made that simply are not correct. Approaching a field condition with a limited viewpoint decreases one's ability to learn from one's errors. Mistakes should be considered part of the learning experience. They present opportunities to broaden one's knowledge base and to better understand how the sewer system as a whole behaves.

Using the monitoring system to guide sewer maintenance and pattern identification to characterize hydraulic response yields substantial return on the utility's investment. The system also aids in the constant battle with the general public, which often perceives utility crews as overstaffed—one laborer working and five watching. Those seasoned in the science of sewer maintenance realize why crews comprise six people; much of the time on the surface is spent searching for the problem, which may not be obvious even if sanitary or combined system discharge is evident. It is in this regard that sewer management systems serve the workers and their reputation as much as they assist the utility as a whole in efficiency.

Here the critical path approach comes into play. In this example, a flooding complaint is received. An emergency work order is generated as a result of the field complaint and referred to the designated crew, whose first stop is the computer room, where the nearby instruments are interrogated and the most recent patterns are drawn. The levels and flows of the surrounding points indicate the approximate location of the cause of the flooding. Levels are high and flow negligible behind a blockage; in front of it, levels are unusually low and velocities may be unusually high. These situations are easily seen on a pictorial chart and their location quickly determined. Naturally, it is difficult to repair a flooding manhole directly. The electronic monitoring system can pinpoint those upstream or downstream manholes from which the maintenance can be carried out. This minimizes the amount of time spent searching at the street level, economizes the efforts of the field crews, and, incidentally, demonstrates efficiency to the public, as all the crew is active.

Chart reviewers should be alert to patterns that should trigger concern. Remember that a developed baseline understanding of the historical dry-weather flows is necessary in order to make these observations. Unusually high levels for dry-weather conditions are the first clue that there is a line blockage or collapsed sewer line. It can also indicate a sudden introduction of grit or silt into the line, which leads to artificially high level readings although the aggregate amount of water being conveyed does not increase. Such depositions generally require line cleaning and remedial action to determine the cause of the deposition. Reductions or reversals in velocities also indicate problems that require correction. Here, the level may appear the same in the line; the trigger instead is the velocity. Bypassing or collapse may be at fault.

Sudden or otherwise unusual periodic pattern changes that appear to cause stress in the system are easily identified once the normal periodic pattern is determined. Leaky tide gates, tide inflow, groundwater infiltration, and related occurrences tend to be periodic over hours, days, and weeks. Tide-related problems cycle with known tide rhythms and are easily identifiable. Groundwater infiltration must be matched to the localized hydrologic cycle for wet and dry periods in the drainage basin. Pattern identification and variances from normal steady-state conditions provide the triggers for these notable alerts. Here, the cycle of disturbance, not the continued disruption, generates the computer pick. The dispatch of the maintenance crew should be timed to correlate with the next predicted occurrence.

Plant hydraulic or pump station actions as a result, or for the protection, of the maintenance crews must be considered with the computer picks. Maintenance crews may complain about flooding and other difficult conditions during dry weather. These complaints may appear unfounded until documentation is provided by the charts. Many times, actions in the treatment works produce conditions in the influent sewers that cause problems for field personnel. The original design hydraulics of the sewage treatment works precluded such occurrences; however, with years of modifications on both sides of the influent gate, conditions change. The monitors can provide this information. As noted many times previously, both the sewage treatment works and the sewer infrastructure are changing entities, continually being modified and enhanced. Monitoring system documentation can show how to modify activity or maintenance cycles to accommodate field personnel or provide a more optimal scheme based on the field crew's findings and reporting. This enhances the expansion and upkeep of the sewers and the process improvements at the treatment works.

The monitoring system's intangible benefits to the maintenance areas generally have no exact dollar value but can be just as important as the tangible ones. A positive public image is certainly needed to readily accomplish work. There is a learning curve, as noted above, in adapting to the system. The payoff is rapid, real, and one that provides a higher-quality work environment on the street than would exist without the equipment.

8.9 ELECTRONIC EQUIPMENT DIAGNOSIS

This is the proper place to introduce equipment diagnosis. The previous section discussed basic maintenance diagnosis. Patterns may be misinterpreted when the instrument itself fails. This discussion presents a methodology for determining the issues that surround equipment failure, minimizing its effects, and ultimately resolving the problem.

Any man-made device is subject to failure. The highest-quality devices of any kind, in any industry, fail at some point for any number of reasons. Here, delicate electronic equipment is placed in one of the most hazardous environments known. Its many mysteries mean that all the planning in the world cannot possibly predict every incident that could lead to instrument failure in a sewer. Manufacturers test for as many potential failure issues as possible. Downtime of equipment in the field is nonetheless inevitable and should be anticipated in the system planning stages. By addressing the situation from the outset and by expecting a percentage of failures, the utility agency or service provider can plan for these events and realize maximal uptime for the monitoring network.

Every manufacturer calculates the mean time between failures (MTBF) of its equipment. The more reputable manufacturers either provide this information to customers or publish it in their product literature. The statistic may be in some form other than MTBF but estimates the time before an instrument fails. Periodic maintenance is a contributing factor. Manufacturers supply tables of recommended maintenance intervals and calibration schedules that should be adhered to. Because the equipment resides in an undesirable environment, maintenance schedules are more rigorous than those for standard process equipment and are vital to maximum performance and longevity.

The MTBF is the length of time that the device is expected to function before a failure of one or more of its components. A manufacturer may calculate this interval in any number of ways; the interest is not the calculation method but the value it yields. It is necessary, however, to determine whether this value was determined in laboratory settings or in actual sewer environments. Laboratory settings are fine for initial analysis, but the devices are expected to perform in the harsh climate of the sewer and should be tested in them. MTBF analysis is also an aid in equipment selection.

Periodic maintenance schedules are generally calculated in more pristine conditions than a sewer; the manufacturer should be asked how the maintenance requirements were generated. The more reputable manufacturers do extensive testing in sewers and under a range of conditions. The first step in equipment diagnosis is to ascertain whether or not periodic maintenance requirements were met by the utility. (Note that the cost of periodic maintenance can be estimated in the simplified cost-to-purchase/cost-to-own model based on the frequency of confined space entries.)

Once the periodic maintenance issues are addressed, the malfunction of the devices can be ascertained. The intermittent recording of the parameters, gaps in the data, or other peculiar aspect of the trend indicates problems that must be segregated from periodic maintenance lapses and equipment failure. Although there is much variability in causes of failure, some generalizations can be made concerning trends related to maintenance and electronic failures.

The problem of grease coatings on the velocity sensors as a result of suspended matter is addressed by periodic maintenance. Such coating is common in heavily populated areas and industrial locations, where greases are prevalent in the waste stream. These coatings generally are not visible in the compromised light of a sewer infrastructure but may cause failure of velocity and level probes exposed to the flow. Characteristic patterns include zero readings at the instrument. The problem is readily resolved with a light cleaning of the sensors. As noted above, solids deposition of all kinds causes equipment failure as often as it does structural deposition problems.

Electronic failures tend to yield erratic or unstable readings where the normal pattern would indicate a more continuous or predictable steady state. Zero drifts and other issues that cause undefined changes in the steady-state pattern are often a result of electronic failure. A zero drift is the movement of a standard pattern up or down a scale from the calibrated zero point for no apparent reason. In the case where the equipment can be detached from the sensors, a simple replacement of the electronics package can make the diagnosis. A defective electronics package is easily diagnosed in this manner if the trend returns to the historic presentation and pattern when the package is replaced. If the manufacturer is to make the repairs, include the subject chart with the return authorization and the package to expedite diagnosis of the cause of the failure.

Much of the expertise generated in equipment diagnosis from the charts is based on the understanding of characteristic dry-weather responses and the variations generated. This is accomplished through longitudinal examination of the information and the ability to discern patterns that call for action. Segregating instrument problems from actual infrastructure occurrences can be a formidable task. However, the more this is accomplished through long-term examination, the more information can be documented and made available to all parties.

8.10 ENGINEERING PLANNING

Accurate long-term data for planning and engineering studies can be a tremendous asset in providing the highest-quality services to ratepayers. Equally as important, the same data can be utilized to protect the watersheds in the most cost-effective manner. Actual behavior patterns of the sewer infrastructure are more valuable and reliable guides than are desktop estimates or the original design theories. The ability to assess actual sewer conditions with the latest technology is one of the greatest payoffs of engineering planning. Making the most of the data is a function of the skill and experience of the engineers.

The largest immediate use for data is to determine the transportation capacity of the sewer infrastructure. As noted previously, the system changes and grows over time. The current system may be quite different from its original structure. The ability to watch and study the response to a variety of storm events, utility-driven conditions, and steady-state form provides valuable information, laying the groundwork for the most appealing solutions to the problems facing the system engineers. The advantage is that solutions can be tuned to the circumstances; engineers need not rely on broad commercial solutions that may cost more and deliver fewer benefits.

Increasing regulations and higher fines for violations demand research of greater depth than ever before expected. Initial studies generally indicate that particular responses at certain points of interest require more immediate attention than others do. Live data, or extended fields of accurate historical information, supports this type of decision making. The determination of outfall capacities, modeling data for calibration, sizing information for bidding and construction costs, complexity estimates, and timescales for programs are all functions readily addressed by interpretation of extensive data sets.

8.11 TIDE AND TIDE CONDITIONS

Many watersheds around the world are affected by the tide cycle. Tides can wreak havoc when their levels exceed estimates and preparations are not in place. Monitoring information can assist in planning for and working with the tide cycle.

In any infrastructure system that has openings to a tidally affected watershed, knowledge of the tide cycle is critical. The tide charts generated by various national and international sources are only approximate calculations based on general formulas. The actual highs of each tide cycle are influenced by many factors that vary by the individual watershed. It is important to realize that tide estimates do not take into consideration the effects of storms or other loadings that affect the aggregate level of the watershed.

The first piece of information that monitoring provides is the aggregate level increases for tidally affected streams and rivers as a result of loadings and runoff due to storms. In bay and ocean areas, the aggregate affect is negligible. However, in more confined watersheds, the difference in high tides due to storm runoff can be dramatic and can adversely affect storm and combined sewer outfalls. Pump stations can also be adversely affected

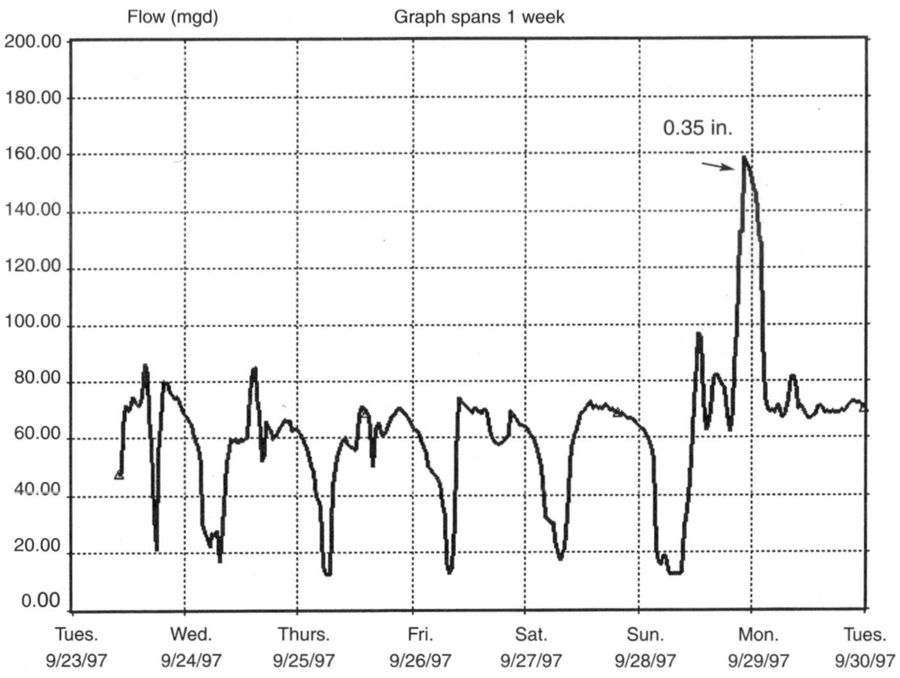

Figure 8-4 A fairly repeatable pattern showing the effects of a rain event. Note the change in pattern from the steady state.

by these conditions, as they must remain on and work harder to keep the water out. Combining rainfall intensity and duration information with historic tidal data enables more accurate estimation of the aggregate levels per storm event.

Tides affect structures in a variety of ways; some of the relevant aspects can be found in the individual discussions on structures. A tide model for combined and storm outfalls is discussed in a later section. It is sufficient to realize here that actual monitoring can more efficiently track tide levels than can estimates or mathematical simulations. Additional factors that increase tide levels can also be tracked as a part of the tide's influence on a drainage district. An example of using monitoring data in tidal situations is included at the end of this chapter.

8.12 CUSTODY TRANSFER

Custody transfer is the movement of sewage from one community to another for a fee. Responsibility for the secure passage of the volume of water is transferred from the sending community to the receiving community. The area in which this transfer takes place is called the *point of connection.* At the point of connection is a device, agreed to by all the involved parties, to measure the flow per the agreement. Historically, this device is a Parshall flume or other primary element. These primary elements are a constant source of controversy.

The implications of custody transfer reach well beyond the accuracies of the flowmeter that measures the volume to the sewage treatment works. It involves an entire body of litigation among all the parties that interface to make the transfer successful. The data derived from the point of connection are critical in this process and are the very core of the cost for which the customer community is billed. If there is any area in which the instruments must be accurate, it is the custody transfer point. A full discussion of custody transfer appears later. The concern here is for the accuracy of the flowmeter.

The custody transfer being presented here has a governmental flavor. This is usually called *billing for profit, billing networks, intercounty connections,* or another name referring to an agreement between two governmental agencies. With the introduction of private sewage operators, many of these agreements become more business-oriented. In this text, the term *custody transfer* is used consistently.

In days gone by, a primary element was the most practical method of performing flow measurement. Under laboratory conditions, the Parshall flume is certainly accurate enough for this purpose. However, in the changing infrastructure environment, the conditions may have adverse affects on the flume's ability to remain unsubmerged. If the flume becomes overtaxed during storm events, in this case, the velocity slows or stops. However, because the measurement is flow calculation versus head rise, the meter registers more flow being received than actually is conveyed. Downstream connections that may not have existed at the time the flume was installed now may adversely affect the device. Hosts of other problems fill court records of interagency disputes. The point is that in the age of modern electronics, the Parshall flume or any other type of primary measuring device may no longer be the best solution for such measurements.

The ability to register velocity accurately in a primary device and prove the data's integrity is one of the many answers to the disputes that arise around primary elements. Many of the other issues can also be resolved with the introduction of electronic devices to the transfer point. The flow of data, and the historic record that can be generated if the devices are installed properly, also assists in generating a fair bill for such arrangements. In the case of a Parshall flume, it is possible to determine submergence using level-only devices. This accepted practice is described in many engineering textbooks. However, the velocity is still not verified and this becomes a source of litigation in billing disputes. Demonstrating that the velocity is within the recommended range of James Parshall's design can be performed readily using electronic monitoring.

There are many methods for incorporating custody transfer sites into the monitoring network. Generally, this site is the first point of interest along a critical path. However, depending on the interceptor run, such sites may feed along some critical path. Therefore, these points of interest must be identified and known to the system operator. Once their locations are established and the primary elements are identified, the process of system integration can begin. Be certain that the manufacturer or installer certifies all the installations for accuracy and provides documentation that can be presented to the customer communities. Careful periodic calibration and verifications should be performed at these locations to maintain the accuracy of the site and a high level of confidence in the data.

The recording of flows at these locations should also be handled differently in each locality because of their relative chance for legal exposure

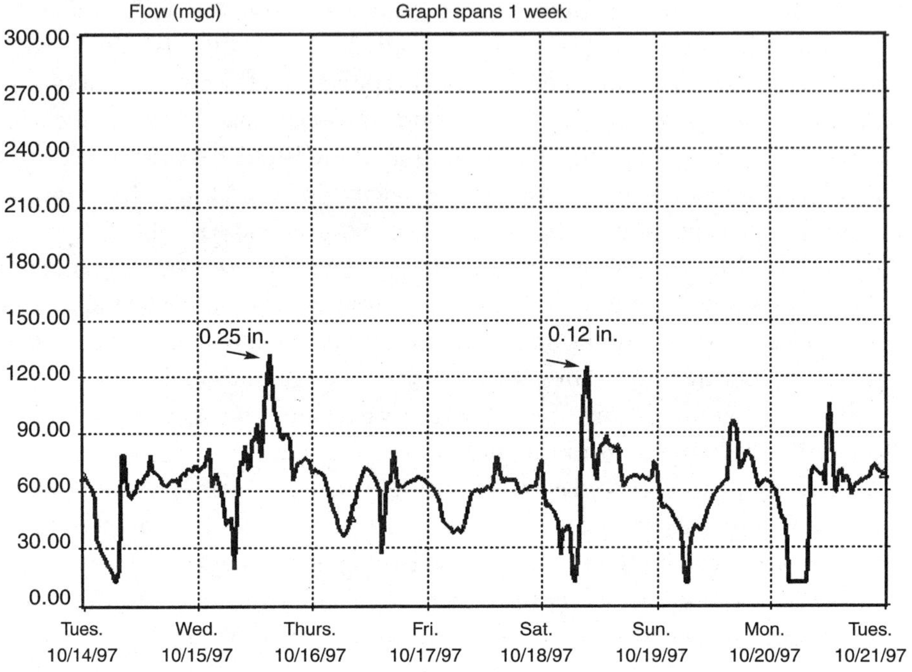

Figure 8-5 Rain sequence causing perturbation of the steady-state pattern. Note the pattern changing with the rain events.

to the interested parties. The system should be able to generate such records quickly, along with calibration certificates, time since last calibration, and related information that may be of value. Even if the system is temporary or partial, data communications of some sort should be included at these vital locations so the parties of interest can see the fairness of the bills.

8.13 RIVER GAUGING STATIONS

River gauging stations are always points of interest, but they may not lie along a critical path. The information they generate is quite valuable, especially when combined with other systemwide data; it makes possible full correlations of the watershed conditions with system events. River gauging stations can be in tidal or nontidal areas. They can be as simple as a float determining the level of the stream or as complex as samplers measuring a variety of chemical compositions, taking samples, recording levels, and transmitting the information to a central location. River gauging stations complement the network at large.

The original river gauging stations measured flood levels and raised an alert when the water body entered flood stage. They have since evolved considerably, and with good reason. Their location was critical long before the development of modern sewer monitoring systems. In many cases, these stations developed independently of any infrastructure monitoring. Their incorporation brings valuable information to the system at large.

The data, when integrated with the sewer monitoring network, can provide loading and runoff information about the entire watershed. This contribution is independent of the sewer infrastructure or the contribution that the sewer makes. Potentially, information can be generated concerning relative factors involving loading lag, rainfall intensity, and effects to the watershed in which the sewer information can assist. These variables depend on the number of stations, their placement, and the information they carry. Integrating the river level staging into the infrastructure database can provide hydrological information essential for regulatory compliance.

8.14 STORM VARIATIONS AND SPATIAL RELATIONSHIPS

The ability to generate large databases concerning storm events and their effects on the sewer infrastructure allows these events to be categorized and referenced as required by various interest groups. The need to segregate storms by their spatial relationships is an issue of great importance, as the dispersion of precipitation may vary greatly during such events. Spatial variations assists in determining storm paths, general trends, and related characteristics that are useful in all phases of operations, planning, and engineering.[3]

A storm's spatial relationship is its distribution of precipitation across a drainage district or a series of districts. Some storms contain cells of intense precipitation that concentrate in specific areas. Others spread evenly

3. T. Day and J. Nicolo, "Rainfall Information Through Automation: Data with a Purpose" (paper presented at the Urban Wet Weather Conference, Quebec City, Quebec, Canada, June 1996).

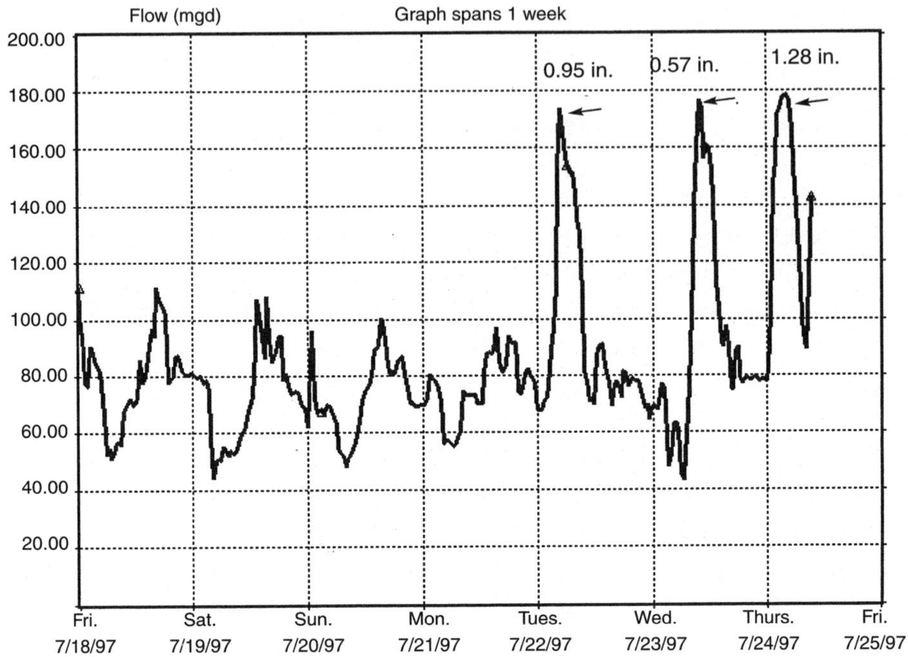

Figure 8-6 Repeating rain events. Note that events 2 and 3 occurred within 24 hours and may have produced excess stress on the sewage treatment facility.

across the entire area. Patterns can be noted in the generation of the localized hydrologic cycle for paths of storms and the zones of intensities that occur as they travel across the region.

Tabulation of storms and their residual effects on the sewers provides the history for patterns that change as a result of alterations to the landscape, climate shifts, or other conditions. Such information, trended for long periods and analyzed correctly, can indicate flooding areas, sections requiring further investigation or development, and areas where there appears to be excess or latent capacity in the sewer system. It can lead to the shifting of flows from one zone into another, maximizing the system efficiency. This can be achieved only if storm trends and their spatial relationships are completely understood. The correlation of storm data is performed through the analysis of short- and long-term rainfall data.

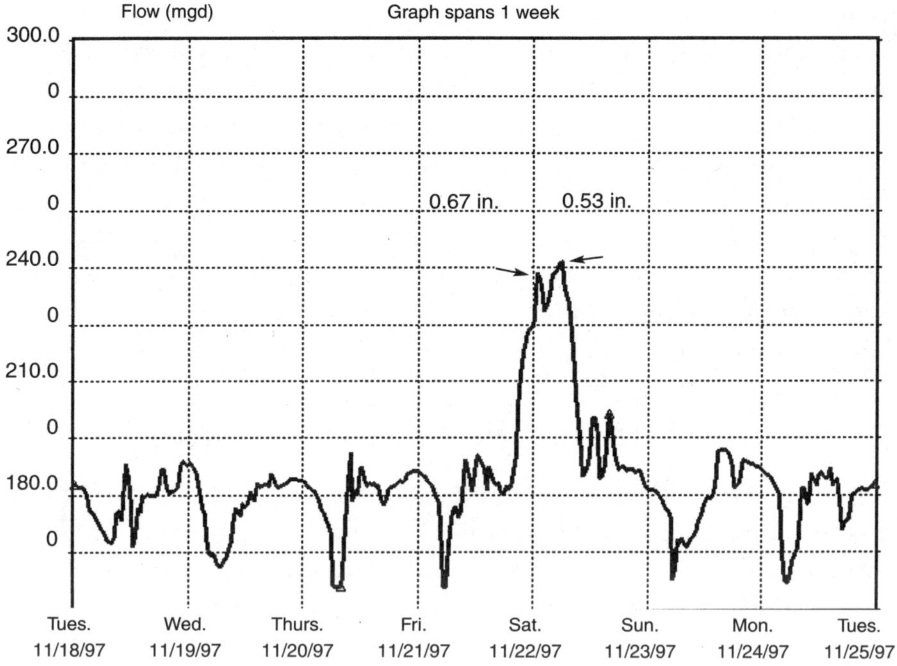

Figure 8-7 This multiple event caused excess loadings to the STW. Note that the instantaneous flow is 180 MGD when the expected flow would be less than 20 MGD.

8.15 ALIGNMENT WITH THE SEWAGE TREATMENT WORKS

Most utilities maintain a great database of information on the hydraulic response of each critical path leading to the treatment works. Because the treatment works is the largest hydraulic control point along any critical path, its personnel should understand the potential effects on the sewers of the plant's hydraulic activities. Throttling the influent gates can send ripples throughout the system and cause problems if notification is inadequate. The evidence of such hydraulic action is recorded through the sewer monitoring information.

The alignment with the treatment works also has immense potential effects for treatment and control, a bonus for the drainage district. If activities are coordinated, the sewers likely can be used for storage during rain events, which can, in turn, increase the plant's ability to perform activities such as load shedding and shaving peak demand. By withholding sewage in the collectors and allowing the plant to meter the amount of water it takes in, peaks may be lessened or eliminated. Also, the number of pumps needed throughout the facility may decrease when the flow is efficiently controlled.

Other positive opportunities may emerge if the sewers and the plants align. In this case, retrieved data is used to better measure the latent capacity within the sewers, which can serve as storage, thus maximizing the plant's operation.

8.16 TRACING ILLEGAL DISCHARGES

The careful evaluation of charts and computer picks can help the utility operator monitor for illegal discharges by users hoping to avoid paying the plant processing fee. Sewage treatment plants offer many services to companies who deal in wastewater removal, including those required to process their water before discharging into the sewers. Although the integrity of these companies is rarely in question, a monitoring system can act as a watchdog for violations.

Tracing illegal discharges requires proficiency in chart interpretation given how less than reputable operators of portable waste facilities usually avoid treatment plant fees: They identify a manhole in a location that does not attract attention and illegally discharge collected wastewater into the hold. Depending on the line, the discharge may cause a spike in the chart data. If the spike appears once or twice, the interpreter may simply discount the point as erroneous and ignore it. However, if the spike repeats for a week or more in the same time period, then investigations to locate the source of the discharge can commence. The illegal operator may not be aware of flow metering within the infrastructure, nor that an influx of several thousand gallons can easily be detected, depending on the time settings and the skill of the interpreter.

This method for monitoring for illegal discharges can be extended to industrial discharges and other points that have the potential to introduce illegal or otherwise undesirable wastewater. The placement of the instruments and the skill applied to their interpretation are major factors in the success of this endeavor. The skill is readily developed through the continuous and regular examination of the data and through knowledge of the pattern that each point of interest typically generates.

8.17 BIOMONITORING

The subject of biomonitoring is addressed in a separate chapter. Biomonitoring is the potential next step in the study of the infrastructure. Its potential for system understanding exceeds that of flow measurement, as it is the study of the quality of the wastewater being transported. Biomonitoring is in its infancy but will grow as the equipment and systems designed to perform in the sewer environment mature.

The applications of data obtained through biomonitoring are increasing as efforts to protect the watershed gain momentum. Cellular phone systems, used in alarming and reporting, have generated a new level of surveillance effort in sensitive recreational areas and fishing regions by allowing equipment to be placed and moved without the need to install direct utility service. This provides discharging information and other water quality data that assist in enforcement, protection, and regulatory areas. The ability to tell if swimmers can safely enter a watercourse is critical to their protection. The ability to detect toxins and other undesirables has direct applications to shellfish and fish habitats in more sheltered areas. Biomonitoring is fast becoming a mature and better understood area of infrastructure management.

Biomonitoring data also can assess toxic and nontoxic discharges from industrial customers, residuals from farm and farm products facilities, and any business that draws and returns water from a watershed district. Protection of the environment is a result of the ability of system operators to translate data into facts relevant to protection program development and implementation. In the majority of cases, simply presenting the unabridged facts to an industry or business is enough evidence for them to wish in assisting these efforts. However, the data must be unabridged and factual, with emphasis on pragmatic protection of the associated watershed and proactive prevention of industrial discharges.

Determining the chemical content of the waste stream can be critical if it will produce residual effects that can adversely affect the environment. One of the issues of concern for any watershed area is the return of nutrients to the district, as would naturally occur if there were no population concentrations. Removing too many nutrients could have a detrimental effect on fish and plant life. The cyclical reintroduction of nutrients to the watershed began long before people domesticated the landscape; it must continue if the local ecology is to flourish.

A major application of biomonitoring is to monitoring for toxins and other undesirables that may adversely affect the treatment process at a sewage treatment plant. The protection of the plant is vital to the protection of the watershed in general. Undesirables, which may affect the growth of the microorganisms that are the backbone of the biological process, can be detected and remediated or at least mitigated, to maintain process efficiency. More uses for biomonitoring may be found as refinements to the treatment plants' biological processes are researched.

The tools and techniques of biomonitoring are developing constantly as understanding the quality of the water becomes as important as the quantity of the flow in the system. The learning curve of sewer infrastructure system managers with respect to the functions and benefits of biomonitoring is accelerating. As their understanding increases and the products mature, biomonitoring will provide new knowledge about both the quantity and the quality of water in the watershed, information that can be applied to the watershed's protection.

8.18 ASSISTING THE LEGAL DEPARTMENT: DEFENDING THE PUBLIC INTEREST

It is difficult to defend public investments against claims made by groups for flooding and other damage from the water system. Such claims are often made when little or no private insurance covers the affected properties. Insurance companies also assess utility operators for such damages when they occur. The majority of claims are legitimate and reasonable; a percentage is not. Fraud may be attempted because of the inability of the agency to defend itself to the satisfaction of a local legal entity. Many fraudulent claims, however, can be overturned with the help of sewer infrastructure information.

A properly deployed rain gauge system can provide reliable rainfall data for a drainage district. A claim made for flooding due to excessive rainfall can be evaluated if the rain gauge network can describe both the general storm event and the area of concern. If the data indicate that storm damage was possible, then remedial action to protect the drainage strict from future similar events can be recommended. If the claim appears fraudulent, the utility uses the data to show that the storm would not have caused the alleged damage and that another source of the cause of damage should be sought. If the claim is fraudulent, the case likely will be dismissed. Even if an out-of-court settlement is made, the amount of exposure should be substantially less than if the data were not available.

Flooding complaints due to overtaxed sewers are a continual headache to all utility operators. Claims for basement flooding and other damage occur with increasing frequency as population centers become more congested. Many times these claims cannot be refuted simply because not enough is known about the hydraulics of the system to understand whether the conditions described could actually occur. A permanent monitoring system can provide information about hydraulic conditions to mitigate these claims. If the sewers actually exceeded their capacity, remedial actions can be implemented to protect the public interest. Ideally, data demonstrate that capacity was sufficient and that flooding could not have occurred.

Monitoring data is a tremendous assist to legal units defending the utility's interest. Much of this is discussed in the funding and justification section in this book. It is presented here to show the technical side of generating the data and making it available. The presentation of the data in a format understandable by laymen helps justify the system and increases its value to the utility.

Applying data from the sewers has far-reaching effects on indirect cost savings. The ability to reduce legal claims using the data can not only justify the cost of the system and assist in risk management but also directly reduce costs associated with legal actions. This concept is discussed at length in a later section. The value is in data that describes actual conditions.

8.19 COMPUTER MODEL SIMULATION

There are a suite of products for simulating conditions within the sewer infrastructure. Generally, these planning tools are complex and specialized and require a high level of proficiency to operate. Interpretation of the data likewise requires a high level of expertise. But these simulation models cannot produce data without live information from the sewers. Computer simulation programs absolutely require calibration and maintenance data to function properly. In short, sewer monitoring data is needed.

This subject is addressed separately here because of the relative impact that live data has on each step of a computer simulation. It is believed that, if set off from other subjects, the fine points of the discussion will make better sense and be applied more effectively.

Computer models simulate only known conditions or those where historic trends are integrated. A hydraulic computer model is a numerical simulation of hydraulic actions within the sewer infrastructure that assumes a known set of parameters and responses. The model rapidly projects future responses; this body of projected data is used in planning and engineering. The boundaries, initial conditions, and operating parameters of the simulation are derived from the live data generated by sewer monitors. The accuracy and integrity of the live data determine the accuracy and integrity of the results of the computer model. If the most accurate computer model is provided with compromised data, the results will be skewed and potentially unusable. It is relatively straightforward to verify the numerical accuracy of a simulator. The variable becomes the calibration and maintenance data required to make the system work.

Calibration data supply the initial parameters for the model. All models require parameter sets, and the modelers themselves have particular requirements for putting these simulations in motion. The calibration data is also used to verify the integrity of the first generated simulation points. Without calibration data, it is difficult to ascertain where the initial conditions should be placed. If they are placed inaccurately, dependent calculations will also be inaccurate through no fault of the simulation.

The second type of data required by computer simulations is maintenance data, which is used to provide long-term verification of the model's projections. Such information is required to determine whether floating point calculation errors, overloads, or other mathematical conditions are making the model unstable. It also tracks changes within the infrastructure and in upstream and downstream conditions that would affect the actual conditions but which were not necessarily programmed into the model in the beginning. Maintenance data ensures the integrity of the model results and provides confidence in the simulation schemes to be utilized in planning and engineering studies.

The data for a model is generally provided in ASCII text format and stream-fed into the simulation. Here, the modeler must ascertain that the data is in the format and scale that best serves the model. If the HGL is utilized, the data must be in elevation or some form of an elevation. If the model makes this conversion based on known benchmarks, then the data can be directly read. The software used to process the field data should be able to make the conversion to the format as required by the hydraulic model.

Live data from the field are critical for the startup, short-term, and long-term accuracy of any numeric hydraulic simulation. The data generated from the field must therefore be of the highest integrity and in a form that the model can use. Without this data, hydraulic models cannot accurately simulate the critical paths of interest nor can confidence in the projections be ensured.

8.20 TIDE MODELING: DATA APPLICATIONS EXAMPLE

The example presented here combines the requirements for good data and an understanding of hydraulic modeling, especially when prompt information is needed to protect the plant and the process within it. The example also demonstrates the requirements for hands-on system knowledge by experienced personnel and a concept of the localized hydrologic cycle. The example can be reproduced quickly and, when properly calibrated, is repeatable to model with an accuracy of gallons on million gallons. It is not meant to take the place of any long-term hydrologic simulation model but rather to demonstrate the critical nature of monitoring data and the ability to quickly translate it into a working information base.

Here is the scenario: A sewage treatment works in a tidally affected combined sewer system complains of high flows at times of high tide. The sewer unit performed an inspection of all the tide gates and found them to be sealed and operating properly. However, the plant is registering these high, dilute flows during known dry-weather periods. The flows are qualified, so the reporting is known to be accurate. Also, the work of the sewer department was checked and is not in question. The problem is to find the cause of the high, dilute flows and present it to the operations personnel for remediation.

The example makes some working assumptions. First, the tide gates are adequately maintained and no other infrastructure maintenance problems would cause this situation. Next, flows recorded at the plant gates are accurate. Finally, combined sewer overflow (CSO) tide stations generally have an overflow weir for discharging excess flow to the watershed.

The high flows are correlated to high tides. Once the high-tide cycles are correlated, data is generated so the tide at each CSO—or at least those that have unobstructed discharge points over the tide gates or dam—is available. The height of the tide is the issue. Remember that a tide cycle chart is valid only for one point along the river, bay, or ocean. Adjustments must be made for any structure up- or downstream of the target point. The tide height must be known at each CSO, not just at the reference landmark. The tide therefore must be measured using instrumentation.

It may be cumbersome to place instrumentation at each CSO outfall if none exists. For this example, it is assumed that the CSOs have monitors and data are available at each point of interest along the critical path. In actual circumstances, groups of monitors recording the levels on both sides of the tide gates are necessary.

A grid is developed with all the CSO outfall elevations taken at some survey landmark. This provides a common scale for all the outfalls along the point of interest. The high of the tide cycle is translated to the elevation correlation either at the survey landmark or at another point with which the utility is comfortable. A hydraulic grade line is projected at each high-tide cycle down and along the grid based on the data taken from the monitors such that the maximum height of the tide is shown against the grid. This reveals where the high tide has the potential of entering the system. It is clear to see, as the line drawn from the high-tide maximum lies above the minimum elevation of the outfall opening. If the line is above the minimum elevation, the tide may be introduced into the regulator and reach the treatment plant. Remember, this is only a projection of an HGL and thus is subject to the interpretation of the data at the site.

A 24-hour chart is then generated for the site. The chart is used in this case to capture two tide cycles. A different time scale can be utilized if there is positive evidence that tidewater was not introduced to the combined system. Otherwise, it makes sense to capture at least two tide cycles. It has to be understood at this point that there is level on both sides of the tide gate or dam of interest. There will be a tide cycle that appears diurnal. Translate the elevation of the bottom of the overflow to the corresponding grid on the chart. The point where the tide cycle crosses that line indicates the beginning of the inflow of tide to the sewer system. The cycle will wane and recross that threshold. The period between the two crossings indicates the time and duration of the inflow event. Record the maximum height that the tide cycle rose to over the minimum point of the overflow.

The correct way to calculate the flow is to use a weir equation and integrate over time because of the rise and fall of the tide. This is impractical and slow. Such analysis is best left to known mathematical models for long-term simulation. The goal here is to obtain operating answers in a relatively short period of time. Therefore, do a cross-sectional area computation calculating the width of the outfall opening, which is fixed, and multiply that by the maximum height recorded by tide introduction. This is the height of record. Remember that this is a one-time maximum and that the entire cycle must be taken into consideration. Therefore, for purposes of estimation, reduce the area by 0.8 so the difference between the rise and fall can be accommodated. This is an arbitrary reduction that may have to be adjusted slightly to the individual utility. This provides a cross-sectional area for use in a flow rate calculation.

Returning to the chart, determine the duration of the inflow. Round off to quarter hours, but keep the figure in hours. This is an estimate. Assume that the velocity over the dam back into the system is around 3 ft/s or 1 m/s, depending on the unit of preference. Use the continuity equation and develop a rate of flow in ft^3/s or in m^3/s. Convert the seconds to the number of hours that the chart indicated inflow. This is the approximate amount of inflow as estimated based on the high-tide cycle.

A table of the latest sewage treatment works dry-weather flow is needed. Take the last 10 to 20 days' worth of dry-weather data and develop an average of the dry-weather flow that was expected. Do not use data from days suspected to have experienced tidal inflow events nor affected by the aftermath of a storm event. Subtract that value from the value found the day that the tidal inflow was reported. This is the difference suspected to be tidal inflow. Calculate and add all the inflow points as the data permit. Compare the calculated inflow values with the subtracted difference. The two values probably correlate.

Note that this model did not take much time to construct. It was a fast use of the data that generates results quickly and allows for decisions to be made quickly concerning remediation. It can be applied to any number of circumstances. The ability of the collector system to remediate the tidal inflow potentially saves the sewage treatment plant considerable funds in processing costs. It also proved that there was, in fact, tide inflow and that it had to be corrected.

8.21 COMPLIANCE PROGRAMS

The issues surrounding compliance programs are intricate and dependent on the individual circumstances. Compliance programs are instituted for many reasons. Some programs are legally mandated as a result of legislation or through local regulatory boards. Some programs are required to meet permit or other guidelines that have large impact on vast drainage districts. Generally, the decisions made by such agencies are based on little field interaction. This causes many potentially difficult or otherwise troubling goals, some which may require mitigation or litigation to clarify and delineate. An excellent tool with which a utility can document, mitigate, or otherwise litigate the compliance programs is through the use of sewer infrastructure monitoring data.

The ability to justify a position, prove an effort, or otherwise disprove a condition may lie with the ability to monitor a point of interest. Monitoring the point and generating data about the location and its responses can substantiate a utility's claim about a condition. The goals set by regulators are often difficult or much too costly to meet. The data generated can provide evidence that the relief claimed by the agency is justified and necessary for its operation.

The ability to translate this data into a clear financial picture is also critical in the costing of a compliance effort. The costs of tasks appear minimal at first but escalate rapidly when actual results rather then estimated outcomes are demonstrated through monitoring information. The outcome of and eventual payback for the compliance effort may be weighed using test locations. The results of actual testing and their associated costs can be substantiated and presented to the regulators. Such defense and the ability to provide a counterproposal based on actual conditions balances the amount of remedial efforts required for a given task. The application of the data set is critical in presenting a balanced proposal justification.

Compliance programs require funds above the resource pool originally set aside. The ability to monitor and understand the sewer system and to provide counterproposals to the suggestions for remedial action relies on data from the field. An infrastructure monitoring system is one of the best methods for obtaining the data required to handle compliance actions. In any compliance activity, remember that the data is only one of the tools. The craftspeople involved must build the most economical compliance program with the available resources and those that can be reasonable obtained.

8.22 TEMPERATURE STUDIES

This is a short example of how a simple temperature gradient can be a gauge in determining infiltration and inflow to a collector system. The use of temperature gradients is a clever and inexpensive method of verifying particulars of the water mixing as a result of the different entry temperatures of the various water sources.

A standard thermocouple is inserted with each monitor as it is introduced into the sewer infrastructure to monitor the temperature of the wastewater. The climate dictates that the watershed has a temperature close to air temperature and that sewage temperature is similar to the homes and businesses from which it emanates. The temperature does not vary much over the year from the homes and businesses, which are generally heated and cooled to around 70°F. However, watershed conditions vary with air temperature.

If a sharp temperature change occurs at a point along a critical path, this is a sign of infiltration from the watershed. The deployment of temperature probes along the critical path can readily identify the two points between which the infiltration begins. The temperature change can be a parameter to generate a work order and expedite repairs. This is because there is a definite temperature gradient between the more variable outside water source and the source of the wastewater. Observing the difference in temperature gradients can lead quickly to infiltration issues and their approximate zone of occurrence.

8.23 MAINTENANCE TRACKING AND PREDICTIVE MAINTENANCE

The issues for benchmarking and maintenance tracking are more for making justifications within the system as presented here than of actual operations procedures. However, it is worthwhile to study the potential of a monitoring system for tracking work performed in the system for purposes of predictive maintenance and manpower allocation.

The assumption is that a monitor is in place at the site where maintenance is being performed. Because of disturbed grit and other maintenance by-products, the data will appear garbled, mostly unusable—but this conclusion is incorrect. The unusable portion of the data is the evidence of the work being performed in the system. It is difficult to note the precise start time, but the time that a maintenance activity is finished, at least the confined entry part, is evident in a return to normal steady-state conditions as recorded by the monitor. From such data, approximations of job times for like work can be generated. Manpower allocations can thus be made based on known times of completion.

Such documentation of maintenance activities in the field can be considered vivid evidence of work performed and justification of resource allocations. It also verifies the location and activities of work crews in the event of questioning by regulators concerned with response times and work activities. This hard evidence supports claims that the utility operators may make for programs best suited to efficient operations and streamlined maintenance.

8.24 SUMMARY

There are many essential uses for monitoring data, along with more exotic uses conceived by sophisticated and imaginative reviewers. The ability to utilize in-place equipment allows the surveillance of many more points, along more critical paths, than would be possible using direct visual inspections.

The ability to monitor and observe conditions over a wide range of critical paths and to apply the data to a variety of programs and circumstances makes the cost of the monitoring systems worthwhile. The ability to offset direct costs associated with the operation of the system by using the data pragmatically makes these systems self-sustaining.

The infrastructure monitoring system can be instrumental in controlling both direct and indirect costs. The direction of maintenance activities, the correction of damaging situations, and the protection of the watershed all incur direct costs. Indirect costs are more difficult to justify but may be higher. For example, legal actions can be exceedingly costly; monitoring data that mitigate these costs is, hence, enormously valuable. Just having the system in place and operating to assist in these efforts can yield after several years a reduction of the numbers of legal claims and funds associated with the loss.

Using the systems to support maintenance activities brings another financial reward by finding problems before they escalate into costlier ones. More accurate targeting of maintenance activities generates a better public image and, potentially, more street cooperation from the inhabitants of the drainage districts. Good press is an excellent protection for field personnel and an excellent method of safeguarding positions to prove conditions in societies and postures in political stands. Accurate orchestration of field personnel through the use of monitoring information makes this possible.

Sewer infrastructure monitoring information is a tool that has many uses. However, like any tool, it is only as useful as the hand that holds it. The skill set necessary to best utilize the tools is developed through experience. The eventual application of the information is limited only by the imagination of the personnel who find uses for it.

CHAPTER

9

Enhancements to the Sewer Infrastructure Monitoring System

9.1 INTRODUCTION

The monitoring system installed in any sewer infrastructure measures flow, level, and other parameters that operators believe are necessary to the operation. This chapter is about these parameters.

Some of the material, at times, leads the discussion from application to theory. This is the reverse of what may be expected. Other parameters and conditions are modified from sewage treatment plant operational requirements and included based on trends for data requests. Monitoring systems found throughout the world, such as compliance programs in the United Kingdom and Scandinavia, already monitor for some of the parameters or have these functions. Some projects currently being planned include many of these parameters because they represent the next level of sewer infrastructure management. The ability to determine water quality and to base control of the system on actual conditions is the future of monitoring technology.

Biomonitoring was briefly discussed in the preceding chapter. This chapter covers the subject as it relates to water quality and to applications for both the watershed and the sewage treatment works. Determining the chemical makeup of the waste stream or the quality of the water can assist both pretreatment and treatment functions. Specific devices perform this monitoring; this chapter described the generic forms of these devices. There is a parameter called *cost to treat*, which is the amount of revenue needed to process a volume of wastewater; this varies with the chemical makeup of the waste stream. Knowing the information needed to calculate a cost to treat in advance of entry to the sewage treatment works may significantly improve process control and population safety along the critical path of interest.

The subject of biomonitoring takes on additional importance with respect to the protection of watersheds. Tracking and monitoring for undesirables that may be discharged assists not only in protection efforts but also in determining which safety measures are required to protect the watercourse. Monitoring beach areas protects bathers and, more broadly, the local tourist industry. Surveillance of watersheds helps protect drinking water from contamination that may be difficult, if not impossible, to treat. Systematic monitoring for such conditions raises the alarm earlier than random sampling does—which means the problems, in turn, are rectified earlier in their development.

The functions of monitoring and control need to be discussed first in terms of alert and alarm conditions. Identifying these conditions provides warnings of both incipient problems and those already occurring. An alert is defined here as a notification that an adverse condition is about to occur. An alarm is defined here as a notification that an adverse condition exists and requires immediate attention. There are methods for determining alert and alarm threshold boundaries. Alerts and alarms are meant to protect the watershed and surrounding property from discharges. Risk management and other financial incentives thus pertain to the determination of thresholds.

Alerts and alarms are activated by protocols. The reasoning applied on their development is vital to preventing false notifications.

The ability to store and relieve the sewer system based on real-time setpoints is the concept of real-time automatic control. Real-time control of the sewer system is not widespread nor readily accepted. Real-time control systems are currently being heavily researched and investigated. If properly planned and implemented, they have the potential to greatly improve sewage treatment functions such as dilute flow, prescreening and flow storage, and bypassing. The ability to throttle and control the flow volume can assist in plant load shedding schemes and in reducing peak demand. Automatic control is difficult to plan and the equipment is expensive, but it is

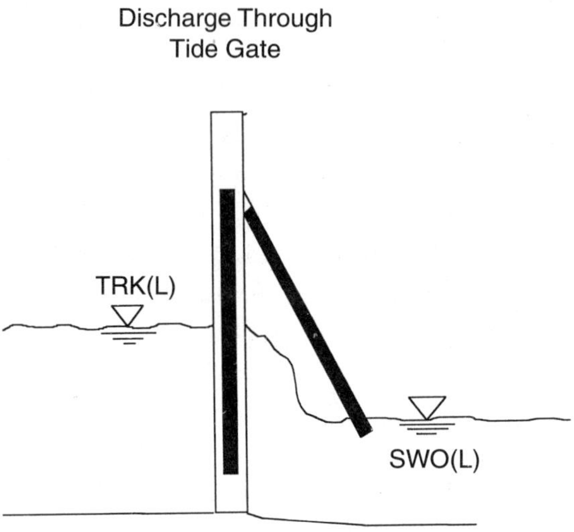

Figure 9-1 Tide gate discharge from combined sewer.

a reasonable alternative to the addition of new sewers, especially when space is limited.

All these system enhancements help in the management of the sewer infrastructure. Tasks that can be performed remotely yield tremendous cost savings in manpower movement. The more information in-place equipment can generate, the greater the financial benefit that can be realized and the greater the safety of the watershed.

9.2 BIOMONITORING: THEORY AND CONCEPTS

Biomonitoring involves the examination of wastewater for chemical and biological components that can affect health and safety. Much of this discussion addresses these compounds and the measurement of chemical potential from a relational standpoint. Biomonitoring focuses on the effects of the waste stream on the environment and the population. The microbiology of the waste stream is of greater interest to sewage treatment plant personnel than to sewer infrastructure management professionals.

The purposes of biomonitoring are many and not limited to the waste stream. Biomonitoring techniques can be applied to water quality monitoring of the watershed, points along the watercourse, wells, and other drinking water sources. Sewers and their contents affect all these areas. Biomonitoring is required for both the waste stream and the areas around the

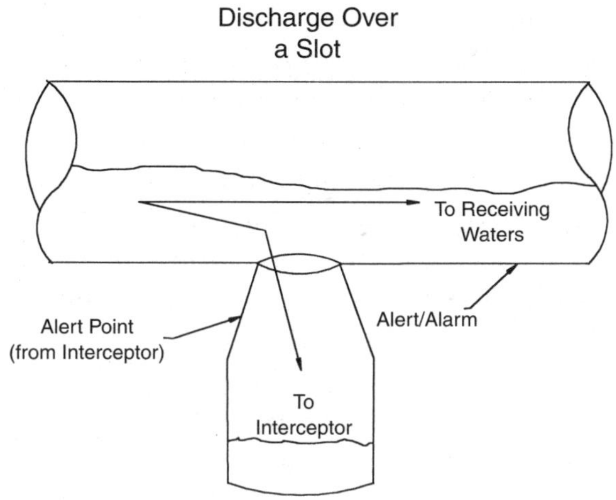

Figure 9-2 Discharge configuration for a slot or other drop device to a lower intercepting sewer.

contained flow in order to protect one from the other. Therefore, the target of the instrumentation may vary.

Water quality information, even if only a sampling, benefits the system operators. The content of the water may be more threatening than is realized. The risk potential can only be assessed if samples are analyzed either by direct instrumentation or through laboratory analysis. (The latter approach may yield too little information too late.) Biomonitoring documentation can be used in court when the utility prosecutes repeat violators, which also makes the effort worthwhile.

This discussion includes the methods of sampling, the parameters for samples, and the applications of all this to the service of the drainage district.

9.3 BIOMONITORING: DIRECT PARAMETER MEASUREMENT AND SAMPLING

Two generally accepted forms of measurement apply to biomonitoring in the infrastructure: direct measurement and laboratory analysis of samples. One is not necessarily better than the other, given modern electronics; selection is based on what fits the situation. Also, there are limitations of the more commercially available equipment to perform tasks that are functions of the type of biomonitoring applicable to the circumstance. Therefore, when choosing the measurement approach for a point of interest, all of the applicable analysis must be performed. The blending of these concepts is presented in this section as well.

It is always important to remember the requirements for local or national certification and the accepted parameters and methods for determining waste stream concentrations. The accuracy of devices, while useful, is less important to regulatory agencies then the method whereby the resultant data is analyzed. Monitoring equipment can always be used for internal purposes. Certain procedures, equipment, or methods of verification may not be permitted by regulations. Consult with regulators before going forward with a compliance-driven biomonitoring program to be sure it meets their specifications.

Direct parameter measurement is the fastest-growing segment of the analysis field. It involves the introduction of probes directly into the water for continuous measurement. Each probe monitors one of a variety of parameters, most being indicators of chemical functions. These can be measured using relational operators in the same way that flow and level are often measured. Some probes are equipped to measure several independent parameters. This economizes on space at the measurement location and saves on installation costs.

Concentrations of chemicals and compounds travel in layers, as does velocity. In the case of chemicals, these layers are called *plumes.* A plume is the concentration of a compound of interest that moves and mixes into the water of study. The plume tends to disperse throughout the water body until it blends semihomogeneously, much like saturation. However, as it is introduced, the plume resembles a velocity layer in that it appears to have distinct boundaries. If the probes are located just ahead of a discharge point, the instruments may not be sensitive enough to read the dispersion of the plume. If the plume is traveling away from the sensor, the instruments may not see it at all. This is why placement is so critical in direct parameter measurement. The potential for sensors to miss a strong, directed chemical plume is the disadvantage of both direct measurement and sampling.

Placing sensors after heavy turbulence or mixing carries the best potential for success. In this case, the dispersion should be great enough to reach the measurement point. If this simply is not possible, the characteristics of the chemicals being scanned for should be understood so optimal placement of the instrument can be achieved. Relevant characteristics include weight relative to water, dispersion and saturation in water, solubility, denseness, and colloidal potential. Many times it makes sense to do a one-time field study to determine the characteristics likely to matter at the location of interest.

Sampling is the act of withdrawing a metered portion of water that, under a defined set of custody chains, is sent to a laboratory for analysis. Sampling can be performed by either a grab or an automatic sampler. A grab sample is obtained by manually filling a prepared bottle with a metered amount of water. This is often done in sewage treatment works to monitor plant performance. This is also the technique used in industrial waste enforcement when there is a suspected toxin in the watercourse. Another use is in regulatory studies, where one-time samples are obtained in this manner.

Automatic sampling is a more regulated method of timing or synchronizing the removal of samples from the waste stream or water body of interest. Many sewage treatment works use flow-paced sampling to guide chemical usage for maximum efficiency. Flow-paced sampling is the relative change of the sample size withdrawn during an adjusted period of time to reflect a more scaled representation of water and its contents being introduced to the process. The specimens must be removed manually and delivered to a laboratory for analysis.

Automatic samplers deliver two types of samples: composite and discrete. A discrete sample is one where a metered amount of water is delivered to an independent bottle for each time period of interest; each bottle's contents is analyzed separately. In a common arrangement, a carousel of 24 bottles is filled, one bottle per hour, over a 24-hour period. The better samplers can vary the time of collection and record when the last bottle is filled. In a composite sample, all the metered samples are delivered to the same vessel, thereby mixing them. In both cases, the bottles are removed and taken to a lab for analysis.

The temperature and cleanliness of the bottles is critical to the quality of the samples; there are specific temperature and cleanliness requirements for many laboratory tests, including setup chemicals and baseline preparation. Quality control of the samples is part of what is known as the *chain of custody*. The actual chain and its regulations vary by jurisdiction, but wastewater samples must be kept at a temperature that inhibits degeneration by the microorganisms. Bottles must be prepared as required by the lab performing the analysis. Transportation of the samples, if the distance between the point of interest and the laboratory is significant, is another factor that adds to the expense and the tediousness of using samplers. However, automatic sampling is accepted in most governmental jurisdictions, and protocols have been established in this area.

Samplers come in three varieties. The first is the standalone powered sampler, in which the samples are placed in a refrigerated environment. A time interval and sample size is programmed, and either the carousel or composite, as described above, is completed for removal and analysis. These devices tend to be permanent fixtures and are AC powered. The second variety is the portable sampler, where the bottle or carousel is packed in ice and is battery powered. This type is much smaller than the refrigerated device and can be hung in manholes or stations. The better devices use peristaltic pumps, which use suction rather than pressure to withdraw the metered sample from the water; this approach does not introduce air to the sample. The third variety is the dipper or flow-through sampler, which removes an agitated sample directly from the waste stream of interest. The distinction is that the waste stream passes directly through the device. Dipper samplers are rarely used in infrastructure management because they are limited by the size of the line in which they are attached.

Laboratory analysis is expensive and time-consuming. It should be used only when the value of the data obtained justifies the expense of its generation. Even if the utility has laboratory facilities or existing contracts for such work, the time and expense requires careful consideration and planning. This is not an endorsement for any other form of analysis. It is only a caveat so the utility is not broadsided with unexpected expenses and delays in the transmission of data from the field to the plant.

A variation of the automatic sampler is the introduction of direct parameter measurement to the sample taken. In this case, arrays of parameter measuring devices are attached to the sampler. The sampler head has the computerized ability to discern and record the various signals generated from the reference probes used. The better samplers feature alarm triggers so notification can be issued in the event of threshold violations. These devices include a special bottle from which a sample of the water in which a threshold was violated can be drawn and analyzed. Such a trouble bottle can be a great asset in industrial waste management, especially when specific compounds are sought. This hybrid sampler has tremendous potential value in the industry and should experience widespread usage as interest in direct water quality and control increases or is regulated.

Direct parameter measurement and laboratory sampling are accepted by most regulating bodies. There is definitely a growing regulatory acceptance of laboratory sampling and more operational acceptance of on-line monitoring. The hybrid approach is gaining popularity and should enjoy a considerable share of placement in the systems being installed around the world, especially in nations active in the protection of watersheds.

9.4 PARAMETERS IN BIOMONITORING

A variety of parameters are of interest to the utility and, potentially, to the regulatory agencies. Each of the major classifications is discussed so there is a baseline for understanding the examination as intended. The parameters are discussed from the simplest to explain to the most complex. The sewer infrastructure viewpoint is emphasized.[1]

Temperature is a parameter relevant to the operation of the wastewater plant, but it also has significant meaning to the sewer infrastructure operator. Domestic sewage tends to approach house temperature. In most winters in most places, the sewers continue to flow regardless of the surface temperature. This is the first significance of temperature of the wastewater and its contribution to sampling and analysis. Plotting the temperature gradient among points along the critical path is an inexpensive and efficient method of determining inflow and infiltration. This tends to work best in cold weather, although warm-weather applications also exist. Temperature also has an effect on pumping, as the maximum density of water occurs at 34°F, when it is still a liquid. Such information can assist in assessing and controlling power costs in pumping the wastewater in areas of load shedding and peak shaving.

Dissolved oxygen (DO) is the amount of oxygen dissolved in the wastestream. The measure of DO indicates the ability of the watershed to sustain aquatic life. It is also a quality measurement for wastewater processing. DO can change readily in a sample if appreciable biological activity or iodine is present. Laboratory-prepared bottles and reduced temperatures curtail the amount of DO consumption for a specified time period or until the sample is again prepared for analysis. Minimums of DO are set by regulation to accommodate local aquatic life and to make sure the wastewater is processed properly. Depletions and excesses indicate problems in the wastewater or the watershed requiring immediate attention.

1. All references to exact parameters in this section are from *Standard Methods in the Examination of Water and Wastewater*, 19th ed.

There are many methods for determining DO in a waste stream. Membrane electrode probes can be utilized and are available through many reliable manufacturers. Membrane electrodes yield a relational rather than direct measurement. However, high-quality instruments are calibrated to perform this relational measurement with precision, and they generally do provide accurate readings to the satisfaction of the utility and regulator alike.

Biochemical oxygen demand (BOD) is the potential or measured difference in dissolved oxygen from a set time zero to the completion time of analysis through incubation. BOD is an accepted water-quality measurement technique and is used extensively in wastewater treatment. It is a discharge permit parameter in many places, even though the method of determining BOD can be difficult to manage. The difference in dissolved oxygen is due to the oxygen's consumption by the microorganisms in the sample. It is necessary to understand the accepted values for DO in watercourses and waste streams to properly understand BOD characteristics. There are two forms of BOD analysis. Nitrogenous BOD is the uninhibited consumption of DO in a sample for the incubation period. Carbonaceous BOD (CBOD) is the exclusion of the nitrogenous component of the water or waste stream through chemical inhibition in the incubation period.

The usual BOD incubation period is five days. This selection is based on the early British observation that no river appeared to take longer than five days to reach the ocean. The ultimate BOD test lasts 30 days and is much more accurate. The problem with both of these analyses is the time frame; either way, it takes days to get results. BOD analysis also requires special seeding of microorganisms and controlled incubation for accuracy. In reality, the same sample incubated under the same exact conditions can produce different BOD values. However, BOD is still an accepted parameter and used fairly consistently across the industry.

Two devices may solve the problem of real-time BOD monitoring: the fermentation monitor and the curve estimator. The accelerated BOD monitor actually accelerated fermentation of the wastewater by its own independent process. On-line BOD devices are quite accurate, although not yet accepted by major world regulatory bodies. They are expensive and bulky and probably more applicable to plant process than field activities. They are accurate despite compressing the usual 5- or 30-day fermentation period into 5-minute increments.

The second solution is the curve estimator. The dissolved oxygen tends to form a deterioration curve that may or may not follow an estimation path. The curve estimator measures two DO points and matches them to known BOD deterioration patterns. These devices are much less expensive than the fermentation devices but have a greater margin of error. Also, like the fermentation devices, their results are not accepted by regulators. However, they are portable and can work in the field if BOD measurements are required.

Suspended solids are the amount of particulate matter contained and moving in a water sample. They are generally measured in milligrams per liter (mg/l). There are two kinds of suspended solids: particulate and dissolved. Particulate suspended solids are those that can be strained and weighed in a laboratory. Dissolved solids do not strain out under normal conditions. The combination of particulate and dissolved suspended solids is referred to as *total suspended solids* (TSS), which is a parameter of interest. TSS, whether organic or inorganic, is a characteristic of both the watershed and the waste stream. Because of this variability, few, if any, reliable devices are available to measure solids directly in the water of interest; this is performed in the more controlled environment of the laboratory. TSS measurement is generally performed and measured in units of turbidity (or NTUs). Turbidity measuring devices generally are not referenced with other water parameter measurements. Equations relating turbidity to density are not generally accepted, even though most instrument manufacturers install both features on their instruments.

The concept of pH is of substantial interest as it is generally a parameter required for permit reporting. This chemical term represents the acidity or alkalinity of water. It is the measurement of hydrogen ions in aqueous solutions. The value is exponentially expressed because of the small values incurred. The inverse measurement, pOH, uses the hydroxide ion instead of the hydrogen ion. Clear water has a pH and pOH of 7, the center of both ranges. Here is an example of the logarithmic function of pH: A solution of pH 8 has 10 times the hydroxide ions of a solution of pH 9 and 100 times the hydroxide ions that a solution of pH of 10 possesses. This parameter of interest can be measured in any watershed or waste stream.

The most common laboratory method of measuring pH, the electrometric probe method, uses two electrodes to generate a potential difference between a glass electrode and a reference electrode referenced to a standards bureau and a temperature probe such as a thermocouple. Such potential difference relationships are generally linear. Using glass liners also makes the probe assembly fairly fragile, but many solutions to the problem facilitate open measurements in water and waste streams. The measurement of pH is related to the presence of industrial chemicals and other undesirables that could have negative effects on the watershed, treatment process, and workers in the facilities.

Another parameter that can be measured by the same device as pH is the oxidation reduction potential (ORP). This measurement was originally geared for freshwater measurements, but interest in applying it to the wastewater process and waste streams is growing. ORP is a measure of the potential for reduction of oxygen, or redox, reactions. It is akin to the potential for oxidation, observed when cut apples become brown or rust forms on a nail. ORP probes generally utilize platinum electrodes (or in some cases graphite) and utilizes the millivolt scale for measurement. Wax-impregnated electrodes are less sensitive to electrode poisoning than the platinum probes. The devices require constant attention and foul quickly. No accepted manner for the presentation of ORP values has yet been defined. This makes the millivolt scale amendable to whatever interface is appropriate for the application.

The conductivity parameter appears in manufacturers' literature as a water analysis feature. It is defined here as the ability of water or the waste stream to conduct electrictry. Conductivity is usually measured in millisiemens per meter. Some countries convert this to micromhos per centimeter, *mho* being the inverse of *ohm*, a measure of resistance. Natural conductivities exist, and by examining them, the presence of such chemicals and compounds of interest can be estimated. Obviously, the exact compound is not discernable from the measurement alone, but changes in conductivity certainly can point to relative changes in known concentrations of chemicals in the waters.

The measurements are made with a variety of conductivity cells, selection of which varies by manufacturer. Many types of cells are equally appropriate for use in water analysis.

All the parameters described here are of significance to the sewer infrastructure management team. Their relative concentrations and presence in various parts of the infrastructure can have substantial effect on sewage treatment plant operations and loadings to watersheds if a discharge is experienced. It is not necessary to be knowledgeable about the intricate details of each measurement unless there is some underlying professional or personal interest. A sewer utility operator *does* need to be aware of the presence and of methods for identifying them in the system. The ability to provide this information as required to regulators and to identify the characteristic presence in both the sewage and watershed can make significant contributions to protection and efficient operations.

9.5 PRACTICAL APPLICATIONS OF BIOMONITORING

The ability to utilize the data generated from biomonitoring activities can significantly enhance the operations of any sewer infrastructure, improve transport characteristics, and contribute to protection of the watershed. Portable telecommunications assists in placing devices in locations that several years ago could not have been monitored.[2]

The protection of watersheds, especially in cities and areas of heavy industry, is the predominant application of biomonitoring data. Biomonitoring may be the first line of defense in the event of a chemical spill or other problem within the plant environment. As an example, monitoring for temperature at a hydroelectric plant can protect the fish population by detecting failures in the cooling systems and preventing or limiting hot water discharge from the sewers to the receiving waters. A device as simple as a temperature monitor is enough to provide this information. In the case of discharges from a food processing plant, rapid changes in BOD can signal a failure in the plant's internal processes, which in turn can cause problems at the treatment plant. A staged BOD device of any kind can alert for this failure so the utility can respond as fast as the food processing plant can. Remember that the first priority of the manufacturing facility is internal and notification of the utility may be delayed. Properly staged equipment can assist in the response to such issues and concerns by alerting the sewer maintenance and sewage treatment plant operators so action may be taken.

Monitoring for industrial protection can also defend against businesses that violate regulations protecting the environment. If questions arise concerning the integrity of a business, the monitors can assist in answering them. Monitoring for changes in the waste stream and drawing the sample into a trouble bottle can provide the evidence necessary to prove or dis-

2. T. Day, "Biomonitoring Equipment and Techniques" (paper presented at the Industrial Pre-Treatment Conference, Indianapolis, Indiana, July 1996).

prove the alleged violation. Such potential for evidence is important in the legal system as part of its effort to bring violators into compliance.

The change in suspended solids in a watershed area can be the result of rainfall washdown, debris from illegal discharges, and other influences of varying complexity. The characteristics of each must be noted, as the chemical content will be different in the recorded measurements. Changes in the pH and conductivity in watershed areas can indicate washdowns of pesticides due to changes in terrain that affect the watercourse. Failures in the sewers running close to the watercourse can be recorded and quickly remediated.

A big challenge, especially to the growing leisure industry, is the protection of bathing beaches and recreational fishing areas. Many communities depend on tourist trade for their survival, including those in coastal areas, tourist and resort islands, and camping areas. Monitoring for discharges that may negatively affect the water supply or the ability of paying visitors to engage in recreational activities protects the financial investment of the governments of these tourist areas. Monitoring also protects the aquatic life of these waters.

Samplers, including flowmeters and standalone devices equipped with probes for this analysis, supply the necessary information. Setpoints for notification of adverse conditions can be programmed, potentially initiating remedial action before damage is inflicted on the area of surveillance. For example, after an alarm, the simple act of placing booms at outfalls may be enough to protect these areas from damage. Biomonitoring may allow a plume to be recognized before it is dispersed, allowing the source of damaging discharges to be more readily traced.

Biomonitoring, depending on local interest and regulatory structure, is potentially the fastest-growing portion of sewer infrastructure management. The ability to trigger notifications based on changing conditions and to warn of and document violations assists in the protection of the watershed and preservation of the integrity of the treatment works. Biomonitoring serves as a kind of insurance for the resources invested by the local utility in protecting its watershed assets for present and future generations.

9.6 ALERT AND ALARM ALGORITHMS

The ability of a monitoring system to issue alerts and alarms based on conditions within the infrastructure is useful and potentially quite valuable. Algorithms provide the notification limits required to generate warnings about problems within the sewer infrastructure and to instigate actions for the protection of the health, safety, and property of the drainage district. The process of developing algorithms enables the utility to incorporate its own values in the equations so the alert and alarm values can be understood in terms of local conditions.[3]

An alert is notification of a situation that may negatively affect the site and, possibly, the critical path. An alarm is notification of a current condition that is causing negative effects and requires immediate attention. The alert is meant to precipitate remedial action to prevent an alarm condition. The goal is action to prevent damage to the drainage district.

This discussion develops a simple set of algorithms that may be directly implemented. The algorithm requires rain gauges and communications devices so information can pass to a central computer system. The actual methods are less important than an understanding of all the parameters needed to generate warnings as needed to protect the infrastructure and adjoining watershed.

The installation or retrofitting of the rain gauge network should be correlated with a duration and intensity of precipitation that correlates to the beginning of a typical storm event. This provides a trigger or setpoint by which a storm can be designated. As an example, a utility operator does not want a lot of high-level alarms during a storm event, when the sewer is expected to be high. The question is what constitutes the beginning of a storm event. For this discussion—and for actual use, if desired—any rain of 0.10 inches recorded within a 24-hour period represents a wet-weather day. (Events yielding less than 0.10 inches in 24 hours and 100-year return-interval storms were eliminated from this analysis.) This figure represents an intensity but no duration that is of use to a dynamic alarm system. Based on the analysis using the localized hydrologic cycle as a base, a storm of 0.10 inches or greater could be determined by recording 0.05 inches of rain in a 0.5-hour period. These values provide, for this discussion, the upper and lower limits for development of the alert/alarm algorithms.

An endpoint of a wet-weather event next needs to be described to bracket the occurrence of precipitation. This boundary condition is necessary to provide filtering for the alert/alarm algorithms. Here, a timing deadband or moratorium on alert/alarm generation is employed to

3. T. Day, "Alert and Alarm Algorithms for Dynamic Wastewater Collection Systems," WEFTEC '96 Dallas, Texas, October 1996.

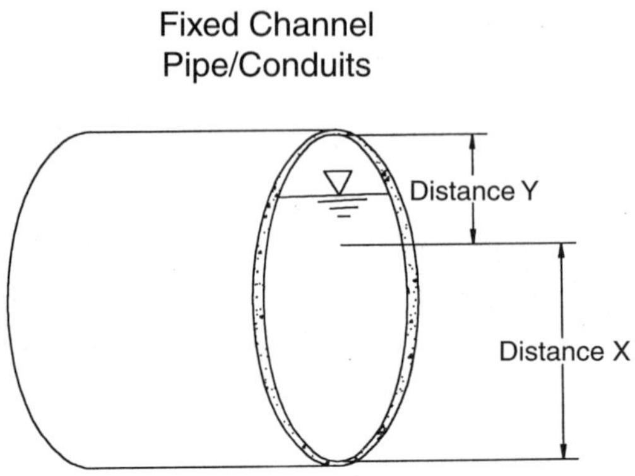

Figure 9-3 Generalized alert/alarm demarcations for algorithm development.

smooth the irregularities in storm patterns. A moratorium of 12 hours past a notification sequence provides the other boundary condition. The 12-hour moratorium was derived from observation of plant flow patterns after storm events and time needed for the collector system to return to steady state or equilibrium. It gives the system time to drain and the plant to process the water stored in relief sewers and other holding points.

The case of small systems is discussed briefly here. It may not be necessary to have the alerts pass through the central computer if the system is small and if it is the preference of the utility operators. Here, a rain gauge can be added to the installation directly as an input, enabling any comparison for rainfall to be made locally. This is impractical in large systems but attractive for small ones. Because most datalogging/RTU/PLC devices have paging/alarming functions, the necessary coding is readily performed. A decision such as is related to local conditions is best left to the utility operators, who base it on product availability and the integrity of the system integrator.

The base system is now established for describing an alarm condition for a wet-weather event. In-place components and support for the generated conditions requiring action have been outlined. The majority of this discussion pertains to large systems as they are difficult to implement. Applications of the algorithm provided to small systems will become obvious to the system integrator and those utility operators who truly are aware of the conditions within their own system.

The following algorithm is based on conditions particular to sanitary sewers and other conveyances where there are no diversion structures of interest. It is usually a pipe or other conveyance with no other unique properties. In the case of a pipe or other conveyance where the structure is fixed:

- The alert condition consists of $X\%$ of conduit potential (diameter or maximum height, level sites only)
- The alarm condition consists of $(X + Y\%)$ of conduit potential (diameter or maximum height, level sites only).
- An alarm condition consists of velocity within the conduit falling below X ft/s or known particular limits and the depth reaching $X\%$ of conduit potential (level and velocity).

The task is to sample the various locations for velocity and level and to understand, usually through temporary monitoring, the hydraulic characteristics of each point of interest. It is through these unique characteristics that the X and Y functions make sense. Percentages are fine, but in the cases of small pipes, an inch range is probably more useful. The idea of extending percentages to the algorithm development is to add versatility as needed to particular conditions to the application and the concepts being considered.

In combined sewers and storm relief systems, dams and other types of fixed weirs are points of particular interest as well as easy places to implement monitoring and alarming devices. In the case of dams, where a tide gate is fixed or not operational:

- The alert condition consists of level $(X)\%$ of the height of the dam.
- The alarm condition consists of level $(X + Y)\%$ of the height of the dam.

X is a predetermined storage level within the trunk sewer or relief system. In this example, $X = 85\%$, or 85% of dam height. $Y = 15\%$, or the additional height that the system will bear before discharge is seen. In tidally affected cities and towns, accommodate the maximum tide height *before* inflow to the system occurs. This is a case where local knowledge and studies assist in fine-tuning of the algorithm values and are imperative in their development.

The structures in a sewer system are unique and have their own particular requirements. For example, a tide gate can either discharge or receive inflow, depending on the conditions that affect it. Therefore, the algorithm for a structure requires development and explanation along the way. In this case, a tide gate is a variable parameter and must be treated accordingly.

The gates open based on tide versus trunk level. The differential determines the point at which the tide gate opens. A differential factor allows each gate to maintain some closure between these two settings. There is also usually a point below which the differences in pressures between trunk and tide have no effect (say, 6 inches). Therefore, the head rise before the opening point becomes the alert factor and the point of excess becomes the alarm factor. These points vary by location. The algorithm development begins as:

(i) $$Level(Sewer) - Level(Tide) = DIFF$$

where the differential level becomes a decision value. If the trunk head rise is greater than the tide level, the decision value is positive. If the differential is 0, then the alert condition should already have been activated. If the *DIFF* value is negative, the alarm condition exists and central operations should be notified after a deadband is reached. Thus, the positive values and zero represent alert conditions, and negative numbers represent alarm conditions. The task now is to define alert and alarm conditions.

Clearly, the alert condition is when the trunk and tide equal each other ($DIFF = 0$) above some minimum, which is:

(ii) $$Level(Sewer) - Level(Tide) = 0$$

This indicates that the system is about to discharge, a situation requiring immediate response. The concern is that the trunk and tide are not also zero or at some minor level at which action would not happen anyway. As noted above, the offset must be site-specific. Therefore, an offset to the elevation is needed that raises the decision value generated from the head rises above the zone of no action. This offset level is added, accordingly, as:

(iii) $$Level(Sewer) + OFF = T$$
$$Level(Tide) + OFF = S$$
$$DIFF = D$$

where *OFF* is the offset level bringing the elevations above the neutral zone, *T* and *S* are the new values that include the site-specific sensor offset information for normalizing the elevations, and D is the normalized differential based on the contribution of the T and S values. Equation 1 can now be expressed as:

$$(iv) \qquad T - S = D$$

for which the following case/condition statements can be expressed based on the generated decision values:

$$(v) \qquad \begin{array}{l} \text{if } D \geqq 0 \text{ then alert} \\ \text{if } D < 0 \text{ then alarm} \end{array}$$

The next variable is the float/weight assembly on the moving gates. This assembly maintains some storage based on the unique site characteristics and conditions. This variable is defined here as W, as it is particular to each site. W determines the range by which the trunk can exceed the tide prior to the system discharging. If the weight system does not provide any storage (as determined by site inspections), then the W value is zero and does not affect the decision value. If necessary, W can be added to the T variable for the final algorithm of:

$$(vi) \qquad (T + W) - S = D$$

No lower limit is currently assigned to the alert boundary. An arbitrary lower limit can be selected—6 inches or 95% of the tide height, whichever is more applicable. This arbitrary value will suffice until trials can be performed.

To summarize, the following developed algorithms apply to an active tide gate:

$$(vi) \qquad (T + W) - S = D$$

$$(vii) \qquad \begin{array}{l} \text{if } D \geqq 0 \text{ then alert} \\ \text{if } D < 0 \text{ then alarm} \end{array}$$

The algorithm above provides a lower alarm limit, an alert condition, and an alarm condition that can be readily coded. The high tide algorithm applies at sites with structures as well. Generalized algorithms are not possible in this discussion due to the nature of wet-weather conditions; however, the subsequent mathematical expressions suffice with the proper modifications as would be necessary for the individual system.

The process of algorithm development attempts to cover all of the parameters that are expected in the development of alerts and alarms for the sewer system. The algorithms may be intricate and require modification for the individual system, but if all of the potential issues concerning alerts and alarms are considered, those that do not apply can be left out. It is better to leave out some aspect that is known and considered than to overlook some aspect and realize it later.

9.7 AUTOMATIC CONTROL

Automatic control is just starting to receive mainstream acceptance in the United States. It has been extensively studied in Europe, the Scandinavian countries, and China. Automatic control provides the ability to utilize the latent capacity within the sewers, level the treatment process, and, potentially, provide power savings for the treatment works. Automatic control of the sewer system, in some form or another, can be a viable option if the equipment is available and the return on investment is great enough.[4]

The main purpose of automatic control is to move sewage from an overtaxed portion of the sewer to a less utilized portion. Another purpose is to use large sewers that appear to have capacity for storm or event storage to level out the peaks and valleys in daily plant flow. These options show significant potential for the future of infrastructure management in densely populated areas. The use of automatic control equipment to move volumes of water through sewer systems is an attractive possibility in long-term planning and permanent improvements to the sewer infrastructure, and it may provide an alternative when construction or additional infrastructure is being considered to best utilize latent capacity in the system.

Two types of control are discussed in this section: static and automatic. Static control is the use of dams and weirs to store the volume of water in the latent portions of a sewer system. Latent capacity is the underutilized capacity of a sewer infrastructure that could be accessed with the proper control scheme. A hydraulic study is undertaken of the critical path of interest. In the majority of sewers, a portion of the critical path could be uti-

4. T. Day, "Real Time Control in CSO: Maximizing Flow with Automation" (paper presented at the Urban Wet Weather Conference, Quebec City, Quebec, Canada, June 1996).

lized if flow could be backed up into it for some time period. This latent capacity can store water underground until the treatment facility can accept it. The capacity is utilized when a series of weirs or dams are staged through the critical path, withholding the volume within the sewer system based on the computed time of storage and eventual release to the treatment works.

Static control has the advantage of failing only when the capacity of the dams or weirs is miscalculated or when structural failure occurs. If designed and implemented properly, the control should withhold the calculated volume for the expected time period. The dams or weirs require only the usual periodic cleaning to prevent solids from collecting on the structures. When systems are carefully planned, solids deposition is minimal and the need to remove it coincides with the maintenance schedule. The downside to static control is the inability to adjust to any given condition and to accommodate a range of volumes. This is not necessarily difficult but certainly desirable, as the plant process is adjusted to accept the throttled volumes. The last issue is if the system is misdesigned, which is always a concern based on ongoing and long-term system studies. There is little flexibility in increasing or decreasing storage capacity to provide system balance. The monitoring system can report the effectiveness of static controls and how they affect the total volume and system balance.

Automatic control is the addition of standard process control equipment to the sewer system to throttle flow or to store flow volumes in anticipation of more optimal treatment times. Leveling the flow peak also levels the demand for pumps and other process equipment within the plant facility. Automatic control is attained by modifying devices designed for standard process control work to withstand the harshness of the sewer environment. Flow and level monitors become the governing devices in the operation of the automatic control system. Automatic control is often referred to as *real-time control*—the system responds to conditions as they occur.

The first use for real-time control is the movement of flow from one part of the drainage district to another. This can be performed by installing gates and perhaps pumps controlled by a device programmed to make decisions based on its inputs, such as a programmable logic controller or a minicomputer with industrial inputs. The flow is redirected from the more heavily taxed portion to the less used portion. The gates can be inflatable dams or standard gating systems. Pumping is not often performed because of the excessive electrical costs, but the option exists.

Inflatable dams are gaining attention for their reliability and versatility of use in automatic control schemes. Such devices are large airbags of canvas, similar to sail canvas, that inflate with a small compressor and either release or retain water in a sewer. The compressor need not be large to keep the dam inflated. Level sensors on each side of the dam are responsible for monitoring; they have setpoints for action as determined by the system designer. A compressor must be mounted in either an underground chamber or a surface cabinet, with local utility services for communications and electric. Because inflatable dams are like big balloons, debris and large floatables bounce off them, causing little or no damage; timbers and other large debris that may be carried in larger sewers are simply deflected or released by the dam. A sailmaker or other craftsperson skilled in canvaswork can repair the dam when required.

Automatic control is also used for storage of wastewater until it can be processed more cost-effectively by the treatment plant. In many schemes, the large sewers, such as combined systems, have capacities that permit the storage of flow volumes, preventing discharges into the receiving waters before the treatment plant processes the sewage. The goal is to throttle the flow to the plant in order for the sewage works to minimize power costs and keep peak demands at a minimum. This approach has been employed in several large combined sewage systems around the world with a range of success rates. In combined systems, tide gates are automated to withhold the water based on a process algorithm, releasing the minimum to the watershed.

Combined systems are ideal candidates for the automation. Often some regulating device is already in service to retain flow volumes in the system. At the time of construction of these systems, there was no way to know the levels that would be achieved in advance. The result was development over time of capacities that vary from the original design specifications. Unless level or flow monitoring is installed, no one really knows the actual capacities of the system. Automatic control can assist in accessing more capacity in combined sewers that appear fully utilized.

Biomonitoring and automatic or real-time control can be combined in interesting ways. Water quality monitors have been installed at automatic control locations to determine the quality characteristics of storm flows. The research appears to show that if the water can meet regulatory requirements for effluent quality, the storm flow can simply be screened for trash and other floating debris and returned to the watershed. This possibility is worth continued investigation. It offers the opportunity to substantially reduce the cost to treat. Obviously, flow volumes that do not meet water quality standards must be returned to the treatment facility. However, in the case of long wet spells, snow meltdowns, and similar climatic occurrences, real-time control systems coupled with biomonitoring information can route much of the dilute flows back to the watershed. However, the size of the system in which the automatic control is to be implemented must be considered. It simply may not be cost-effective, no matter how promising the desktop studies appear, to implement this technology in small systems or small pipes.

Many flow-measuring devices today have inherent control capabilities that add real-time control in small sewer systems. For a minimal cost, the contact closures that appear as options or standard features can be altered to perform specific control aspects as desired. In this case, automatic control may be a viable option if there is significant tangible benefit to the system at large. Small packaged control solutions also can assist in this effort. Consider how much it would cost to implement automatic control compared to the cost of constructing additional infrastructure to achieve the desired result. If the cost of real-time control seems to be significantly lower, consider the possibility further. If the costs appear to approach those of new construction, then the dig should be considered. As noted earlier, automatic control is an expensive solution and one that requires extensive maintenance for proper operation.

The methods described here outline how to construct one site. The instrumentation of one location allows the cascading of other locations using similar technology or more applicable solutions. The test site requires a dedicated controlling device or a flow-measuring system with a series of contact closures and high/low limits. It also requires the ability to communicate with a central location, to facilitate monitoring of the conditions. Real-time systems for the sewers generally require remote control, as it is desirable to control the system from a distance; However, for purposes of this demonstration, it may simply be best to allow the system to function on its own. By keeping the example simple and general, it will better allow a utility to adapt the concepts to their particular requirements.

A site survey should be performed along all critical paths of interest. The transport capacity of all paths should be observed and, according to the best judgment of the system operators, one point selected that appears to lend itself to testing. Electricity and communications should be available. Determine whether the location is a sanitary sewer, a major interceptor into which the flow is to be diverted, or a combined sewer in which discharges are to be mitigated. Inflatable dams are best for the sewers into which flow is to be detoured. In the case of working tide gates, the packaged hydraulic systems that throttle the gate are a good solution. Inflatable dams should include a means of measuring the pressure within the bag. Tide gates should include a means of moving the gate and determining its position. Several packaged solutions are available that include this hydraulic function. It is advisable to choose packaged solutions when they are available, as the manufacturer can then render assistance in the application.

The site selected and the proper technology targeted, the on-site construction can begin. Whether the test involves flow shifting or storage, goals should be established up front so the results can be weighed. In the case of flow shifting, temporary monitors should be added to the area that will receive the shifted flow, to determine the actual results. In the case of storage, the latent capacity and the currently used capacity should be calculated and the results substantiated with monitoring data. In the case of storage, calculate how long the volume needs to be held within the infrastructure before the sewage treatment plant can receive it.

Communications should be installed at the sites so central maintenance can monitor them and determine if they are responding as anticipated. At this point in the demonstration, it does not matter if the storage or shifting is achieved; what matters is that the equipment responds to the environment. The verification of the effectiveness of the technology can come later. The first shakedown period involves adjustment of the deadbands—ranges of measurement within which no action takes place—in order to minimize system bounce that is a result of jitter, a back-and-forth crossing of thresholds that causes the equipment to turn on and off needlessly.

The effectiveness of the automatic controls can now be calculated as a function of the data measured and processed. In the case of load shifting, it needs to be ascertained whether or not the technology shifted sufficient quantities of the volume to the intended portion of the sewer infrastructure. If the goal was storage, the volume of storage achieved and the time of storage can be calculated. If the targets were met, then maintenance requirements can be evaluated.

The fact that a program does not meet its goals is not necessarily a cause for concern. The demonstration program is also a test of the rationale used to develop program goals and the reasonability of their attainment. Missing a target in the demonstration usually indicates that another option or target was missed in the first screening. Perhaps the flow was shifted to the wrong portion of the sewer. It may be proven that the latent capacity that appeared in one portion of the infrastructure was due to upstream conditions or to the unanticipated effects of another critical path. Perhaps the wrong site was chosen for the evaluation. These evaluations provide information that will likely be useful the next time around.

Calculations should be performed immediately at the time of hydraulic action to determine the electrical cost of moving or storing the flow. This cost must be included in the cost to treat. If there is no significant movement in the cost to treat, then the real-time process has potential value. If the cost to treat escalates, this indicates that automatic control is not cost-effective. Power factors are a consideration in the case of storage; the shaving of peak demand and a concomitant reduction in peak penalties can in itself provide significant cost justification for real-time operations. This is why the time of storage is so important to the evaluation program; there needs to be enough storage time to remove peak demand on equipment already taxed by a storm event.

The limitation or elimination of discharges to the watershed must be considered before the maintenance options are screened. This has value not in potential cost savings nor in mitigated resource use but in the ability to provide watershed protection. There are cost reduction and screening methods to achieve this but, in general, there should be no cost limit associated with watershed protection other than what is reasonable for a ratepayer to afford. Technology should not be judged solely on its cost. Cost should be viewed as a screening method of segregating that which is practical from that which would pose excessive burdens on the ratepayers. Therefore, always consider the potential impacts on the watershed of any control scheme to determine if the level of protection derived from the automatic control merits further investigation.

The best method of determining whether the control scheme has a positive effect on environmental protection is through a combination of automatic sampling of pollutant loadings and observation. As part of the total analysis, before and after control options based on lab samples from a more typical storm should be combined with observation of the amount of visible debris being introduced or reduced to the watershed. If the potential for the introduction of pollutant loadings is significantly reduced and can be measured in the sampling, the screening for automatic control should be continued. If the effort appears to make no difference, the use of real-time control may not be justified. However, the safety of the watershed should always be of paramount consideration, and if the control strategies may contribute to this goal, they should be pursued.

The last issue in the screening process is maintenance, potentially the largest cost in the real-time control scheme because of the nature and application of the associated devices. These schemes introduce sensitive control equipment into environments of extreme hazards for which they were not designed. Because all real-time control systems are custom designed, the ability to integrate them to the point of interest is critical to the success of the program. Maintaining these custom devices is especially difficult as a result of all the integration required. Also, the equipment may be inactive for long periods due to seasonal variations, dry spells, and other climatic occurrences.

Consider maintenance performed during the period, including work crew activities. Downtime of the devices and equipment installed should also be recorded. Inflatable dams tend to get punctures from floating timbers and other debris, requiring patching. Hydraulic equipment decays in the corrosive sewer atmosphere. If the system is maintained by existing crews, workers must receive pertinent training. If the plan is to contract out the work, the cost of the maintenance contract should be factored into the assessment. The cost to keep equipment operational can far exceed the actual capital outlay for its purchase if it is incorrectly applied, improperly maintained, or simply not appropriate for the job.

Maintenance costs should be compared to benefits as a first-line screening. Does the expense justify this approach to achieving the water quality targets? Additional evaluation should be made of the cost to treat as well as potential benefits to the watershed. This process yields an overall picture of the benefits achieved, resources required, and action generated by the automatic control test. The utility is in the best position to make this evaluation as the benefit factors or total return on investment, for concepts such as water quality sometimes have little basis in cost. Benefit factors are goals and benefits and need to be evaluated on results.

The screening of the results and costs of real-time control allow further evaluation based on experience. The solution picked for evaluation may turn out to be a poor fit with the particular infrastructure, but the experience of working with that piece of technology enables more informed analysis of others. Gaining this knowledge is more valuable than striking the right solution on the first try. Technologies can be more reliable screened or redeployed to other parts of the sewer infrastructure.

Inflatable dams can be installed in combined sewer overflows and used to regulate the discharge to the receiving waters. This requires sealing off tide gates, in order to make the dam the regulating device. Care must be taken that the dams actually deflate during intense storm events to provide relief to the system; otherwise, manhole covers may be blown off along the critical path. The evaluation study checks on this mechanism. Assuming that latent capacity in the system is sufficient, strategically installed inflatable dams in the CSOs may provide substantial protection to a watershed that was previously exposed. This reduction of exposure does not incur either a cost or cost savings per se, but it is the goal of every water quality program.

Another potential effect of flow control devices in a critical path is the regulation of flow to the sewage treatment works. Many storms are concentrated in cells and affect portions of the watershed, taxing particular critical paths based on the track of the event. The installation of inflatable dams can allow the flow of an overtaxed critical path to enter the facility while storing or otherwise throttling the flow in the path on which the storm had the least effect. This is a method of reducing the peak demand on the sewage works and more evenly dispersing flow across the plant process, and it can help lower the cost of sewage treatment and produce less stress on plant physical equipment.

CSO and stormwater compliance programs can greatly benefit from real-time controls strategies that are custom matched to compliance objectives. Underground construction is always expensive. Installing off-line storage tanks and other diversion paths to the sewer infrastructure is also costly. The addition of automatic controls for regulation and maximization of the structure increases the value of the fixture to the utility. The cost to install real-time controls at the time of infrastructure installation is always relatively small. Maintenance costs must be considered, but if the automatic control is designed as part of the infrastructure addition, these are often reduced.

The majority of real-time control implementation is installed in existing sewer infrastructures. It is harder to predict the maintenance requirements of retrofitted equipment than those of equipment installed at the same time as the sewer itself. The ability to anticipate maintenance demands allows designers to mitigate the required effort in the planning stages.

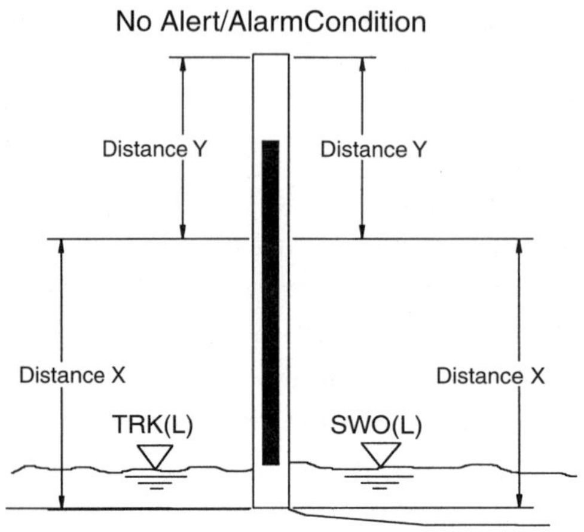

Figure 9-4 Baseline demarcations for dam or tide gate.

Real-time control, by the definition presented here, is not limited to the inclusion of structures to divert or store flow. Automatic control can be synchronized with pump station activities for maximum benefit. Depending on the utility's scheme of operation, the pump station can drain and fill the off-line storage, the relief sewers, or another structure of choice. It can be aligned with the sewage treatment works to orchestrate the flow, leveling the contribution to the plant. The concepts associated with real-time control should not be limited to any one application; they can be applied to all aspects of the infrastructure that can be controlled, throttled, or otherwise managed.

The example of deep tunnels is a case in point. Utilities implement deep tunnel technology, described in earlier chapters, for purposes of off-line storage. Pumping the flow in and out of the tunnels is an excellent example of a real-time process. Not only does the flow need to be monitored as it enters, occupies, and exists the tunnels, it must also be controlled in the sewer making the final delivery to the treatment works. The capacity of deep tunnels far exceeds that of the main sewer to the treatment works. If the flow is delivered too fast back to the main sewer, the overtaxing issues resurface. For this reason, the deep tunnel real-time control must be integrated with the automatic control of the system at large.

Automatic control is a subject of intense debate, and few successful examples exist. Combined with biomonitoring, automatic controls may assist in reducing the cost to treat. This helps the treatment plants in their operations. However, it is an expensive technology to implement if retrofitted, and it is costly to maintain in every situation. It demands a full permanent monitoring system for proper implementation. It requires utility service and extensive maintenance schedules. Despite the difficulties associated with automatic control, it may be one of the only options open to large utilities facing crowded underground conditions or those in high-population areas.

The screening approach taken in this technology evaluation assists in determining whether or not the devices of interest will cost-effectively meet their targets. As an additional issue, the evaluation of watershed protection, which really has no price, must be included in the evaluation. It makes sense to implement automatic control when modifications are made to the sewer system. However, the evaluation is still riddled with questions that only the operating utility can answer as part of its analysis.

9.8 COMPUTER SYSTEMS

The final step in developing monitoring and control capabilities involves the computer system. In the not-too-distant past, highly trained professionals were required to manage such systems. Today, minimal training—the equivalent of a few evening college courses—generally is enough. Software still remains an issue, but this is quickly being resolved. It can be difficult for a utility to sort through the variety of products available in order to choose the best system for the task at hand. The discussion in this section addresses choosing the technology on a needs basis, cutting through the hype of the modern computer world.

There are as many manufacturers and integrators of computer systems as there are methods and modes of overall system operations. In all, only a few companies manufacture accepted computer operating systems for these devices. Because of this, the hardware platform tends to be level, varying only by the quality and the options available on the market. Such products drive the development of third-party software, for which user preferences develop because of the learning curve required to master it. Through all the maze of products and services, it's important to determine which system is best for the task at hand.

The first evaluation is simply to discover which software is available. If the manufacturer's software is targeted for use, specifications are fairly clear and straightforward. Simply purchase a machine matching or exceeding requirements, and purchase the software service agreement as well.

For small systems, this is likely the best solution. If the monitoring system is large and third-party software solutions are necessary, software vendors must decide which machines and operating system platform optimize their product. In this case, it may make sense to let the software provider perform the hardware installation.

Local area networks in some form probably exist in the building housing the utility administration. Therefore, getting the data to others should not be a technological challenge; the true challenge lies in setting up the system such that the standard polling and data handling do not interfere with standard data requests. Interference can occur, especially in heavy data traffic on well-utilized local area networks. Therefore, a targeted personal computer or networkable storage device featuring large data storage capacity should be reserved for this data. Personnel should be allowed only to copy and view data on this system; any other function should be performed on a copy of the data downloaded to individual machines and not on the network data storage device.

It is essential to provide system security for the data and firewalls so there is no direct interrogation of field equipment without the express permission of the responsible utility system operator. Since incorrect operation of the monitoring devices tend to produce work orders regardless of the intent of the party attempting the interrogation, it makes sense to know who is interrogating the devices so changes that appear in the data can be explained. For example, it is possible for someone to poll a field device and inadvertently erase the memory. On the next scheduled system scan, the interrogation computer finds the device has no data and issues a work order for repair. Nothing is wrong with the device, but data handling was sloppy.

It is useful to know the approximate size of the data files that will be handled by the system. This can be estimated by multiplying the size of a sample file from a single device in a single day by 365, then multiplying the result by the number of sites from which data is expected. This is the estimated file size of one year's worth of data. Disk space today is not the problem it once was. However, knowing how many hard drives are required to harbor the information is part of good planning.

The number and types of sites determine the file structure and labeling methods. Manufacturers of instrument software have automatic labeling schemes, useful for small systems, that include a site code and a date stamp. On large systems, this approach may not be adequate and a separate labeling system should be developed. If the utility has an existing internal protocol, it may serve well. Otherwise, one should be developed whose methodology will be useful to the system operators.

The size of a sewer system will be a factor in the size of the computer system needed. Small sewer systems tend to encompass relatively few field instruments and usually need relatively small computer systems. Large sewer systems have many diverse points along a variety of critical paths. Such systems generally require a large and complex computer system as well as a staff member dedicated to managing the data and handling the day-to-day data requests. There are no fixed delineations for small or large systems other than their relative complexity.

The data communications solution must be addressed at this point. Assume that the technology for the monitoring system has been chosen. Now the interface to the computer system must be considered. If the system is small and the interface equipment simple enough, then the choice should be straightforward. Manufacturers usually offer solutions that interface with most standard protocols and their input will fill any portals not covered from the packing literature. Tone modems for telephone lines are very straightforward. Cell phone interfaces for commercial applications are slightly more complex but still can be installed readily. Interface cages are available if more than one modem is required. For small groups of modems, only one personal computer is required. For large systems, a data communications front end should be considered. This is simply a dedicated personal computer that has the task of extracting the data from the field devices and moving it along to another processing area. Even on mainframe systems, personal computer interfaces are now the norm.[5]

At this point in the discussion, the computer system can be chosen. As noted, a single personal computer may be adequate. If space permits in the computer, install two tone modems that can back each other up. At current prices, this is a smart investment in data quality and consistency.

If a large system is required, then a networked solution may be the best. The need for full mainframe systems is declining as the power of the personal computer increases and technical requirements decrease; the recommended platform is thus a string of PCs. As noted earlier, the information systems department often has already dictated specifications for

5. T. Day and C. Bradstock, "Moving from Old Computer Systems to New Platforms," WEFTEC 1996, Dallas, Texas, October 1996.

personal computers and network functions. Most manufacturers of process software can work with these specifications, resulting in optimal performance.

A central data repository is essential. This central repository should be accessible on a read-only basis to virtually all staff. Only the system manager and a designee should have the ability to delete a file from the system. If a data file is considered permanent, it should not be alterable; this precaution preserves the integrity of the data set. This rule should apply even if the data is incorrect. This sounds odd, but the ability to demonstrate to an outside party that the data constitutes a permanent record that cannot be altered can serve the utility well, especially in lawsuits. To maintain data integrity, the files should not be alterable in any manner once placed in the repository.

The consideration for the network hierarchy and protocol should also include the ability to accept data from temporary field devices. Such devices are often overlooked when large systems are being developed. Questions concerning how to label the data and where to put the files emerge later. Therefore, always keep in mind temporary devices and related files when generating system requirements.

The personal computer platform means that minimal training is needed to generate a maximum return for the system operators. In the recent past, weeks or months of training was needed to learn to operate cumbersome proprietary mainframe systems that were also expensive to run. This is not the case today, when only a few night school classes are needed to develop basic proficiency. A few more classes builds expertise that can assist in day-to-day operations. This means someone with initiative and drive can be trained to manage even large computer systems in about a week. This is a great benefit to the utility operator.

9.9 SUMMARY

Enhancements to the monitoring system increase returns on the investments made by the utility. Whenever a feature can be added that provides information about the infrastructure, the number of confined space entries and related travel time between sites and the central facility declines. Even if monitoring data is manually collected, the data set is considered more reliable than data sets derived from direct observations.

Biomonitoring is the determination of water quality and of the effects man-made pressures exert on this quality. The measurement of chemical parameters or the indication of potential reactions generally is informative enough so that reliable assertions about water quality can be made. Biomonitoring supports protection of the watershed by identifying the presence of hazardous substances before they are introduced into the receiving bodies. It assists the utility operators by supplying early notification of undesirables that not only would be harmful to sewer workers but also would upset the biological treatment process. Reaction agents within the waste stream may indicate other hazards or the potential for loadings that would overtax the biological treatment works. For example, excessive BOD depositions from foodstuffs manufacturers are not necessarily hazardous but could cause nutrient-dense material that upsets the microorganism staging at the plant facility. The ability to determine waste stream characteristics such as pH and temperature also assist in protecting the biological treatment process.

The components of biomonitoring include direct monitoring and sampling. Direct monitoring is the preferred method of providing notification of potential hazards. Sampling is the accepted method; many schemes for generating a sample for analysis are available. Sampling always requires that a laboratory perform the analysis. A combination of these approaches allows a sample of suspect water to be drawn while directly monitoring for unusual contributors to the waste stream. In watershed areas, samples may become court evidence when a suspected polluter is prosecuted.

Interest in biomonitoring is growing quickly. The ability to monitor water quality has great potential to support watershed protection. Direct measurement of the microorganism makeup of wastewater and other parameters yields immediate results, as opposed to sampling and its attendant time lag.

Monitoring systems that feature automatic polling and data collection can convey an additional benefit through alerting and alarming algorithms. The discussion included an algorithm development process that defined the limits and guidelines of alert and alarm conditions. Determining the beginning and end of a storm event, turning on and off the alarms, and understanding the difference between alerts and alarms all assist the utility operators in system management.

An alert is a notification of an impending hazardous condition. An alarm is a notification a current hazardous condition requires immediate attention. The devices that generate these notifications must be shut down during storm events to avoid false alarms and turned on again after them. The setpoints, limits, ceilings, and other parameters assist in providing returns on the investment by extending an additional level of protection to the watershed, the drainage district, and the community it serves. Alerts and alarms also allows for locations of trouble to be readily identified, including the potential site, its critical path, and the times that the situation began and ended. These times may be retimed for regulatory purposes. Rapid response times give regulators confidence in the handling of problems and assurances to ratepayers that the funds for the system were spent wisely.

Automatic control systems, also known as real-time control systems, are under intense study in many areas of the world. They may constitute a significant asset by making use of latent storage capacity within larger sewers. The ability to identify latent capacity and to design systems to utilize it can affect plans for additional infrastructure construction. The use of automatic control systems also can smooth out the peaks in power consumption by pump stations and level the loadings to sewage treatment works.

Automatic control can be coupled with biomonitoring to produce a system potentially able to return essentially clear runoff to a watershed. This substantially saves on the cost to treat and limits the conveyance of clear water to a treatment plant. Obviously, this approach must be approved by regulators; screenings for trash and other debris are necessary. The capability, however, could be quite beneficial to both the treatment works and the sewer infrastructure.

Computer systems are fairly straightforward today in comparison to several years ago, thanks to the evolution of the personal computer. Today, a personal computer has more power than mainframe systems did not long ago. It is difficult for professionals versed in current computer technology to distinguish many of the mainframe specifications of yesteryear from the personal computer specs of today, except today's machines have more hardware components per footprint.

Most utility information systems departments follow standard patterns employed by most manufacturers of software and supplies. It is thus relatively easy to match products to the computer platform of choice.

It is usually necessary to limit access to the data and field devices for security purposes. Field devices should not be polled by anyone outside the utility's system management groups; this rule helps ensure data integrity and device performance. If work order generation is based on site responses, strange or unexpected incidents in the field can be explained without undue investigation. The data should reside in a permanent repository from which it can only be copied. Other data functions should not be available to the general user.

System enhancements benefit the utility. Additional monitoring functions make the equipment more useful and provide improved service and protection to the communities and the watershed. These general enhancements are a start in understanding the particulars that each utility requires. They make the system more valuable, providing a greater return on investment to the ratepayers it ultimately serves.

CHAPTER

10

Funding and Justification of Sewer Infrastructure Management Systems

10.1 INTRODUCTION

There is probably no more important aspect of sewer infrastructure management than funding. Seeking funding requires that all aspects of the program demonstrate, in detail, its projected value. Those promoting the implementation of a proposed program must be able to anticipate questions that will be asked by outside parties. Without approval from those who control the funding, the program goes nowhere.

Questions must be asked after all the good reasons that support the endeavor are placed on the table. Those involved with a program always know why it should be implemented regardless of cost. They often overlook arguments against the program until some outside party asks pertinent questions. This section explains the basis for asking such questions as part of the development program and suggests ways to fund the project as deemed necessary.

Many of the benefits that justify a sewer infrastructure program have no visible return on investment. This can be stated in another manner: Money is not always the best justification for the program. Money is the focus of this discussion, but it must be understood that dollar issues may not be the best criteria for justifying a program. One of the best reasons to implement a sewer management program may be the potential environmental protection that it enables. Protection of watersheds and receiving waters certainly yield financial benefits due to reduced water processing costs. However, the inherent human responsibility to preserve natural areas from pollution is usually justification enough.

The concept of spending resources for environmental protection is not new, though it sometimes gets sidetracked with all the emphasis on outsourcing and cost cutting in today's world. The ability to offer real and tangible guardianship of natural reserves in itself is a program selling point. Everyone has seen the devastating effects of pollution on the environment. Showing in no uncertain terms that investment in sewer management systems will directly and tangibly protect the environment is a tremendous justification. It should also be the ultimate goal of every environmental program. Protection of the environment is a duty and its preservation for future generations a responsibility.

Several financial issues are manifest nevertheless. The first and most logical is the challenge to do the best job for the lowest cost. The goal is not in question; only the implementation is. Another issue is the ranking of the tasks on the drawing board. Important projects get funding based on funding allocations particular to the utility. There is also the second-guessing of the motives and goals of the projects by those who will be fiscally responsible for the programs. The program managers and coordinators generally put their reputations and, potentially, their jobs on the line for the success of funded endeavors. Financial people approving frivolous jobs may find themselves in another line of work. Ratepayers press to maximize their investment in the utility. These reasons all contribute to the importance of cost justification to program approval and implementation. Funding is also the reason so many options are counterproposed when a program is placed on the table for review.

Privatization is outsourcing, either by direct sale or agreement, services of a government-owned utility. The subject tends to be controversial because of misconceptions and a general lack of understanding, inadequate planning, and resistance to change. It is presented here in order to dispel the myths and allow a better understanding of what outsourcing is and does.

In convincing funding sources to support sewer infrastructure management systems, two benefits must be demonstrated: direct cost savings and indirect cost savings. Direct cost savings can be accurately computed.

Indirect cost savings are harder to substantiate. This is a matter of concern because indirect cost savings can be the biggest asset of the program. Properly packaging and presenting such savings, making them appear credible, is one of the keys to successful funding applications. The terms *cost savings* and *return on investment* here are used interchangeably even though their technical definition differs.

Financial people generally prefer that a project be self-funded or paid for by another source. If this can be shown, the program generally is approved. Even if just a portion of the program is funded by another source, governments are more likely to support the effort.

A program that is legally mandated can also expect funding, but how much and for what part of the program are open to question. The issues are applicability of the program to related areas and the continuance of the programs once the court-ordered part of the endeavor is complete. The programs have to be designed and constructed so that there is return not only on the legally mandated activity but also on as many others as possible. Obviously, the legal mandate is based on the constructs of the court order and the basis from which it was developed.

The issues surrounding funding often prove more formidable than the technological issues of equipment and services. The justification materials should be clear enough that financial people can readily understand why the program was developed, what its needs are, and exactly what benefits are expected. Depending on the financial status of the utility, most programs that can show substantial benefits have an excellent chance of passage, even in difficult economic times. Shared or outside-funded programs have an even greater chance of being approved. Legally mandated programs will proceed, but only with assurances of continuance, benefits to additional parties, and compliance with all legal requirements.

Justification goes hand in hand with funding. The ability to sell the idea of the program to the non-user is critical in that it generates approval and subsequent funding. The tighter the financial restrictions, the more difficult the justification. The more resources are secured from outside the agency, the better chance of approval of the project. Financial people are not necessarily tyrants. They simply have to be made to understand, in a nontechnical way, why sewer management programs are in their best interest. This is the key to approval.

Outsourcing is always an issue and one that prompts controversy. Privatization is the mode of the day and cannot be ignored. Choosing which functions should be outsourced, if any, and which should remain in the utility must be carefully considered. A thorough examination of the options must be documented in the justification package. This constitutes one of the best defenses of a decision to keep all work in the utility agency. It also allows for a complete examination of the agency to determine its competence and what is required to improve it.

A good justification plan, whatever its form, serves the utility by opening access to funding pools and thus allowing programs to move forward.

10.2 THE INITIAL CONCEPTS: OVERVIEW

At this point, all the technological and logistical reasons for the program are understood. The system designers are familiar with the product market, the equipment and maintenance offerings, the potential uses of the data, and the providers of the support services. The effort now is to move the program out of the concept phase into funding, approval, and buy-in from administration.

Three big questions must be asked when a utility decides that sewer infrastructure management is necessary: (1) why this program should be implemented, (2) how the program will be funded, and (3) what benefits in services and real return on investment are expected. The approval or denial of any program is based on these questions.

Program designers can identify many reasons to approve a program. They do, however, tend to rely on technical information that few reviewing parties are competent to understand. Financial and management people understand bottom-line issues, results, benefits, and returns; they care less about the methods, technology, and inner workings of the sewer infrastructure system. Answering the questions listed above helps program designers address the interests and concerns of the reviewers who have the final say in funding or approving the program.

10.3 THE INITIAL CONCEPTS: REASONS FOR THE PROGRAM

Financial managers ask practical questions in overseeing any expenditure because individuals and committees responsible for development funds must answer for their utilization. Expenditures must produce tangible benefits. Therefore, financial and other oversight groups need to understand, in layman's terms, the program in its entirety.

This leads into the first issue in program approval, which is the reason a program should be initiated. There are many reasons why a program should be approved, initiated, and completed. First and foremost, there is the ever-increasing awareness of environmental protection and the need to preserve the watershed for today and for future generations. Such protection efforts are paramount and, as noted above, a duty and responsibility. Nations that have not protected their natural resources, or not reclaimed them, experience water emergencies and watershed endangerment. Such conditions are precipating searches worldwide for new technologies to alleviate these pressing conditions. Protection efforts thus are critical to preventing the loss of and need to recover the environment. In any program, the goal of preservation and protection of the environment should transcend all financial reasons, political interest, and public pressures.

People involved with the day-to-day operations of any water utility are always interested in working more efficiently in order to increase environmental protection. One reason for this dedication is the conscience and good intent of the workers. Another is the threat of outsourcing and privatization. A third reason is the possibility of budgetary constrictions as a result of declining population, revenue losses, or political issues. Other reasons include political and regulatory concerns that directly affect workers and the way they perform their tasks. In many cases, all of these reasons are present. Internal motivation drives the program forward and is vital to the internal program buy-in. Without this buy-in, the best program will fail. People make up the heart of any system, and it is only through people that systems can succeed.

There needs to be a central reason, a tangible purpose, for the implementation of any sewer infrastructure program. The suggested first step is to write out and examine all program goals, from maximization of the field forces and protection from outsourcing to preparing for outsourcing. The important thing is to establish a target toward which the program is aimed. A program preparing for outsourcing capital construction projects takes on a different form than one whose purpose is legal defense. It is only through establishing a tangible goal that the path of the program can be charted. Program goals must conform to the utility's governing policies and practices and be consistent with the goals of the utility. Without this alignment, the program cannot succeed.

10.4 THE INITIAL CONCEPTS: FUNDING THE PROGRAM

The funding of a program is of critical importance. Every utility has money to spend for programs, even utilities that are considered insolvent. There is a constant intake of funds as a result of sewer rents and other methods of revenue generation. Money is always being reinvested into the utility's infrastructure as a result of this intake. The need is to show the priority and importance of the desired program. This may appear difficult, especially to those who have experienced serious budget constrictions in the past. However, the truth of the matter is that money is certainly being spent, invested, or otherwise utilized for specific programs. Obtaining funds is only a matter of showing the importance of a sewer management system and the benefits that can be achieved.[1]

The task is to demonstrate the need versus the return to the people who make the approvals. As noted in previous sections, there may not necessarily be a monetary return on investment, but the program may ensure that the utility's duties and responsibilities for environmental protection are carried out. This argument has credence even with financially strapped utilities. The ability to show the direct and indirect returns on the investment, along with the program target and its alignment with the utility's general policy, can generate the interest necessary for funding. The actual goal of the program is an important link in making this area complete. This is why the definition of the program is so important. This is also why the program must be explained in terms the nontechnical reviewer can understand. The completion of this step leaves only how the funding or the program will materialize.

It makes sense at this point in the justification process to show the funding source. There are three funding sources to consider when designing these programs: internal funding, external funding, and a blend of these two. Internal funding is monetary support from within the organization. Repayment can be based on return on investments, both direct and indi-

1. T. Day and J. Nicolo, "Funding and Justification of Infrastructure Management Systems in Difficult Economic Times," WEFTEC '96, Dallas, Texas, October 1996.

rect, which justified the funding required for the program. In the case of legally mandated programs, there may not be a potential monetary return; the program may simply require direct funding. However, any sewer management system should provide a direct return on investment that can be readily identified using the definitions provided in this text. If there is no mandated return on investment, the system designer must be creative in designing the program to take in many areas of concern and potentially address areas not specifically targeted.

Mandated return on investment is an area that should be considered. As an example, assume a mandate or consent decree is issued that requires a utility to monitor and report on time and duration of discharge at six CSO points in the drainage district. Here is a task that incurs no direct return on investment as mandated. However, the program can go forward if concepts such as sewer system maintenance, predictive maintenance, environmental protection, and transport capacities studies are rolled into it. The two maintenance issues noted here provide a direct return on investment by potentially reducing the number of inspections of the combined sewer location, which is performed by and derived from monitoring the data. Environmental protection generally yields no direct return. However, there is an indirect return by the protection the monitors extended for discharging systems. Solids screening generates another indirect return on investment by reducing litter and other floatables on the watershed banks; this trash must be removed by litter crews. If the maintenance aspects are coupled with the monitoring, then there is a direct return on investment with the predictive maintenance aspects, and planning of work crews. Also, knowing the amount actually discharged from the point allows the utility to determine if upstream studies should be performed for transport capacities or other issues.

Note how quickly the uses for those six standalone monitors accumulate, even without communications features. Just by being in place, they support watershed protection, maintenance improvements, and other benefits not possible before their deployment. If each of these features were calculated by cost, the return on investment would become clear. Such justification is what financial people and administrators need to make go-ahead decisions.

Custody transfer, a subject of considerable interest, is the change in responsibility for a volume of sewage as it passes from one jurisdiction to another. This source of revenue is a constant cause of arguments between utility agencies when limits are placed on the flow between jurisdictions. It may be wise to consider what revenue can be realized after expenses.

The design of most wastewater treatment plants either allows for expansion or has inherent reserve capacity. Most municipal utilities do not generally lease out this capacity, nor do they attempt to expand the plant unless there are specific reasons to do so. Additional revenues from custody transfer locations can finance these expansions or maximize that which is underutilized. Internal audits and the careful setting up of the account system are required. Most of the account system can be set up using the installed sewer infrastructure management system. Therefore, another revenue source usable in justifications is funding a related project.

10.5 THE INITIAL CONCEPTS: THE EXPECTED BENEFITS

The final part of this discussion addresses the expected benefits of the monitoring system being procured. Expected benefits are not really a matter of how novel the engineering is, but of how pragmatic the application of the technology is. It is important to show how the system will be practically utilized, along with the plan for its short- and long-term upkeep. The many facets of this type of program should be examined individually.

A financial manager reviewing the program now knows why the program is required and how it will be funded. The next logical question, and one of critical importance, is what fiscal benefits will be obtained. This is difficult for most technical people to explain to nontechnical people, because their notes, sketches, and ideas revolve around the program's technical aspects and logistical concerns. It seldom addresses the bottom line. As an example, remote data communications is an option on any permanent monitoring system. If a technical person were to make that justification, it would rest on the basis of convenience, getting data whenever required, alarming capabilities, and other frill factors. This does not generally appeal to financial people. Technical people have a pragmatic understanding of the cost savings involved; their problem involves presentation.

Look at whether the project can be justified on the basis of a reduction in the number of confined space entries as a result of electronic monitoring allows reallocation of manpower to emergency, reconstruction, and proactive work. Here is one way to phrase it: "Currently, *X* number of visual inspections occur per period with confined space entries. This can be reduced by (*X-Y*) because of constant monitoring and alerting." Point out that alerting and alarming can protect against property damage and associated lawsuits. "Last year, there were *X* flooding occurrences, which cost the utility *Y* funds. This program is geared to address these complaints and reduce related claims. Just the ability to defend against frivolous claims should reduce costs by 10%."

A major issue that financial people always wish to see addressed—but which seldom is—is the short- and long-term maintenance of the devices. It is wise to show what mechanism is in place to ensure that the benefits previously outlined will continue. The plan has to be thought out and in line with current policies and procedures. If there is a trend to outsource all repair of a specialized or technical nature, include the estimates for these contracts and long-term agreements. Show that the devices will be repaired in a particular time frame or the vendor will replace the device to continue monitoring of the location. The importance of this is to show that the system is being managed and maintained to the best of the staff's ability. Any investment, including a sewer monitoring and management system, demands a continuation plan. The plan should be included as part of the overall system benefit. This is proper, as it shows that as much risk as possible has been removed from the proposed program.

A method of justifying potential benefits is the use of peer examples. Most reputable manufacturers of sewer management systems maintain lists of current clients who have agreed to share with potential clients their experiences with the company's products and services. The input they provide can be of great value to the program when utilities share similar issues and concerns. The potential client can learn a lot from experienced clients and avoid problems as a result. These experiences can be included as part of the justification. The ability to tell a success story that supports the system designers' arguments angurs well for the success of the installation in the sewer infrastructure.

There is generally some internal bias in every utility concerning the complete uniqueness of the subject utility's organization. It is understood that every utility is unique and that every system has issues that cannot be found anywhere else. However, there are enough similarities across utility agencies to support the comparison of core issues as a way of measuring a program's potential for success. Of course, the choice of reference is important. It is not wise to choose for the model a utility that is much larger or smaller than the subject system. It also does not benefit the justification to make associations with dissimilar areas, unless the potential for rationalization can be made. Common sense must prevail in comparing the reference and the client. If this is not understood and managed, the bias will remain, even though it will only stifle the program.

The use of the system can be demonstrated by applying the material presented in the previous chapters. The ultimate use by all parties of interest of the data generated from the monitoring program is in itself a justification and benefit. The case for the system can be made by the hard data generated by the monitors, even if it shows deficiencies in the infrastructure. If this is the case, such data actually works positively by showing that constant monitoring and proaction would prevent such conditions from accumulating.

It is always advisable to show bottom-line benefits of the proposed system in an easily understood way. The interjection of technical jargon and complicated hyperbole does not impress financial managers; it serves only to bore and insult them. Bottom-line, concrete, results-oriented benefit is the only concept on the mind of financial managers and administrators. The system designer should always refer to the targets and resources that will benefit from the program in simple terms.

10.6 PRIVATIZATION: THE OVERVIEW

Privatization is discussed at this juncture for a variety of reasons. Justification of sewer infrastructure management often involves this issue. The topics surrounding privatization must be understood so that acceptance or rejection of the proposal can in turn be understood. Without this understanding, the potential benefits are missed and the usefulness of the project to the utility is never fulfilled. Privatization matters thus should be integrated with the management program design.

Privatization is the ownership or operation of a formerly public service utility by a profit-making company. This concept began when local governments attempted to cut the rising costs of their water utilities by contracting out seasonal or periodic services, thus reducing the number of staff needed. Over the years, privatization grew into an industry of its own. In the water field, it may be the fastest-growing movement in the world. Entire countries have privatized their water services in an effort to lower costs, reduce government liability, and imporve services to the citizens. The

United Kingdom simply privatized all water utilities through a stroke of legislation. Many countries all over the world look to the United Kingdom for the privatization model established there. The relative success of privatization may be a matter of opinion, but nonetheless, it was done and it flourishes. Many other countries, especially in Eastern Europe and the Far East, are looking at privatization as a way to manage not only new construction but existing facilities as well. The trend is global and the examples many.

The privatization of sewer infrastructures has been slower than that of treatment facilities for several reasons. As noted above, infrastructure generally has received little attention and much of it is now in need of extensive repair. It is difficult to price the work in order to balance what part of it, if any, should be in the hands of a private operator and what should remain under direct utility agency control. The United States is certainly one of the nations struggling with the questions of who will fix them and who will pay for the repairs. Local government must identify solutions for both long- and short-term concerns. Much of the federal monies earmarked for renovation programs has been spent, so structural repairs and system expansions are unfunded. It is only recently that privatization has emerged as a potential solution. The idea holds great appeal to American sewer infrastructure management teams.

10.7 PRIVATIZATION: CLASSIFICATIONS

The privatization of entire sewer systems at the same level that treatment works have experienced is comparatively limited. The only nationwide example appears to be in the United Kingdom. As a result, the definition set is not yet well established. This is an effort to describe the general potential scenarios as would be assigned for privatization of the infrastructure, as the concepts and definitions are still evolving.

There are two approaches to privatization of the sewer system: outright sale and management. Outright sale of the system is difficult to price but less complex than management. This is because there are so many potential management scenarios that confusion and misconceptions are common regardless of the amicability of the agreements. Each scenario is worth discussion.

To sell outright a governmental asset is difficult from the standpoint of pricing. It is difficult to price a length of sewer line, a pump station, or a lateral. The costs to replace these infrastructure facilities is known; however, replacement costs and sale costs are considerably different. When a house is sold, it is not priced by replacement value; its purchase price is based on demand for similar properties and on the tax base. The same is true with the purchase of a sewer infrastructure. On the other hand, it is expected that the sewer will have to be replaced or repaired at some point in its life cycle, but it is not expected that a house will have to be replaced. Also, the value of house can be compared to those of nearby properties. This is certainly not the case with sewer infrastructures. Therefore, many problems arise due to the lack of experience in these transactions. The biggest issue is determining the asset that is to be conveyed in the sale.

The first segregation to make is the boundaries of the asset. If a pump station is conveyed, the station, wet well, structure, and adjoining fixtures are conveyed as well. The land, or at least the rights to the land, must be conveyed as well. The next step is to classify the parts of the asset as fixed or replaceable. Land, for example, is a fixed asset; pumps and motors, which must be replaced periodically; these are called replaceable assets. The total asset is the combination of the fixed and replacement assets.

Fixed-asset pricing is difficult in government sales, as there generally is no basis for setting the price. Even exclusive communities have sewers and pump stations that are no more or less valuable than those in a tenement neighborhood; they all perform the same function. Zones where infrastructures reside do not convert readily to any other function. Therefore, the fixed assets are priced by agreement per the life of the system, necessity of the station or location, and usefulness of the facility. Land, however, still has price and it can still be conveyed. Several attempts have been made to devise equations for costing this out, but there is still a lack of transactions to validate the equation used to calculate the asset value. Therefore, the agreement between the private operator and the governmental agency offering the sale is the only method for pricing the fixed asset.

Replaceable assets are a bit easier to price according to the number of replacements per the expected ownership interval. This cost issue matters to the private owner, which becomes responsible for the replacements. Therefore, the number of expected replacements should be factored into the sale price. It is up to the ultimate agreement of the assessors if the price is accepted. Regardless of price, the station must be maintained and the water has to flow.

The sale of the asset is based on the perceived value of the total infrastructure and the life cycle of the replaceable assets. Agreement on a price benefiting both parties can be difficult, but it is possible to achieve. Remember, in these sales, that the hole in which the sewer sits has value just like the sewer structure itself does. The issue now is the size of the hole, the structure within it, the above, below, and side-to-side underground utilities of others, and the legal ramifications that pertain. For this reason, the widespread sale of sewer infrastructures appears unlikely.

Privatizing utility management may be more popular than privatizing the infrastructure because it is more flexible; management agreements can be custom designed to suit all participating organizations. It allows the local government to maintain control and ownership of the asset while the day-to-day activities and long-term replacement, construction, and abandonment plans are left to another party. Combined arrangements are possible, as when employees remain in civil service but are managed by the private operator. No assets are transferred, only the responsibility for the system's upkeep and support. This scenario has wide appeal and can accommodate many of the concerns people have about privatization.

The terms and definitions for issues surrounding privatization of the sewer infrastructure will become more mainstream as interest in this form of outsourcing grows. With government agencies striving to maintain compliance with regulators and deal with environmental pressures, the trend to outsourcing functions of or all of a sewer system will grow as well. The privatized management approach appears to hold the most promise for acceptance and may suit more utility agencies than outright sale.

10.8 PRIVATIZATION: OPERATIONS

The goal of any sewer infrastructure utility is to provide the maximum service at the most reasonable cost. The goal of the privatizer is to ensure the most reasonable cost for providing the maximum service. This may appear to be a word game, but it is truly not. The distinction is that any profit realized by a utility is put back into the infrastructure or returned to the overseeing government, whereas in the case of private operations there is reinvestment in the infrastructure, but stockholders and investors require dividends as well. The goal is always to provide maximum service. The question is who receives the balance of the revenue.

Private operation of a sewer infrastructure differs from public operation, at least superficially, only in the ability to procure and maintain technology and services. Private operators are less likely to encounter the pricing structure to which governments are accustomed because of their ability to generate a timely contract and a payment. Private operators also have less interest in performing nonperiodic tasks in-house. Periodic and recurring tasks are structured according to the available resource pool. However, those tasks that are not performed in-house are usually outsourced in a more competitive manner in private entities than they are through governmental agencies. Fewer contract parameters must be met with respect to terms, conditions, and eligibility than in the standard government certification process. For this reason, the nature of doing business with the privatizer is vastly different than with government but yields the same results.

The focus now returns to management systems. Realizing revenue on these endeavors involves performing the task more efficiently and at greater cost savings than the governmental agency did. The ability to generate contracts for repairs by private entities occurs faster and without the usual government bureaucracy. Certainly the utility, as an agency, can perform these functions, but the government would have to restructure it to attain the same latitude and incentive programs offered by private management.

The privatizer has to outperform its competition. This extremely competitive market requires the careful examination of every expenditure. The competition among companies is extremely tight, with talent and expertise being traded and utilized throughout the business. It is therefore imperative that the privatizer be able to fully document to the customer utility that they are providing the ratepayers the best, most cost-effective solutions. Many people involved with utility work do not see the tremendous and often stressful competition among the outsourcing companies. This competition generates the incentives to do the job in the most economical manner.

One method of maintaining a cost-competitive edge is the use of modern sewer management systems. Constant long-term monitoring and reduction of the effort expended for physical inspections and travel provides for a more efficient response to issues as they arise. It also shows attention to and care of the infrastructure, as noted in previous chapters. The tools of automation, the use of which speeds repairs, and extends the lifespan of the replacement assets. Extended lifespan also means the utility has more leeway to research replacement technologies that are more cost effective and easier to utilize.

The implementation of modern management techniques in the sewer infrastructure allows for a better cost analysis of the total asset and hence greater insight into what should or may be privatized, and the associated and projected costs for it. The cost set for the assets is critical because, as they are public, no previous costing likely was performed. The lack of precedent means that extra care must be taken in the implementation of the systems and the interpretation of the data they generate.

Outsourcing can be of tremendous benefit to agencies that have no desire to privatize but are having difficulty competing in the marketplace. Portions of services can be outsourced. A policy can be established that any task requiring specific skill sets should be outsourced. For example, electronic repairs can be outsourced because they require expensive diagnostic equipment and technicians. It may make sense to maintain the field crews as instrument exchangers and contract for specific device repairs. The ability to repair equipment quickly and efficiently is a key to efficient sewer monitoring operations. Shedding the need to continually upgrade a repair shop with diagnostic equipment and to train technicians can yield substantial cost savings for the privatizer. Also, equipment fails on an intermittent basis, and usually under the worst possible conditions. Emergency repair contracts can get devices back on line quickly, with substantial cost savings if the repairs are minimal.

10.9 PRIVATIZATION: THE FUTURE

The privatization of the sewer infrastructure and related management systems appears to be on the rise. More experience with and confidence in private operations can only benefit the ratepayers of a sewer district.

The reorganization of the management aspects under the private operators can be achieved quickly and cost-effectively. Responsible bidders make an effort to provide services in excess of the specifications—an extra benefit to the utility and the ratepayers. Also, privatized management of sewer operations can reduce the fears of long-time workers with respect to issues such as job security and benefits.

A key to the ability of private entities to competently operate or own sewer infrastructures is their ability to utilize the management systems and equipment already in place. Full use means maintaining the devices and understanding the hydraulic response of the sewer system so the best protection can be provided to the drainage district and surrounding communities. The protection of the watershed is best achieved with monitoring, advanced data communications, and the skilled people who interpret the measurements. Private operators can use these tools and techniques to the benefit of all.

10.10 THE CASE FOR CUSTODY TRANSFER: OVERVIEW

Custody transfer generates funds for the agency regardless of its management. Custody transfer, as described previously, is the change in responsibility for a volume of sewage passing from one jurisdiction to another. For this responsibility and the processing of the sewage, the receiving agency collects revenue. This revenue is one that can have great return to the agency if properly handled. The following material can assist the utility in establishing the base for a custody transfer program. The discussion focuses mainly on the logistics of setting up such a program, including data handling, billing, interagency liaison, and responsibility for matters such as points of connection and liabilities.

Any agency or utility can benefit by establishing a custody transfer program. Even if the effort requires the expansion of plant capacity, the use of the system to convey interagency sewage can provide a welcome source of revenue. Also, for the communities that join the system, it can mean relief from any of the issues of sewage processing depending on the arrangement. It certainly can be a shift of focus to not have the responsibility for a treatment plant. The service is bought wholesale and marked up to the ratepayers. All parties benefit, depending on the perspective of the participants.[2]

The issues concerning sewage treatment plant capacity are introduced only generally here, for the purposes of discussion. Treatment plant capacity in both wet weather and dry weather must be understood, as must the customer's requirements. A third party should review the transport capacities of the customer's system and evaluate the excess capacity at the treatment works. It needs to be established, through a certifiable and reliable source, that the sewage treatment plant has sufficient capacity to handle all of the customer's requirements. The difference between the rated capacity and the actual capacity is the capacity available for sale.

The rated capacity is the wet- and dry-weather flow maximums that the plant is certified to handle while maintaining effluent quality. The actual

2. T. Day, J. Nicolo, and M. Kyser, "Custody Transfer: Lessons in Successful Wastewater Metering and Billing," WEFTEC '96, Dallas, Texas, October 1996.

capacity, known by many names, is the average flow that the plant is currently processing and reporting. The difference is the reserve capacity of the treatment works and can be sold to outside jurisdictions. The reserve has both wet- and dry-weather flow figures in its description.

The customer is the agency wishing to purchase part of the reserve. The reserve capacity to be procured can be either as a complete connection or in parts.

The customer utility needs to fully understand the amount of flow and the quality of the sewage that is to be conveyed to the treatment works. As noted above, a third party should measure the parameters of the flow. If a treatment plant is being shut down, then the records of flow when the plant was operational may be a good source for this information. The amount of flow being lost through CSOs and other relief points should also be known, as should the condition, time, and duration of discharges, as regulations and changing attitudes eventually may require this flow to be returned to a treatment works for processing. The potential for their connection requirements should be weighed against the sewage treatment works' reserve capacity.

Water quality issues should already be well documented in the customer agency's files. Industrial or other loadings that may produce issues for the treatment works should be noted and handled prior to final implementation of the agreement. Some communities may require specific industrial monitoring and pretreatment. The variability of the chemical content of flow being delivered dictates the requirements for water quality monitoring. Connections that largely handle domestic sewage probably do not require long-term water quality monitoring. Connections that include manufacturing or other extensive services should have water quality monitoring for the protection of the treatment plant and its operation.

Choosing the meter or instrument location occurs after the capacity and loading issues are resolved. Factors of jurisdiction, maintenance, and ownership have to be considered. The matter is more complicated than simply picking a spot and installing a meter. Complications include local laws and regulations and such guidelines as are specified in the interagency agreement.

The physical location and the housing of the meter must be established. The usual choice is a place with an existing pipe connection or that is at the closest point of intersection between the two communities. Proximity minimizes the need for construction that may occur as a result of the agreement. It may lower the cost if a pump station is constructed with exposed piping simply emptying into the treatment plant's collector system. The choice of location is obviously particular to the terrain and existing conduits. A plan, mutually agreed upon, is required to make the connection. At this point, the actual electronics is not an issue. The choice of connection points is a technical choice as well as a logistical one.

The big issue at this point is who actually owns the new facilities. It is generally a good idea that the owner, whoever it is, provide the maintenance to the site; in fact, this should be required. Ownership generally drives upkeep in any area, and the upkeep of site is critical to inter-agency relations. The ownership of the monitoring and sampling portion of the site must be agreed upon as well. This is also best in the hands of the party liable for the maintenance and calibration of the devices. It is suggested that the treatment plant side of the connection be liable for the equipment and its calibration, as it already contracts for the technical support of process control equipment.

Determining the actual point of connection, or where the actual responsibility for the sewage ends for one party and begins for the other, is not as simple as deciding where the meter goes. A flowmeter requires specific upstream and downstream conditions to be accurate. Therefore, there needs to be a point in the system where the actual responsibilities for the sewage shifts. This point should be above or below the flowmetering area rather than precisely at the meter. Such a choice makes all of the flow-measuring site entirely the responsibility of a single party. The maintenance of the adjoining pipe must also be defined; therefore, the nearest upstream or downstream manhole can be a logical choice for custody transfer point. It is recommended that any potential metering location be checked to ensure that there is no crossover or cross-connections.

In sum, the logistics of custody transfer include decisions about place, type of monitoring, jurisdiction, and maintenance responsibilities. Note that there are no technological issues here. It is expected that at least one of the two parties is expert in the technical area as a result of experience with sewer infrastructure monitoring.

10.11 THE CASE FOR CUSTODY TRANSFER: THE MECHANICS

After the negotiations have been completed, installation and startup can begin. Many facets of startup are not relevant to the general sewer infrastructure monitoring network because the issue here is money, not information. Therefore, to limit the arguments related to money, specific issues of accuracy, integrity, and equipment failure should be explored, documented, discussed, and agreed upon based on the best available and affordable technology.

A shared communications system should be installed so both parties can examine the flows being recorded, as is their right. The agreement may specify that only the party performing maintenance can retrieve data directly, as a security measure. In this case, access to the central computer system should be established such that the excluded party can observe the final billing data at will. With this access, any questions or concerns about the data can be addressed promptly rather than delayed to the time of billing. The sooner a problem is identified, the sooner it can be resolved to the satisfaction of all parties.

The data should be stored in such a way that no alteration of the files is possible once they are written to final memory. The generated value is either used or not used in form; it should not be altered or massaged once generated. It should not be possible to change the parameters for scaling once the program is running. Such a change should require both parties to witness and agree that this is necessary for the accuracy of the device. Data communications lines should be used only for interrogation and program loading.

The previous discussion now makes its own case for the choice of technology for flow measurement. Several recommendations are made here; one is to use velocity equipment. At all times, the ability to determine velocity should be paramount. Accurate velocity measurements are reliable and viable with today's modern electronics. The velocity estimation routines, as performed by flumes and weirs, have substantial shortcomings when the upstream or downstream flow profiles are disturbed. Profile disturbance is a familiar problem that is quickly remediated with velocity measuring devices. Remember that the reason for the development of flumes and weirs in the first place was to determine the velocities of a known set of head rises. In days gone by, few reliable methods were available for this measurement. This is no longer the case. The use of any flume is expensive and complicates the logistics of the measurement point. If there is a downstream restriction, the level in the flume rises artificially. Such a rise in level is translated by a level measurement device as a high flow. This causes inaccurate billing. A velocity device provides reasonable evidence of a blockage by showing either reduced velocity or a skewed reading.

The chosen equipment should be approved by the manufacturer for direct billing cycles. Some instruments measure too coarse a data set for this purpose. Parties involved may have experience with technologies on the market; this should be applied to making the best choice. If new construction is planned at the location, the device can be worked to best advantage. If existing locations are being retrofitted, equipment choices tend to be more or less based on the size of the connection and the environment in which the device must operate. The manufacturer should be consulted for the best model and modifications required to accommodate the installation and retrofit.

There is always an issue of accuracy of the device of choice, and the method for proving such tolerances especially in the billing situations discussed here. Measured flows in billing locations should not exceed the flow accuracy of the majority of the best available technology. Generating 2% accuracies may be expensive but, depending on the amount of flow at the location, it may be necessary. It is probably sound engineering to reexamine the reliable flow technologies for such accuracies and the rationale that they can produce this number. The adaptation to achieve this value may require additional surcharges to the measuring devices. The cost of the meter and its support must be weighed against the amount being billed generated as a result of the flow.

Calibration of the location can be handled in any of several ways. The instrument should be third-party certified at some manufacturer's recommended interval. This guarantees the integrity of the electronics and provides confidence that the most reasonable approach is being used in the device of mutual choice. The manufacturer should certify the technicians installing the device. It does not matter if the service is on contract or provided by a utility's technical maintenance force. It does matter that the manufacturer assures that the methods, tools, and equipment employed are appropriate to the recommended procedures. Calibration issues are also extended to the technicians performing the installation. The manufacturer must certify that the technicians charged with this task understand the device's optimal installation, operation, and maintenance procedures. Any calibration or service to the device should be witnessed by both parties unless otherwise agreed. A certificate of calibration should be provided to both parties. Certification should be renewed at the manufacturer's recommended interval unless there is agreement that the cycle should be shortened. The cycle should never be extended beyond the manufacturer's recommendation.

Equipment and installation requirements being set, the next concern is for electronic and site maintenance. All electronics must be maintained properly to operate reliably. Prescribed maintenance schedules that should be followed to maximize the life cycle of the device. Responsibility for maintenance should rest with those who do the calibration. This approach yields familiarity with the site and its idiosyncrasies. Even if the entire task is outsourced, understanding the particulars of a location assists in ensuring the maximum amount of uptime for the electronics. Maintenance of the equipment should be performed only by those individuals or companies certified by the manufacturer to do so.

If it mistrusts the accuracy of the flow measurements, one of the sides may wish to place another flowmeter at the point of connection. This is always a major headache because the readings are never exactly the same. The topic is presented here to describe methods of discouraging or coping with the desire to install multiple flowmeters at one location.

One reason that the flows will not match is the time slice and sampling period of the data. Unless the two meters are of the same style and type, their timing will always be out of synchronization. The result is flow readings that appear slightly different because the data were captured at slightly different times, though the compiled data will probably be the same. It is therefore unadvisable to use dissimilar measurement devices at the same site. It is a good idea to establish a protocol in the written agreement for handling flow measurement disputes.

The next issue is billing the customer. The usual approach involves the same type of billing sequence used to bill ratepayers for standard residential or commercial service. If this is the case, then the commercial billing cycle should be examined and reevaluated for applicability. The flow values for the new custody transfer customer can be added directly into this sequence. The question is how to do this. Problems arise because the data for the rest of the billing sequence are generally received from another source. The current billing software may have to be modified. However, the data generated will probably conform to the standard computer platform, as this was specified at the time the sewer infrastructure monitoring system was installed. Therefore, the problems are not as great as one might fear.

If there is no possibility for introducing the commercial account into the standard billing sequence, there are two potential solutions. The first is to generate from the computer a bill based on the field data and then, once a month, have the accounting department hand-generate a bill. The second is to set up an internal system whereby the bill is sent directly to the customer from the computer at operations headquarters; a copy is sent to the accounting department for use in resolving problems that arise. The only problem may involve getting the data from one computer system into the format used by the computer in the accounting department.

One issue to consider at the time a bill is generated is the amount of flow received at the plant or point of connection over the agreed amount. The billing should be able to flag and note all times that the flow was over the agreed volume if that was in the agreement. This is generally an issue when customer agencies initiate new compliance programs and send more flow than expected to the treatment works. Keep in mind that storm water is much more dilute than domestic sewage and contains more trash and other strainable solids as well. The plant needs the biosolids in the process, not necessarily the trash or dilute flows. Therefore, it needs to be decided if a penalty for flows beyond the agreed parameters should be calulated if the condition exists. Such dilution can be verified through biomonitoring and sampling programs.

The situation of biomonitoring and sampling should be included in any custody transfer agreement. The use of BOD billing can be a source of dispute if not properly addressed. BOD analysis requires an organic seed to complete the test. The use of different or nonhomogeneous seeds can cause the BODs of split samples to vary greatly. A split sample is the even division of a metered amount of wastewater for analysis at separate laboratories. Split sampling and biomonitoring procedures should be follow predetermined protocols. Overall, it is unadvisable to split samples and perform separate tests. Even with homogeneous seeding, the same sample may produce completely different results as a result of transport delays, the rate of material decay, and the unique composition of wastewater in general.

Little biomonitoring is generally performed at the custody transfer location. Sampling is performed periodically and the results analyzed for consistency with terms of the agreement. It may make sense to collect water quality data at a custody transfer location, as much to protect the plant as to trace polluters. The desire to locate and prosecute polluters is universal among water utility agencies. Therefore, sampling and water quality monitoring at the custody transfer location can provide an additional benefit by identifying general areas that would be the potential entry point of polluters. Such information can shrink the area of search for this condition and expedite remedial action.

A working log of all activities performed at custody transfer stations should be maintained. In the event of litigation, it makes sense to be able to retrace the steps taken to accomplish each task. Complete documentation of data interrogation, storage, calibration and repair, certifications of the technicians or contractors, and all other relevant issues leading up to the various tasks can resolve many arguments prior to entering litigation over disputed billing. Remember that disputes nearly always center on the flow from which bills are generated rather than on the bills themselves.

10.12 THE CASE FOR CUSTODY TRANSFER: ESTIMATION

All equipment, even that of the highest quality, eventually fails. Anticipating this, a routine for estimating the missed values should be agreed upon. There are two kinds of failures: partial day and full day. At some point in a partial day failure, data for that day can no longer be reconstructed and must be completely estimated. Partial days, depending on the extent of data loss, can be reconstructed based on experience and understanding of the hydraulics. Full-day failures require estimation in their entirety. Parameters should be established to differentiate a partial day's loss of data from a full day's loss. This section also addresses the issue of data rejection due to unrepresented or out of range information records.

The agreement on estimation routines should begin with a definition of a full day's data. In this example, a full day's data are recorded when all of the data points for one 24-hour cycle beginning at midnight and progressing to the next midnight are present. If the full day's data are collected and recorded, no estimation is required. If any data are missed in the 24-hour cycle, the record is considered partial. A partial day's record can be reconstructed based on the previous dry-weather day for the same time period of record within the previous 14 calendar days. The estimation may be performed for up to 6 hours in any given 24-hour period defined as a day's record. The data must be marked as altered and the period of reconstructed data noted. If data are unavailable for a period longer than 6 hours or no suitable dry day in the 14-day calendar period occurred, then an estimate taken from a rolling average of the last 120 days' data substituted for the entire day's record and the file marked as estimated. If 120 days' data are not available, the estimate is based on the last full 120-day estimate, and notification of the situation is sent to the customer agency.

This example covers most of the major issues facing the substitution or replacement of data in billing circumstances. First, the sampling cycle for the data record must be known. The data record is the same as the time slice and sample interval, discussed in previous chapters. An average value is generated per time slice and a number of time slices is generated for a 24-hour recording period. Missing time slices can be counted per hours of recorded time. If the data are missing, a day considered dry is retrieved, times are matched, and the data copied to fill in the day. A dry day is one following at least 72 hours without rain. This accounts for plant operations and storage/relief programs. Using the previous 14 calendar days maintains consistency with the season in which the data is missed. If the data from the site is not available, the day is completely estimated. The 120-day rolling average accounts for highs and lows in the flow being measured. The availability of less than 120 days of data indicates another problem; perhaps the instrument is experiencing a long-term malfunction that requires more than a few hours of maintenance or repair. Such situations should be addressed as soon as possible; all parties should be notified.

Rejection of a day's data is raised here because the rejected data are estimated in the same manner as missing data. Any visual observation or review of data that shows corruption in the electronics or some other concern must be reported and the problem addressed. Missed data is a condition that periodically occurs in data handling, and this must be understood by all parties. It should be directly written into the estimation agreement. The agreement should specify who can reject data because of unrepresented data sets—a possible result of maintenance work at the site as well as faulty recording electronics. Addressing these issues before problems occur prevents many arguments as time progresses.

It is important to specify estimation protocols in advance in order to maintain a continuous flow of data for billing, which leads to a steady stream of income to the receiving agency. In economic and political times such as these, the ability to produce bills with integrity saves greatly in litigation and claims.

10.13 DIRECT REVENUE GENERATION: FUNDED USE OF THE SERVICE

The current political climate demands the fiscal justification of additions to the sewer infrastructure. This is quite acceptable even if the primary justification is solely the protection of public property or health. There is a greater benefit if the service can provide a direct monetary return even if there is a recovery time involved. This section examines direct funding sources other than governmental subsidies.[3]

Direct revenue is generated when funds are obtained from the use of the sewer infrastructure monitoring system directly as a cost item to the system. There are few methods other than custody transfer whereby this can be accomplished; possibilities include providing flow data to private developers, land grant information, rain data to third parties, and historical data for private engineering programs. Traditionally, water utilities have been forthcoming in sharing data with anyone who requested it, regardless of the reason. As budgetary belts tighten, some of these services may be billable. As an example, if a private development company wants to understand the local hydrologic cycle, it will probably hire, for a cost, a private engineering firm. The firm will obtain the hydrologic cycle, for free, from the local water utility and generate a report at a cost to the developer. The water utility's data about the sewer infrastructure and the localized hydrologic cycle thus shows potential sale value.

Excessive claims by insurers wanting information about property damage as a result of flooding may indicate another market for the data. Natural occurrences such as snow meltdown and heavy rains can cause property damage over which the utility has no control. Insurance carriers can purchase utility data in an attempt to defend against the same fraud that the utility itself wants to reduce.

Another potential source of revenue is public-private partnership funding. Such partnerships are especially true in industrial waste and biomonitoring situations. Many times, there may be interest among industries and water utilities in cofunding such projects if mutual benefit can be demonstrated. This mutual interest may be difficult to show, but if there is a common goal or target, costs may be offset by industry contributions. Protection of the watershed where the industry sits may provide substantial

3. T. Day, "Automatic Wastewater Flows for Billing: Philadelphia's Success in Process for Profit," Water Engineering and Management, September/October 1996.

motivation. This is an area that should be reviewed as the sewer monitors are being staged.

Most equipment that monitors the sewer infrastructure was installed at great cost to the ratepayers. The ability to recover costs directly through the sale of monitoring data can assist the budgetary efforts of the water utility and reduce the strain on subsequent ratepayers. These possibilities should be explored wherever possible.

10.14 INDIRECT REVENUE GENERATION: LEGAL LIABILITY SUPPORT

The sewer infrastructure monitoring system may generate its greatest return through indirect revenue. By understanding the potential for indirect cost savings, the potential for a return from the investments is greatest. A present indirect return on investment may have substantial future direct financial impact. Indirect returns can curtail operating and logistical costs and keep risk expenses in check as well.

The most outright example of an indirect return that has substantial and wide-reaching implications is the ability to challenge and defend against frivolous lawsuits. One always wishes to pay fair claims; this is ethical business. But the issues with fraud are numerous, difficult to manage, and substantially increase insurance premiums or artificially and unfairly increase the amount of settlements that utilities and insurance carriers must pay. Successfully defending against fraudulent claims can have a tremendous positive impact on the percentage of claims paid and the amount.

Monitoring equipment can also provide help with legal defense and justifications. Accurately showing the conditions within the sewer infrastructure can limit the exposure of the utility to extraordinary claims and lead to favorable settlements. The limitation of exposure to such instances can have a positive downward pressure on the amount spent on monetary awards, making them more manageable and potentially more predictable. An actuarial study may, in fact, show the potential for such reduction in awards if various parameters could be identified in the litigation process. Such parameters are unique to each jurisdiction, but the value of the monitoring program should be obvious.

An example is a negligence claim for damage due to flooding in a block of homes because of a sewer system overtaxed by heavy rains. Rain data can accurately show the time, intensity, and duration of the event locally and across the drainage district. Sewer monitors can show the levels of the sewer in question to demonstrate its transport capacity. Also, a historical search for rainfall based on the localized hydrologic cycle can be performed. Details that surface during this process include the track of the event, the amount of precipitation that fell per time period, and the volume being transported. If the data supports the claims for relief, a settlement without any court hearings may be possible.

If any of the claims can be disputed as a result of the data obtained, then the potential liability for award can be reduced. If the localized hydrologic cycle shows that the ground was already saturated because of heavy rains, the claim may simply be referred to the homeowner's insurer. If the data from the sewer shows that transport capacity in the system was sufficient, other potential flooding sources, such as a nearby stream or creek, may have to be investigated. Data showing that a storm exceeded the parameters of a design storm may reduce the liability of the utility because the infrastructure was legitimately not designed to carry that load. This may sound like shifting blame, but the ability to assign liability is critical in establishing fairness in the event of loss claims.

These means of reducing loss liability potential yield a substantial indirect return on investment. Indirect return can be estimated through attorneys' records; legal counsel often has an excellent idea of cases lost when monitoring data was unavailable. There are probably as many different methods for presenting the package of information in court cases as there are different circumstances for legal awards depending on the local government and country in which the utility is located. There may be other areas where the legal department and the engineering groups can use this information together in defense of the work of the utility.[4]

Many interest groups, as noted in earlier chapters, may bring suits based on any number of claims, again depending on local and national government. Such lawsuits are generally made from the surface perspective, based on observation of a sewer-related condition that allegedly caused an issue. Such defenses are less time-consuming if information is available about the hydraulic condition of the infrastructure, precipitation, and the localized hydrologic cycle. The engineering department assembles the data; the legal unit uses it to make the case for the utility's good work and its efforts to protect the watershed. Here, the claims for damage or neglect generally are

4. T. Day, "Sewer Infrastructure Management: New Concepts and Tools for Your Collection System," Water Engineering and Management, February 1998.

less frivolous and stem from a one-time observation. However, when there are many witnesses and no data from the utility to show its position, defense is weak. A dead fish near a treatment works may be classified as a fish kill; a deceased duck may raise the charge to a fish and wildlife kill. Such situations can be mitigated using monitoring data, which can be used to show that the deaths were natural and unrelated to the treatment works. This demonstration also improves the utility's public image.

The same pertains to fines and penalties resulting from regulatory actions. If sufficient water quality information is available from the infrastructure management system, showing that there was no issue able to cause a given condition, the regulatory agencies are hard pressed to impose fines. The same data can also reduce litigation required to mitigate the fines. Legal departments are always busy, but utilities need to focus on the business of water rather than defending the business. Anytime the position of the legal department can be strengthened, the utility's exposure is reduced. Support of the legal division assists, ultimately, in protecting the assets of the utility.

10.15 INDIRECT REVENUE GENERATION: PROTECTION AND SUPPORT

Many uses of monitoring data do not involve a monetary return at all, or at least not a return that can be readily calculated. As an example, it is difficult to calculate the value that the monitoring system has in the protection of workers. It is difficult to price the savings from discharges that do *not* wash up on the beaches of a local watershed. It is possible to calculate the historical loss due to beach or shoreline closure, but the actual return is difficult to ascertain. This section outlines sources of return that benefit the environment but do not accrue direct investment returns in dollars.

The maintenance of the system, including the instruments themselves, is difficult to quantify in terms of actual return on investment. The reduction or repositioning of work crews *can* be shown and the payback *can* be estimated via the cost-to-purchase/cost-to-own model. However, these financial returns exist only on paper and not in the bank. Potentially, budget reductions are a method of showing this cost savings. In reality, this simply makes more time available in which the work crews can respond to alerts and alarms. This discussion brings up the cost savings that cannot be calculated, which is the potential to reposition hiring requirements as a result of more efficient manpower utilization. Such repositioning of the utility's work crews, assignments within the work group to other tasks, and the tasks assigned provide sufficient justification for the instruments alone. The number of sewer maintenance workers per miles of sewer line can be substantially reduced as a result of monitoring. This approach must be adapted and applied to the justification process in a manner consistent with the policies of the utility.

The issues noted above can be applied to risk management. Utility risk management departments increasingly rate the potential for incidents and the extent of the utility's liability exposure, then recommend remedial actions to limit the exposure. Many issues concerning sewer infrastructures are determined from histories of complaints and lawsuits. Monitoring data can be used to identify areas of potential hazard and precipitate remedial action, thus reducing exposure, especially where flooding and other damage issues are prevalent.

Risk management and the reduction of utility liability are discussed because the goal, although at times not clear, is always the same: the protection of health, safety, and property within the drainage district. The path may vary, but the goal does not. Reducing exposure to liability as a result of risk management investigation adds a new layer of safety to the drainage district. Such protection does not show a direct, measurable return, but it certainly benefits the public.

The *appearance* of increased protection can be an excellent public relations tool. There is no way to measure the financial reward for positive public opinion, although it is clearly great. If the public and the local industries perceive that their interests are being protected, they tend to offer less resistance to new ideas and are more tolerant when undesirable events do occur. The ability to partner the public in such a manner makes for an excellent public relations tool and one worth advertising. There is obviously no dollar return here, but good public opinion yields a tremendous return in goodwill and security.

An additional benefit of a monitoring system, one that brings all the other benefits together, is in the area of municipal bonds and other forms of public stock. In the write-up in which bond raters evaluate the utility as a whole and provide a rating, assets that provide service and protection to the service area are always listed. An in-place monitoring system that provides alarms in the event of high flows and levels is such an asset. Bonding and stock issues are valued based on the condition of the agency and the performance of its management. Monitoring systems have the potential to add to the assessment and enhance the rating of the utility.

Often, the only viable funding source for the monitoring system may be direct funding by the utility. This being the case, it may benefit the utility to develop an entire public awareness program, using the electronic surveillance information as a public relations tool as well as an engineering instrument. The return generated by imporved public opinion, though intangible, may be greater than any funds that can be received from financial sources. The investments having a direct and indirect return on investment can assist the utility both in its day-to-day operations *and* when issues such as rate hikes and assessments are discussed. Also, street crews will probably experience an easier time and greater productivity through the reduction of public ridicule.

A lot of the intangible benefits of monitoring stem from the protection of the watersheds and the good public relations aspects that the systems can generate. It is important to completely understand all of the areas that the sewer infrastructure monitoring system will affect. Comprehending all the benefits, along with the issues necessary for support, points the way to a full range of justification measures that can be properly documented and presented to reviewing authorities.

10.16 JUSTIFICATION OF TARGETS AND GOALS: PERFORMANCE MEASURES

It is important to understand the measures on which a justification for a monitoring system is based. Unless some performance measure can be shown to gauge the effectiveness of the program it makes little sense to write a justification. But it makes considerable sense to allow the benchmarks that will be used to become the tools of analysis in the evaluation period. These tools and methods themselves also become the verification of the effectiveness of the equipment.

The justification process is one of showing how the program will be of benefit to the sewer infrastructure. This sounds easy but can be difficult without performance measures to integrate with the write-up. Performance measures add a level of clarity to the description of the program's effectiveness. Such benchmarks are seldom followed up on by the reviewing agencies but should be realistic nevertheless. They are a way that the operations division can learn and gain from the system.

The benchmarks and other related areas can be drawn from the uses of the data described above. By tracking results, the effectiveness of the program can be ascertained. Such results can be incorporated with future justifications and programs as needed.

An example of a simple performance measure is the determination of issues within the sewer infrastructure. This was described in previous chapters but is reviewed here because of its importance. The total number of computer picks relative to the total number of successful picks is a percentage measure of data effectiveness and hence of the effectiveness of the program. It is also a benchmark for the prevention of problems that occur in sewer infrastructures. The database generated from the successful picks evaluated to determine relative issues before they caused overflows or other problematic results. The example is taken a step further; the success of the program is now tied to historic data. Determine next how many work orders were generated for a single critical path, then determine the number of lawsuits and claims for that period. Compare the quantity of work orders developed with the monitoring period and the percentage of successful computer picks. Note the difference between the number of claims and the number of actual complaints for the period. The data should show a drop in the number of complaints per critical path of interest. Also, when complaints drop, the number of lawsuits should also drop. This is a benchmark for maintenance efficiencies and one that has performance attached to it.

It is also possible to use monitoring data as a function of time and benchmarking for the maintenance activities. An example is the examination of the data for time a job began and the time it was completed. First, there is the type of problem that the monitoring data detected. The type of job is classified and the work monitored. The start and end times are marked and compared with the work logs. Note that the recorded time interval is possible in blockages, as the block will be freed at some point and the flow return to normal dry-weather levels. If a database exists of the work along the same critical paths with the same size sections, a benchmark for the similar work can be generated. Such benchmarks probably cannot be used as a standard for work performance per se, but they can be used to develop relative times of like tasks. The relative times can be a great tool in work planning, especially in the area of predictive maintenance. The ability to generate this information can be an asset when equipment and services are considered to assist the work crews in their efforts. Such information can also be incorporated in other types of services for other parts of the utility's functions.

Benchmarking and performance testing now have both work orders and the monitoring data as supportive evidence in the documentation program. The data being recorded, stored, and presented in a more amicable format now has direct and indirect return on investment in a different way. The direct return on investment is a result of the cost savings realized from the more efficient method of determining the time and resources required to perform tasks in the infrastructure. The data stands as its own support, as it represents the actual conditions within the infrastructure and can provide the time series data required for further studies and understanding.

The ability to generate benchmarking data from monitoring data can show where much of the work within an infrastructure is being performed. If the average confined space entry time is used, for example, a correlation can be made between times required for maintenance and potential structural alterations to the system. This is described as the total cost of the confined space entries plus the resources used to make the intermediate repairs versus the total cost to perform reconstruction and eliminate the repetitive visits. Before, the choice was a matter of the best judgment of some long or short term engineering study. It was based on desktop transport capacity studies and on the original construction prints, along with site inspections. The monitoring data can now provide the engineer with a history of hydraulic responses based on storm conditions and other climatic events; the resultant evaluation has more clarity and promises greater benefit to the ratepayers.

10.17 SUMMARY

Financing and obtaining approval for any project is half the battle. The resources required to get a project moving are just as critical as the technology itself. Many times, engineers get lost in the technology and forget the necessary logistical aspects of getting the program off the gound. It is important, if not imperative, that the complete use of a sewer monitoring network for the infrastructure be demonstrated in a way that non-engineers can understand.

The use of the data also constitutes the justification for the system. Each potential use as well as those already familiar to the agency should be fully documented in a manner that suits a nontechnical audience. The data also point to milestones that can be documented and used in the next justification effort.

Equally as important as the justification is the funding source and investment type. Revenue generation and direct return on investment make the best sell to a utility oversight group. Where a direct financial return cannot be projected or simply is not possible, the indirect return on investment must be shown. This often has a greater impact than the direct return with respect to protective measures and data gathering.

The ability to defend against frivolous legal claims is one of the better examples of indirect returns on investment. A known percentage of legal claims simply are not representative of the damage incurred from sewer infrastructure malfunction. The ability to provide data that can reduce these liabilities is enough to justify the cost of most monitoring systems.

Returns from a sewer monitoring system are not always monetary. Goodwill and other intangibles are beneficial to all parties concerned. Returns and benefits studies may also may show that the funding source may best be completely internal, with no assistance from outside agencies. The continuation of the health and longevity of the watershed is a duty and obligation. Monitoring the water quality and alarming for discharges is a way of fulfilling that responsibility. Carrying out the duty of protection may be justification enough.

Benchmarking and performance standards are tangible benefits that may not yield a specific financial return on investment but do convey long-term benefit through the generation of databases for average times of task completion, documentation of tasks completed, and implementation and documentation of preventive work performed. These additional data can justify other support programs, assist in related service selection, and provide the future benchmarks by which related programs can be gauged.

Creativity and ingenuity are essential in obtaining funding for the systems proposed. All that has been presented here is methods and tools. It takes people with know-how and drive to create working programs from concepts and promises.

BIBLIOGRAPHY

Burgess, E., T. Day, J. Smullen, and L. Roesner. "CSO Model Development and Verification Study. Paper presented at the North American Environmental Conference, Anaheim, California, June 1996.

Day, T. "Alert and Alarm Algorithms for Dynamic Wastewater Collection Systems." WEFTEC '96, Dallas, Texas, October 1996.

———. "Applying SCADA-Type Technology to Sewer Maintenance." Paper presented at the Urban Wet Weather Conference, Quebec City, Quebec, Canada, June 1996.

———. "An Assessment of State-of-the-Art Flow Measurement Techniques for Sanitary Sewers and Overflows." Paper presented at the U.S. Environmental Protection Agency Sanitary Sewer Conference, Washington, D.C., April 1995.

———. "Automatic Wastewater Flows for Billing: Philadelphia's Success in Process for Profit." Water Engineering and Management, September/October 1996.

———. "Biomonitoring Equipment and Techniques: Modern Instrumentation Applications to Monitoring the Ecosystem." Understanding the Industrial Pre-Treatment Program, Indianapolis, Indiana, July 19965.

———. "Demystifying Doppler Flowmeters in Sewage Flow Measurement." Keystone Water Quality Manager, April 1996.

———. "Demystifying Doppler Technology." Water and Engineering Management, April 1996.

———. "Electronic Surveillance of Sewer Systems Using Monitoring Data for Protect the Environment." Understanding Industrial Pre-Treatment, Indianapolis, Indiana, July 1996.

———. "Modern and Advanced Flow Measurement: Equipment for Wastewater Process and Applications." Chesapeake Water Environment Association, Ocean City, Maryland, July 1996.

———. "Overview of Advances in Flow and Level Measurement Technology." American Water Resource Association, Fort Lauderdale, Florida, September 1996.

———. "Permanent Large-Scale CSO Monitoring Programs." Sewers of the Future, Houston, Texas, September 1995.

———. "Proactive Use of Sewer Monitoring Data to Protect the Watershed." American Water Resource Association, Fort Lauderdale, Florida, September 1996.

———. "Real-Time Control in CSO: Maximizing Flow with Automation." Paper presented at the Urban Wet Weather Conference, Quebec City, Quebec, Canada, June 1996.

———. "Sewer Infrastructure Management: New Concepts and Tools for Your Collection System." Water Engineering and Management, February 1998.

———. "Sewer Monitoring Information for Proactive Infrastructure Management." New England Water Environment Association, Marlborough, Massachusetts, September 1996.

Day, T., and C. Bradstock. "Moving from Old Computer Systems to New Platforms." WEFTEC '96, Dallas, Texas, October 1996.

———. "Rainfall Information Through Automation: Data with a Purpose." Paper presented at the Urban Wet Weather Conference, Quebec City, Quebec, Canada, June 1996.

Day, T., J. Nicolo, and M. Keyser. "Custody Transfer: Lessons in Successful Wastewater Metering and Billing." WEFTEC '96, Dallas, Texas, October 1996.

Day, T., et al. "Alternative Design of a Wastewater Monitoring Network." WEFTEC '95, Miami Beach, Florida, October 1995.

Drinkwater, A., et al. "A Guide to Short-Term Flow Surveys of Sewer Systems." WRc Ltd., Frankland Road, Swindon, United Kingdom.

Giles, R. *Fluid Mechanics and Hydraulics.* 2nd ed. New York: McGraw-Hill, 1962.

Grant, M., and B. Dawson. *Isco Open-Channel Flow Measurement Handbook.*

Miller, M. "Lightning Protection for Submersible Pressure Transducers." Water Engineering and Management, February 1999.

Szunko, G. "A Pollution-Focused Strategy for Sewerage Investment in the North West of England." North West Water, Ltd. Paper presented at the Reducing CSOs Conference, Louisville, Kentucky, July 1994.

Tchobanoglous, G. *Wastewater Engineering.* 2nd ed. New York: McGraw-Hill, 1979.

U.S. Environmental Protection Agency. EPA-600/2-76-243, Project No. 11024 FVQ.

———. EPA 625/k-94/003.

———. EPA/625/R-93-004.

———. 382-B-95-002.

Viessman, Knapp, Lewis, and Harbaugh. *Introduction to Hydrology.* 2nd ed. New York: Harper and Row, 1977.

WEF MOP #FD-20. *Design and Construction of Urban Stormwater Management Systems.* 1992.

WEF MOP #21. *Instrumentation in Wastewater Treatment Facilities.* 1993.

INDEX